AF719103

The Radiance of Jade and the Clarity of Water

The Radiance of Jade and the Clarity of Water

KOREAN CERAMICS FROM THE ATAKA COLLECTION

Organized by
Ikutarō Itoh and Yutaka Mino,
with contributions
by Jonathan W. Best and
Pamela B. Vandiver

THE ART INSTITUTE OF CHICAGO
HUDSON HILLS PRESS, NEW YORK

This publication was issued in connection with the exhibition "The Radiance of Jade and the Clarity of Water: Korean Ceramics from the Ataka Collection," held at The Art Institute of Chicago from November 23, 1991, to February 3, 1992; at the Asian Art Museum of San Francisco from March 2 to April 26, 1992; and at The Metropolitan Museum of Art, New York, from May 20 to July 12, 1992. This exhibition and publication were made possible by the Sumitomo Bank, Limited, Japan.

THE ART INSTITUTE OF CHICAGO
Editor: Dr. Deborah Del Gais Muller
Executive Director of Publications: Susan F. Rossen
Production Manager: Katherine Houck Fredrickson

HUDSON HILLS PRESS
Editor and Publisher: Paul Anbinder

Designer: Binns & Lubin/Betty Binns
Composition: Paul Baker Typography
Printer: manufactured in Japan by Nissha Printing Company, Limited

JACKET/FRONT IMAGE: NO. 15

FIRST EDITION
© 1991 by The Art Institute of Chicago

All rights reserved under International and Pan-American Copyright Conventions. No part of this publication may be reproduced without written permission of the Publications Department of The Art Institute of Chicago.

Copublished by The Art Institute of Chicago, Michigan Avenue at Adams Street, Chicago, Illinois 60603; and Hudson Hills Press, Inc., Suite 1308, 230 Fifth Avenue, New York, New York, 10001-7704.

Distributed in the United States, its territories and possessions, Canada, Mexico, and Central and South America by Rizzoli International Publications, Inc. Distributed in the United Kingdom and Ireland by Shaunagh Heneage Distribution. Distributed in Japan by Yohan (Western Publications Distribution Agency).

ISBN: 1-55595-065-5

Library of Congress Cataloguing-in-Publication Data

The Radiance of jade and the clarity of water: Korean ceramics from the Ataka collection / organized by Ikutaro Itoh and Yutaka Mino; with contributions by Jonathan W. Best and Pamela B. Vandiver.
158 p. 27 x 13 cm.
Catalog for an exhibition held at the Art Institute of Chicago, the Asian Art Museum of San Francisco, and the Metropolitan Museum of Art, New York.
Includes bibliographical references.
ISBN: 1-55595-065-5 (Hudson Hills Press)
1. Pottery, Korean—Koryŏ period, 935–1392—Exhibitions. 2. Pottery, Korean—Yi dynasty, 1392–1910—Exhibitions. 3. Porcelain, Korean—Koryŏ period, 935–1392—Exhibitions. 4. Porcelain, Korean—Yi dynasty, 1392–1910—Exhibitions. 5. Ataka Sangyō Kabushiki Kaisha—Art collections. 6. Potter—Private collections—Japan—Osaka—Exhibitions. 7. Porcelain—Private collection—Japan—Osaka—Exhibitions. I. Itoh, Ikutarō. II. Mino, Yutaka. III. Art Institute of Chicago. IV. Asian Art Museum of San Francisco. V. Metropolitan Museum of Art (New York, N.Y.)
NK4168.6.A1R34 1991
738'.09519'07473–dc20

91-25553
CIP

Contents

Sponsor's Foreword

THE ATAKA COLLECTION is known worldwide as a treasure trove of Asian ceramics. It contains about one thousand individual pieces, of which 793 are Korean and 144 Chinese.

The Korean ceramics being seen in the United States have been particularly admired as the highlights of this unique collection. It presents a comprehensive view of ceramics from the Koryŏ and Chosŏn dynasties, or over one thousand years, when an original culture emerged amidst the many Chinese influences in the country.

The collection was assembled by Ataka and Company. In December 1975, the firm suffered a management crisis. At that time, the question of how it would dispose of its wonderful collection became a matter of concern. The Sumitomo Group, which consists of twenty-one companies, including the Sumitomo Bank, purchased the entire collection to protect it from being dispersed and donated it to the city of Osaka, the cradle of the group. Since 1982, numerous visitors to the city, as well as residents of Osaka, have been able to see and enjoy it.

It is meaningful, from the point of view of international cultural exchange, that this exhibition of Korean ceramics from the Ataka Collection is being held at The Art Institute of Chicago, located in a city that is closely linked with Osaka; at the Asian Art Museum of San Francisco; and at The Metropolitan Museum of Art, New York. Thus, a broad spectrum of people in the United States will have the opportunity to study this unique group of Korean ceramics and appreciate its many virtues. We are pleased and proud that the Sumitomo Bank has been able to contribute as a participant in this project.

Finally, we sincerely hope for the success of this exhibition and express our heartfelt thanks to those individuals in Japan and in the United States who have made great efforts to realize it.

S. Tatsumi

Sotō Tatsumi
President
The Sumitomo Bank, Limited

Foreword

THE KOREAN AND Chinese ceramics collection now housed and exhibited in the Museum of Oriental Ceramics, Osaka, formerly belonged to the trading firm of Ataka and Company. The collection became public in 1982 after an elaborate and unprecedented effort was made to preserve it upon dissolution of the firm. Numerous visitors, from places as distant as Europe and the United States, have become familiar with the collection since its installation in the museum. Most have a general knowledge of Chinese ceramics; for many, however, the visit affords their first contact with Korean ceramics. These viewers often express pleasure upon encountering the collection and discovering the appeal of Korean ceramics. I remember a female potter from a village in the Ukraine who, after seeing the museum's Korean ceramics, exclaimed, "How lyrical!"

Korean ceramics from the Ataka Collection are crossing the Pacific Ocean for the first time to be seen in three American cities. I am deeply gratified by this event. Korean ceramics occupy a unique place in the development of East Asian ceramics, having influenced aspects of Chinese and Japanese ceramic manufacture and having been influenced in turn by the ceramic art of China. This exhibition not only offers Americans a chance to become better acquainted with Korean ceramics as an art form, but also hopefully provides an opportunity for renewed recognition of the rich cultural legacy of Korea and of East Asia in general. The aesthetic qualities of Korean ceramics have been closely linked with the aesthetic consciousness of the Japanese people from ancient times and have exerted significant influence on the formation of Japanese artistic sensibilities; with Korean ceramics as a vehicle, we can introduce one aesthetic microcosm from which a particular aspect of Japanese aesthetics evolved. Finally, what these objects reveal about how aesthetic sensibilities and artistic intentions are expressed in ceramics has profound significance within the broader field of the arts as a whole.

To achieve these goals, the Museum of Oriental Ceramics has lent almost all of the representative works of Korean ceramics normally featured in its permanent installation. Pieces were chosen on the basis of their place in the historical development of Korean ceramics; also considered in the selection process was the evolution of the various technical aspects of ceramic manufacture. I believe that this approach permits a better understanding of the diversity of these superbly crafted ceramic works. To be sure, it is difficult to make an all-encompassing statement about the appeal of Korean ceramics through an exhibition of one hundred and fourteen pieces, forty-four of which are from the Koryŏ dynasty and seventy of which date to the Chosŏn period. Nonetheless, I am convinced that visitors will appreciate the aesthetic sophistication and remarkable level of artistic achievement of the Korean people upon viewing such extraordinary examples of ceramic art.

In conclusion, I wish to express my thanks to those who helped make this exhibition possible: James N. Wood, Director of The Art Institute of Chicago; and, above all, Yutaka Mino, the museum's Curator of Asian Art, who provided so much guidance in handling the details of this project. In addition, I would like to thank the staffs of the Asian Art Museum of San Francisco and The Metropolitan Museum of Art, New York, for their warm cooperation in this joint endeavor. Finally, I offer my deep gratitude to the Sumitomo Bank, which should rightly be called the founder of the Museum of Oriental Ceramics, for its magnanimous support of this exhibition.

Ikutarō Itoh, Director
Museum of Oriental Ceramics, Osaka

Foreword

"THE RADIANCE OF Jade and the Clarity of Water: Korean Ceramics from the Ataka Collection" is the most important exhibition of Korean ceramics to be mounted in the United States. Ceramics are the most widely appreciated of the arts of Korea, a medium in which this nation has made significant artistic achievements and contributions. The selection we are presenting from the Museum of Oriental Ceramics in Osaka, Japan, features outstanding examples from the Koryŏ and Chosŏn dynasties (ninth through nineteenth centuries).

The exhibition has drawn the full support of the City of Osaka and its mayor, Masaya Nishio, along with his staff in the Department of Education. Similarly, in Chicago, we are grateful for the assistance of Mayor Richard M. Daley.

The showing of the Ataka Collection has been made possible through the generous support of the Sumitomo Bank, Japan. I am grateful to its former chairman, Ichiro Isoda, and current president, Sotō Tatsumi, for their early commitment to and continuing interest in this project. The relationship between the Museum of Oriental Ceramics and the Sumitomo Bank and Sumitomo Group is a long and important one. Not only has extensive funding from Sumitomo made it possible to preserve the museum's collection, but it has helped to build a distinguished building in which to house it. Additional support for this project was received from a grant from the National Endowment for the Arts, a federal agency.

I am very grateful to Ikutarō Itoh, the museum's director, for all the efforts he and his staff have expended to make this exhibition possible. I would also like to thank his partner in this collaboration, Yutaka Mino, Curator of Asian Art at the Art Institute, who has worked with great determination to bring this exhibition to the cities of Chicago, New York, and San Francisco. I wish to thank my directorial colleagues at these institutions, Philippe de Montebello in New York and Rand Castille in San Francisco, for the important part they have played in bringing this unique collection to a wide American public.

James N. Wood, Director
The Art Institute of Chicago

Acknowledgments

THE OPPORTUNITY TO see a great collection of Korean ceramics is rare. It is my hope that this exhibition of selections from the Museum of Oriental Ceramics in Osaka will increase our awareness and appreciation of Korean ceramics and Korean art in general. It has been my dream to bring highlights of the collection, which I have long admired, to the United States so that Americans might enjoy the extraordinary forms and beauty inherent in these pieces.

Many staff members at The Art Institute of Chicago helped to realize this exhibition. Among them should be singled out the following: I am grateful to James N. Wood, Director, for his support, and to Dorothy Schroeder, Assistant to the Director for Exhibitions and Budget, for her assistance. I also wish to thank the staff of the Department of Asian Art and, in particular, my associate Della Quinn. Chicago architect John Vinci provided a sensitive exhibition design.

A fine team also collaborated on the catalogue. I am grateful to contributors Jonathan W. Best, Professor of Asian Art History, Wesleyan University, and Pamela B. Vandiver, Research Physical Scientist, Conservation Analytical Laboratory, Smithsonian Institution, for their excellent essays, and to Ikutarō Itoh for his preface and illuminating entries. Suzanne Arata and Jung-kang McClain ably translated Mr. Itoh's texts. The photographs were taken by the staff of the Museum of Oriental Ceramics. The catalogue was carefully edited by Dr. Deborah Del Gais Muller; its fine design is that of Betty Binns of Binns & Lubin Design, New York. Susan F. Rossen, Executive Director of Publications, oversaw the preparation and production of the catalogue, ably assisted by Katherine H. Fredrickson, Production Manager; Cris Ligenza, Publications Secretary; and Elisabeth Dunn, Special Projects Editor. Brooke Hammerle and Dr. Helen Kang also provided assistance.

Last, but not least, I deeply appreciate the long working relationship I have enjoyed with Ikutarō Itoh, Director of the Museum of Oriental Ceramics in Osaka. Much of the credit for this exhibition belongs to him.

Yutaka Mino
Curator of Asian Art
The Art Institute of Chicago

Korean ceramics in the Ataka collection

THE FORMATION OF THE COLLECTION

IKUTARŌ ITOH

THE MUSEUM OF Oriental Ceramics was founded by the city of Osaka in November 1982 to house and exhibit the Ataka Collection of Oriental Ceramics, which had been presented to the city by the twenty-one companies of the Sumitomo Group. The Ataka Collection was begun as one of a number of cultural endeavors undertaken by Ataka and Company, Ltd., which had intended to construct an independent art museum later to house it. The collection encompassed three areas: Korean ceramics, Chinese ceramics, and works by the Japanese Nihonga painter Hayami Gyoshū (1894–1935). Ataka and Company, a *shōsha*, or trading firm, was established in 1904. Ranked in 1975 as the ninth largest trading company in Japan, it experienced a management crisis in that year and was dissolved upon its merger with C. Itoh & Co., Ltd. in October 1977. Two years before, the firm had been forced to sell part of its art collection, the 106 works by Hayami Gyoshū, to the Yamatane Art Museum, Tokyo; thereafter, the issue of the disposition of the East Asian ceramics came to the attention of the Japanese public and the world at large. It was even discussed in the Diet.

Management of the matter was entrusted to the Sumitomo Bank, the largest creditor authorized to handle the business affairs of Ataka and Company. Resolving the situation, however, was very difficult for the following reasons: first, the collection had great value as a cultural legacy; second, the monetary value of the collection was very high; and third, there was a policy in effect to prevent its dispersal or sale outside Japan. In 1980, the Sumitomo Bank decided that the Sumitomo Group would donate the Ataka Collection to the city of Osaka. It was certainly the best possible solution, even though it required that the Sumitomo Group make a financial contribution to implement it. Funds were accumulated over a period of two years, during which time each of the twenty-one companies deposited a certain amount in a bank account held by the city of Osaka. When the total was sufficient to pay the purchase price of the Ataka Collection, the city was to officially assume ownership. At the same time, another donation was provided for construction of a museum to contain the collection. The total contribution was unprecedented in the history of preserving Japanese cultural property. An achievement worthy of historical record, it establishes a model for future cultural preservation efforts in Japan.

The contents of the collection

The 965 items in the Ataka Collection include 793 pieces of Korean ceramics and 144 examples of Chinese ceramics. In addition, there are two works of Japanese ceramics, five pieces of Vietnamese ceramics, and twenty-one other articles in metal and lacquer. Among the Chinese examples are two pieces designated as National Treasures and ten designated as Important Cultural Properties; thus, in terms of quality, the collection of Chinese ceramics can be regarded as one of the best in Japan.

But the main and most important component of the Ataka Collection is its Korean ceramics, which comprises eighty-two percent of the total collection. Among the pieces are 304 examples from the Koryŏ dynasty and 485 from the Chosŏn period. Earthenware is not a major focus of the collection: it contains only four earthenware objects, dating from the United Silla period. The ceramics of the Koryŏ and Chosŏn periods were selected on the basis of chronology and technical excellence. In terms of quality and quantity, the Ataka Collection ranks as the greatest collection of Korean ceramics in the world to have been assembled by a private individual.

The Ataka Collection fully reflects all the techniques that figure in the making of ceramics in the Koryŏ period. Koryŏ ceramists specialized in celadon. They adopted the firing techniques of Chinese Yue ware, perfecting the so-called jade color of its beautiful glaze in the first half of the twelfth century. Jade-colored celadons were decorated in a variety of ways, including the use of monochrome glazing, the application of incised and carved designs, and the employment of openwork and sculpting techniques. The Ataka Collection contains ninety-two examples illustrating these methods. There are 128 examples of inlay technique, which became popular in the mid-twelfth century. The collection's twenty-four pieces of underglaze iron-brown slip-painted celadon, the seventeen of underglaze iron-brown slip-coated celadon, the eight of underglaze copper-red spotted celadon, the four of underglaze white-slip-painted celadon, the two of overglaze gilt celadon, and the three of marbled ware are all rare examples of other decorative techniques used in the Koryŏ period. In addition to the celadons, there are also sixteen examples of white porcelain and ten of black-glazed and iron-brown-glazed ware.

The Chosŏn ceramics are the most substantial part of the collection in both their quality and quantity. Punch'ŏng ware, which developed in the early Chosŏn period, is divided into many types according to decorative techniques: inlaid, stamped, or white slip that can be brushed, scraped, incised, iron-brown-painted, or applied overall (Kor. *punin*; Jap. *kohiki*), and so forth. The 171 examples of Punch'ŏng ware in the collection provide an overview of these types. The majority of the Chosŏn ceramics, however, consists of seventy-three examples of white porcelain, one hundred twenty-seven examples of blue-and-white ware, forty-five examples of underglaze iron-brown ware, and twenty-four examples of underglaze copper-red ware. In addition, there are forty-five examples of white porcelain with inlay, blue-and-white ware with touches of underglaze iron-brown or copper-red, underglaze lapis-blue ware, and black-glazed and iron-brown-glazed ware. With these 314 pieces, we can trace the development of Chosŏn ceramics over the 519 years of the dynasty. Since most of these pieces are considered representative works, the history of Chosŏn ceramics can be constructed on the basis of their highly selective and distinctive character alone.

From the sixteenth century on, the Japanese purchased Korean ceramics for use in the tea ceremony. The Ataka Collection, however, features numerous examples of Korean ceramic ware that were imported into Japan at the end of the nineteenth century as objects of connoisseurship and appreciation (i.e., *kanshō tōki*). Formation of the collection was guided by the aesthetic sensibilities that the Japanese perceive and respond to in Korean ceramics. In its character and quality, that is, the collection succeeds in expressing the deepest understanding and appreciation on the part of the Japanese for Korean ceramic wares.

The collector

The Ataka Collection was put together over a relatively short period of almost twenty years, beginning in 1951, by Eiichi Ataka (b. 1900), former chief executive officer of Ataka and Company. Ataka has the soul of an artist more than of an executive; he plays the piano, has provided financial support to many young musicians, and is one of those whose substantial monetary contributions nurtured the Japanese music world. At the same time, his natural gift for discerning artistic quality resulted in the formation of a splendid collection.

Ataka's passion for collecting was intense. He decorated his bedroom with photographs of works he thought of acquiring and, until he did actually obtain them, would gaze at the photographs nightly and would rest only after considering the various strategies required for their acquisition. In many instances, it took more than ten years to obtain a desired work. Fearful of missing the opportunity to acquire a work of art because he was out of the country, he reached a point where he avoided traveling abroad. He visited antique dealers incessantly, attended auctions, and maintained constant contacts with other collectors. Having observed Ataka's relationships with dealers and collectors firsthand, I can say that I have never known anyone to conduct himself with such propriety. His New Year's activities began with an invitation to various dealers to be his guest at a fine Japanese restaurant; at the party, he would place himself in the most humble seat and request their cooperation in his collecting efforts during that year. Ataka's extreme tenacity was the driving force behind his creation of the collection. When I began to assist in the formation of the collection under Ataka in 1955, I came to realize that the activity of collecting itself can become an artistic act. As I have said elsewhere, "the Ataka Collection was created and guided by the collector Eiichi Ataka, a man blessed with a unique artistic talent. It is not a perfect collection but is itself an imperishable and awe-inspiring work of art."

Ataka's standards of selection were strict. When it came to collecting, he was a perfectionist, not permitting even the slightest flaw. By flaw, I am not referring to physical damage to the object, but rather to a flaw in its artistic value. For Ataka, a work of art had to express a sense of tension. No matter how celebrated a work may have been, he would avoid it if it was at all excessive, showy, or vulgar. The objects that appealed to him most had to have dignity, tranquility, and severity, and had to manifest fastidiousness and restraint. It was as though he were searching for human characteristics in the ceramics he collected. These measures, which can be described as the "Ataka taste," form the basis of the unique sensibility present in all the works in the Ataka Collection. If there is a universal aesthetic shared by all Korean ceramics, it was apparently deeply rooted in Ataka's discerning judgment. He had a passion for collecting, particularly for acquiring Korean ceramics, and it was here that he concentrated his most sincere efforts. If the activity of collecting itself is an impassioned quest, then surely the Ataka Collection should be regarded as the manifestation of a distinctive Japanese taste. It is significant to observe in this context that someone, such as Eiichi Ataka, who wishes to implement an idea, must of necessity possess a disposition that is strongly motivated toward that end.

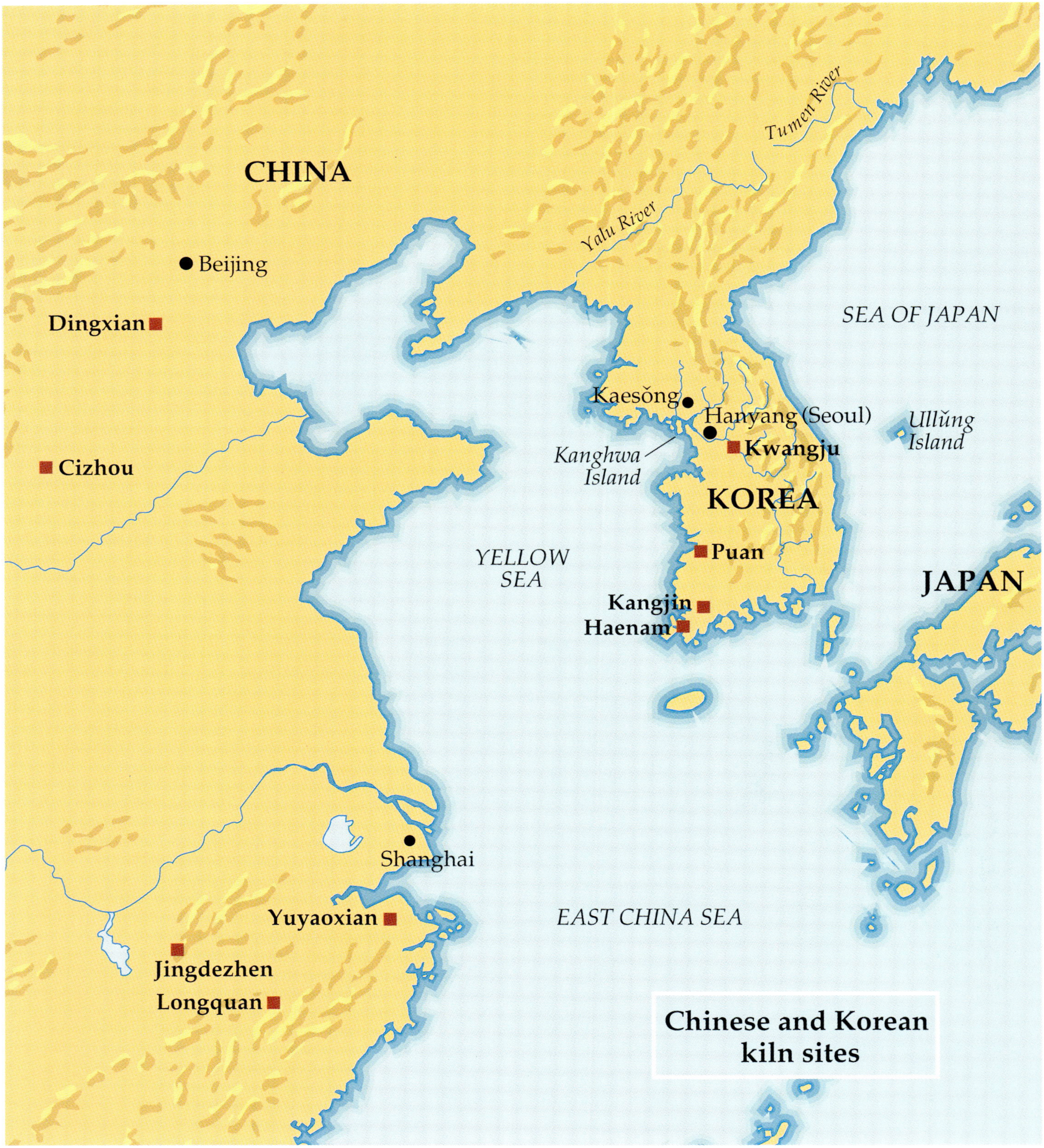
CHINA
Beijing
Dingxian
Cizhou
Yalu River
Tumen River
SEA OF JAPAN
Kaesŏng
Hanyang (Seoul)
Ullŭng Island
Kanghwa Island
Kwangju
KOREA
YELLOW SEA
Puan
JAPAN
Kangjin
Haenam
Shanghai
Yuyaoxian
EAST CHINA SEA
Jingdezhen
Longquan
Chinese and Korean kiln sites

Context of the artistry:

A VIEW OF THE KOREAN PAST

JONATHAN W. BEST

THE ELEMENT OF truth contained in Louis Sullivan's dictum that "form ever follows function" applies with special force, perhaps, to such eminently useful arts as ceramics and architecture. Certainly, the Korean ceramics of the Koryŏ and Chosŏn periods not only reveal in their shapes the purposes for which they were wrought, but also reflect in the artistry of their finishing something of the aesthetic predilections of their users which, in turn, were conditioned by the historical and cultural circumstances attending their production. Created objects are not, however, mere passive mirrors of the past. They participate in shaping the present moment by affirming and reaffirming through repeated use the cultural values and meanings given expression through them. Social confirmation of the sophisticated elegance of a twelfth-century Koryŏ aristocrat's life-style was, in part, derived from the celadon wares possessed and displayed at table and in ceremony. The ideal Confucian simplicity and propriety of a fifteenth-century Chosŏn official's life was invoked by the unpretentious beauty of the white porcelains used in family rites and ritual. Art's influence in shaping the past has been little explored and is, as yet, poorly understood. The role of the past in the shaping of art, however, is both more apparent and more studied. Knowledge of a society's physical environment, socioeconomic structure, beliefs, and political history serves to enhance the understanding and appreciation of its art. This premise underlies the following brief discussion of Korea and its past. The intention here is to provide a general introduction to the historical and cultural context in which the ceramics of the Koryŏ and Chosŏn periods were produced and of which they were a part.

The history and culture of the Korean people have, in no small measure, been shaped by the geography of their land. Bounded on the north by Manchuria, the mountain-spined Korean peninsula juts some 550 kilometers from the Asian landmass to face China in the west and Japan in the south (map, p. 14). From early times, Korean rulers showed their admiration for Chinese civilization and manufactures through the cultivation of commercial and diplomatic ties with the Middle Kingdom, while Japanese rulers commonly demonstrated their appreciation of the culture and products of Korea through emulation, trade, and piracy. The peninsula's physical link to Manchuria served to facilitate, not infrequently to Korea's misfortune, interactions with the nomadic cultures of the Northeast Asian steppe, as well as with the agrarian society of northern China.

Geographical factors also exerted a major influence in the formation of national consciousness in Korea. With all but its northern boundaries fixed by the ocean's waves and measuring on average just some 250 kilometers in breadth, the Korean peninsula has long been readily conceptualized by both its residents and neighbors as a geopolitical entity. Internally, however, the unity of Korea is fractured by major mountain chains. The greatest of these dominates the eastern half of the peninsula and extends from the Yalu and Tumen rivers in the north to the southern coast. Reaching westward toward the Yellow Sea from this massive granite wall are two secondary ranges of historical importance. One of these forms a natural barrier across much of the peninsula's width just north of the Han River valley, and the other angles down to the peninsula's southwestern corner, separating the fertile lowlands of the Naktong River valley and the southeastern coast from those of the southwest. At an early date, these natural subdivisions of the peninsula into northern, southwestern, and southeastern sectors engendered a corresponding sense of pronounced regional distinctiveness among Koreans that has continued to be a potent force in politics and culture down to the present day. Such regional identifications are asserted in spite of the simultaneously held and equally powerful awareness that the Korean peninsula constitutes an inherently fitting national entity.

Traditional Korea was a stratified society, with social position determined by birth. Viewed in the broadest perspective, the population was divided into two classes, the freeborn and the unfree. A small percentage of the freeborn class was composed of the hereditary elite, families that, during the Koryŏ and Chosŏn periods, provided the state's officials, while the bulk of this class consisted of the taxpaying peasantry. The freeborn class also included low-level functionaries in local government, itinerant peddlers, and, during Koryŏ times, Buddhist monks and nuns. The unfree class consisted of slaves, who were the property of others and whose daily labor was determined by their owners. Also part of the unfree class were certain hereditary occupational groups, including practitioners of culturally disesteemed professions, such as butchers and entertainers, and some specialized laborers, such as gold miners and potters, who were the property of the state and whose labor served its needs. The position of individual families within the social structure could change due to alterations in political or economic circumstances, but both the hierarchical structure of the traditional Korean social order and its domination by a hereditary elite remained constant until the nineteenth century.

Over the course of the ninth century, the centralized authority of the sovereigns of the ancient state of Silla had increasingly eroded in areas beyond their traditional stronghold in the southeastern corner of the peninsula. By the century's end, their power was openly challenged by the self-proclaimed monarchs of Later Paekche in the southwest and Later Koguryŏ in the north. In 918, a military coup toppled the king of Later Koguryŏ from his throne, and Wang Kŏn (877–943), who had earlier gained prominence as a general in the deposed king's army, assumed control of his former master's domain and declared himself king of the new state of Koryŏ. Thereafter, through a skillful combination of diplomatic and military initiatives, Wang Kŏn (widely known by his posthumous title of T'aejo, or "Grand Founder"; r. 918–43) brought about the political reunification of the peninsula through the "voluntary" submission of Silla in 935 and the armed conquest of Later Paekche in 936. During the century that followed, the monarchs of Koryŏ sought to strengthen their authority vis-à-vis the established aristocratic lineages through such expedients as the institution of a civil-service examination system for the recruitment of state officials and the inauguration of a new land system that appropriated most of the nation's real estate for the needs of the royal government and the maintenance of its officers. Although Koryŏ reforms of this type, as well as others undertaken later with similar intent by kings of the Chosŏn dynasty, were temporarily successful in elevating the authority of the throne, in the long run their effect was always subverted and traditional Korean society and culture remained dominated by those of exalted birth.

According to the official history of the Koryŏ period, the *Koryŏ-sa* (completed in 1451), from his deathbed King T'aejo enunciated Ten Injunctions to his heirs for their guidance in ruling the state. These admonitions provide interesting insight into the attitudes and beliefs of Koryŏ court society. The first injunction credits the prospering of the state to the beneficence of the Buddhas and yet simultaneously warns against allowing the Buddhist church to meddle in politics. The second injunction speaks of the importance of geomancy (Kor. *p'ungsu*; Chin. *fengshui*) in the selection of propitious sites for Buddhist temples and urges that, to preserve national vitality, temples only be constructed at those sites previously identified by the geomantic adept especially revered by T'aejo, a Buddhist priest named Tosŏn (d. 898). The fourth of the Ten Injunctions is also historically revealing. Its opening acknowledgment of Korea's institutional and cultural indebtedness to China is immediately balanced by the assertion that the Korean people are, in their circumstances and character, sufficiently different from the Chinese so that the ways of the latter should not be slavishly followed. The same injunction also characterizes the Khitan, a seminomadic people of southern Manchuria who were soon to establish the Liao dynasty and eventually dominate a sliver of northern China, as a nation of "beasts" that was entirely unworthy of emulation. Although the importance of Buddhism and geomancy for the preservation of the dynasty is especially pronounced in the Ten Injunctions, the value of Confucian precepts for the maintenance of an effectively ordered state is also acknowledged in the seventh and tenth injunctions.

The distinction accorded Buddhism in particular by the Koryŏ court contrasts markedly with the official policies of the contemporary Song dynasty in China. The many Buddhist sects that claimed adherents in Korea during this period can be divided into two major groupings: those sects that largely emphasized the intellectual mastery of doctrines articulated in particular religious texts or sutras (Kor. *kyo*) and those that stressed the individual's practice of meditation (Kor. *sŏn*) under the direction of a master. The former grouping, known historically as the Kyo tradition, includes the Korean manifestations of the Chinese Huayan, Tiantai, and Pure Land schools; the latter grouping, known as the Sŏn tradition, is equivalent to the Chan tradition in China or the Zen tradition in Japan. Both Kyo and Sŏn Buddhism were munificently endowed by the court and the aristocracy. T'aejo's injunctions to the contrary notwithstanding, Buddhist temples increased in number and in landholdings and wealth, as well as in political influence, over the course of the dynasty. During this era, royal princes became powerful abbots, and priests became intimate royal advisers, while lavish rituals were regularly performed for the state at public expense and for those private devotees who could afford the cost and found satisfaction in the display. The *kuṇḍikā* bottle (Kor. *chŏngbyŏng*; nos. 8 and 22) is a specially shaped Buddhist ritual vessel of ancient Indian pedigree that in Koryŏ Korea was exquisitely fashioned in celadon and frequently donated to temples, it is said, as a merit-producing offering by aristocratic patrons of the faith. The esteem accorded celadon wares by the Koryŏ nobility is also evident from contemporary painted depictions of such wares being used by Buddhist divinities, as in the instance of inclusion of a celadon *kuṇḍikā* in paintings of the Bodhisattva Avalokiteśvara (Kor. *Kwanseŭm*) meditating by the sea (fig.1).

Although Buddhism was the religion of the state and the dominant belief of the age, geomancy and Confucianism also made significant contributions to the ways in which the Koryŏ elite ordered their lives and viewed their world. Geomantic factors were taken into consideration in the siting of palaces and temples, and it was even argued in the fifth of T'aejo's injunctions that his successors should reside in P'yŏngyang, one of Koryŏ's secondary capitals, for almost a third of every year because of the auspiciousness of its location. Similarly, in his eighth injunction, T'aejo used a geomantic rationale to justify his prejudice against the people of southwestern Korea (i.e., the territory of the former state of Later Paekche) as untrustworthy.

Knowledge of Confucianism had been present on the peninsula for a thousand years prior to the establishment of Koryŏ rule, and its principles had long been expressed through the bureaucratic structure and daily ceremony of Korean court life. During the latter part of the Unified Silla period, a civil-service examination testing the candidates' understanding of the Chinese Confucian classics had been inaugurated. In 958, Shuang Ji, a Confucian literatus, who had come to Korea two years earlier as a member of an official mission from a minor northern Chinese state, proposed to King Kwangjong (r. 949–75) the establishment of a Koryŏ civil-service examination system. His recommendation was accepted by the king, who evidently saw in the examination system a way of checking the political power of the most influential aristocratic families. In time, there came to be several different types of examinations in the Koryŏ system, each of which dealt in some way with aspects of Chinese learning. There were even state-sponsored examinations on Buddhist teachings through which priests could acquire official ecclesiastical rank. One indication of the value that the Korean elite had come to place upon a literary education is that by the early thirteenth century—or some two centuries before Gutenberg—a movable metal type had been developed to facilitate the distribution of texts. The earliest surviving text printed in this manner is a work on Confucian ritual published in 1234.

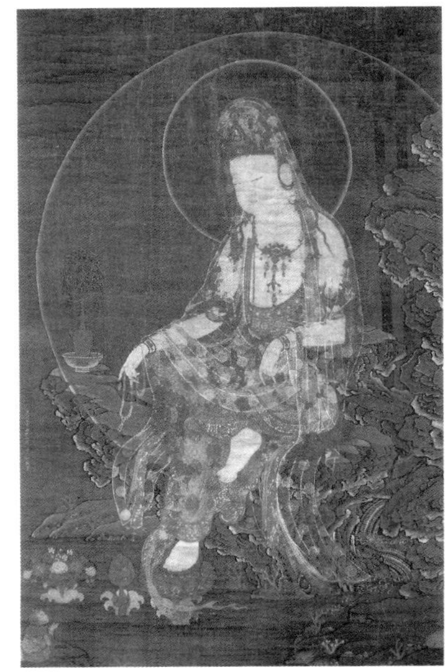

FIGURE 1
Sŏ Ku-bang. Avalokiteśvara Bodhisattva, *Koryŏ dynasty, dated 1323. Colors on silk; 165.5 x 101.5 cm. Japan, Sumitomo Collection. All artists and works of art included here are Korean, unless otherwise noted.*

Late in the Koryŏ dynasty, during the reign of King Ch'ungnyŏl (r. 1274–1308), Neo-Confucianism was introduced to Korea through the efforts of scholars such as An Hyang (1243–1306), who had been attracted to the philosophy while in Peking serving as a member of a diplomatic mission to the court of the Yuan dynasty. Neo-Confucianism, a systematically coherent reformulation of basic Confucian teachings, was developed in China during the Song dynasty and given its most influential exposition by the renowned twelfth-century thinker Zhu Xi (1130–1200). From 1313 until the rise of the Chinese republic in 1912, Zhu Xi's construction of Neo-Confucianism was the officially ordained interpretational perspective to be used in China's civil-service examinations, and thus it essentially constituted the orthodox ideology of the Chinese imperial state. Neo-Confucianism attached importance both to an individual's comprehension of its particular conception of the natural and proper order of the universe and to the crucial role of the ruler in that order. It is not surprising, therefore, that there were many among the philosophy's early proponents in Korea who, in the late fourteenth century, openly questioned the domination of Koryŏ society by Buddhist clerics and a small number of aristocratic families.

King T'aejo's fourth injunction clearly discloses the difficulties of Koryŏ's place in the diplomatic and cultural politics of contemporary East Asia. Koreans were fully conscious of their cultural differences from the Chinese, yet realized that China had long been and would likely long remain a critically important source for cultural borrowing. Immediately beyond Korea's northern border, however, ranged the nomadic and seminomadic peoples of northern Asia. Even though the sedentary population of the peninsula might regard these peoples as "beasts" and unworthy of emulation, as T'aejo said of the Khitans, their military prowess was repeatedly to extract the Korean court's professed allegiance, if not its esteem. At the time of the dynasty's founding in the early tenth century, China was divided into several contending states and the Khitan dominated a large area of Manchuria. In 947, the Khitan ruler proclaimed himself the emperor of the Liao dynasty and soon thereafter, in 960, the Song dynasty emerged as the sole political authority in most of China proper. The combination of Korean prejudices concerning the Khitans and the existence of a disputed border separating their domains quickly provoked tensions between the Koryŏ and Liao courts. These tensions flared on more than one occasion and prompted three major Liao invasions plus several smaller armed incursions into the peninsula between the years 993 and 1018. The Koryŏ court responded to this challenge by, on the one hand, strengthening its armies and, on the other, commissioning the carving of woodblocks for printing the complete Buddhist canon (Tripitaka) in Chinese. Although this monumental undertaking was initiated early in the eleventh century to invoke the aid of the Buddhas against the Khitans, it was not completed until 1087, long after peaceful relations had been established with the Liao.

Between the late tenth and early twelfth centuries, while the Song and Liao regimes remained in contention, the Koryŏ court modified its foreign policy in response to the changing pressures and opportunities generated by these two mutually antagonistic continental neighbors. At some points, Koryŏ sent embassies to both neighboring powers and at other moments just to one; in such trying times, the choice could shift abruptly. Even though the might of the "barbarian" Liao court was sufficient for several decades in the mid-eleventh century to command Koryŏ's undivided diplomatic attention, the Korean elite continued throughout this period to welcome Chinese merchants bearing goods from the culturally honored Middle Kingdom. Among the products of Song China that always found a ready market in Koryŏ were Buddhist and Confucian texts, medicines, and porcelains. Similarly, certain Korean products—including celadons—earned favorable attention in China. One Song scholar, known only by the pseudonym of Taiping Laoren (dates unknown), expressed the highest admiration in his *Xiuzhongjin* for the "secret" color of Koryŏ's distinctive wares. A much more detailed appreciation of Korean celadons was penned by the scholar-official Xu Jing (1091–1153), who in 1123 visited the Koryŏ capital at Kaesŏng as part of a Song diplomatic mission. In the account of his travels, the *Xuanhe fengshi Gaoli tujing*, Xu Jing described a number of characteristic celadon vessel types and noted that in Korea the peculiarly subtle hue of this ware was commonly termed kingfisher-colored (Kor. *pisaek*; Chin. *feise*).

Early in the twelfth century, the Liao dynasty of the Khitans was supplanted by the Jin dynasty of the Jurchens, another seminomadic people from Manchuria. The power of the Jurchen armies, which also succeeded in driving the Song totally out of northern China, earned for that regime Koryŏ's begrudging diplomatic recognition. A little over a century later, however, first the Jin government in 1234 and then the Song in 1279 were toppled by the devastating assaults of the Mongols. The Mongol rulers then proceeded to set themselves up as the emperors of China and designated their new regime as the Yuan dynasty. Korea was also ravaged by Mongol military might, suffering six invasions between 1231 and 1257. In 1231, the Koryŏ court fled from Kaesŏng and took refuge on Kanghwa, a large island in the Yellow Sea some eighteen kilometers away, but less than two kilometers offshore. The royal court and the officers of the military government that at the time was the real political authority for the state were able to live comfortably on Kanghwa, while the population left behind on the peninsula suffered terribly at the hands of the Mongols. This ignoble situation persisted for some forty years; it was during this period that the Koryŏ court again ordered the preparation of a complete set of woodblocks for the Buddhist Tripitaka. This set, over eighty thousand woodblocks in all, is housed today at the Haein-sa (a temple in South Kyŏngsang province). Its carving had two purposes: to gain divine protection against the Mongols and to replace the eleventh-century set originally created as a defense against the Khitans and burned by the Mongols in 1232.

By 1270, a peace had been negotiated with the invaders, the military government in Korea had been terminated, and the Koryŏ court entered an era of over a century of very close relations—including royal intermarriage—with the Mongol emperors of the Yuan dynasty. The close official ties to Yuan China facilitated the extensive exchange of ideas and goods, as evidenced, for example, by the introduction of Neo-Confucianism to Korea at this time and by the recent archeological discovery of a shipload of Yuan ceramics off the peninsula's eastern coast. This era of friendly relations with the Yuan also facilitated the importation of Cizhou wares to Korea, with the result that significant quantities of Koryŏ celadons began to be adorned with underglaze iron (brown/black) designs unmistakably reminiscent of characteristic Cizhou decoration. It is also likely that the decorative use of underglaze copper (red/reddish orange), a technique believed to have been developed in Korea during the twelfth century, was introduced to China at this time.

The rapid disintegration of Mongol authority in China in the late fourteenth century that culminated in its replacement by the native Chinese dynasty known as the Ming caused substantial turmoil at the Koryŏ court. Conflict arose between those who felt that Korea ought to continue to render allegiance to the beleaguered Yuan court and those who favored alignment with the nascent Ming. It was this tension which precipitated the circumstances that brought about the Koryŏ dynasty's downfall. In 1388, Yi Sŏng-gye (1335–1408), founder of the succeeding Chosŏn dynasty, was put in charge of a body of Korean troops sent to support the Yuan cause by attacking Ming positions in Manchuria. Having reached an island midstream in the Yalu River, however, Yi turned his forces back to the capital and ousted the reigning monarch and the officials responsible for initiating the anti-Ming policy. For the next few years, Yi and the group of Neo-Confucian reformist advisers that he gathered were content with the indirect exercise of political power, but, in 1392, this pretense was cast aside, the new Chosŏn dynasty was proclaimed, and the geomantically auspicious site of Seoul selected for its capital.

The change in dynasty brought in its wake major social and cultural shifts. Yi Sŏng-gye (r. 1392–98) and his immediate successors aggressively moved to strengthen the power of the royal government while reducing the wealth and influence of both Buddhism and the noble families that had been prominent during the Koryŏ era. These ends were accomplished in large part through legislation that had the effect of making the Neo-Confucianism of Zhu Xi the official ideology of the Chosŏn state. During Yi Sŏng-gye's de facto rule in the last few years of the Koryŏ monarchy, he had overseen the confiscation of privately owned lands throughout the country. A portion of the lands so acquired was utilized to recompense government officials and the remainder reserved to meet the requirements of the royal state. Similarly, beginning during Yi Sŏng-gye's years on the throne and continuing during the reign of his son T'aejong (r. 1400–1418), increasingly stringent restrictions were placed on Buddhism and many of its properties were confiscated. These various measures effectively destroyed the societal influence of both Buddhism and the old aristocracy, thereby preparing the way for the rise of the new hereditary elite that was to dominate Korea socially and culturally throughout the half-millenium rule of the Chosŏn dynasty.

This new elite, known collectively as *yangban*, consisted of the small portion of the total freeborn population that over time was able to monopolize civil and military posts in the national bureaucracy. Success in the regularly held civil-service examinations constituted the primary entryway to the bureaucracy and, as in contemporary China, required adherence to a Neo-Confucian perspective on the part of the examinee. Official position within the bureaucracy, especially the civil bureaucracy, conferred prestige and financial security within Chosŏn society. Since *yangban* families were exempt from payment of taxes and the corvée, their male children had substantially greater opportunities to obtain the rigorous Neo-Confucian education necessary to succeed in the civil-service examinations.

Neo-Confucianism having been thus privileged by the Chosŏn government, its influence came to pervade elite Korean society and culture. The hierarchical notions sanctioned by the philosophy provided the ideals that underlay the structuring of all social relationships from the intimate circle of the family to the public arena of the bureaucracy. Neo-Confucian values were celebrated through the textually prescribed rites and ceremonies for both the family and the state. The humanism of Neo-Confucianism, coupled with its firm belief in the value of education, resulted in the invention early in the fifteenth century of *han'gŭl*, a simple script perfectly designed

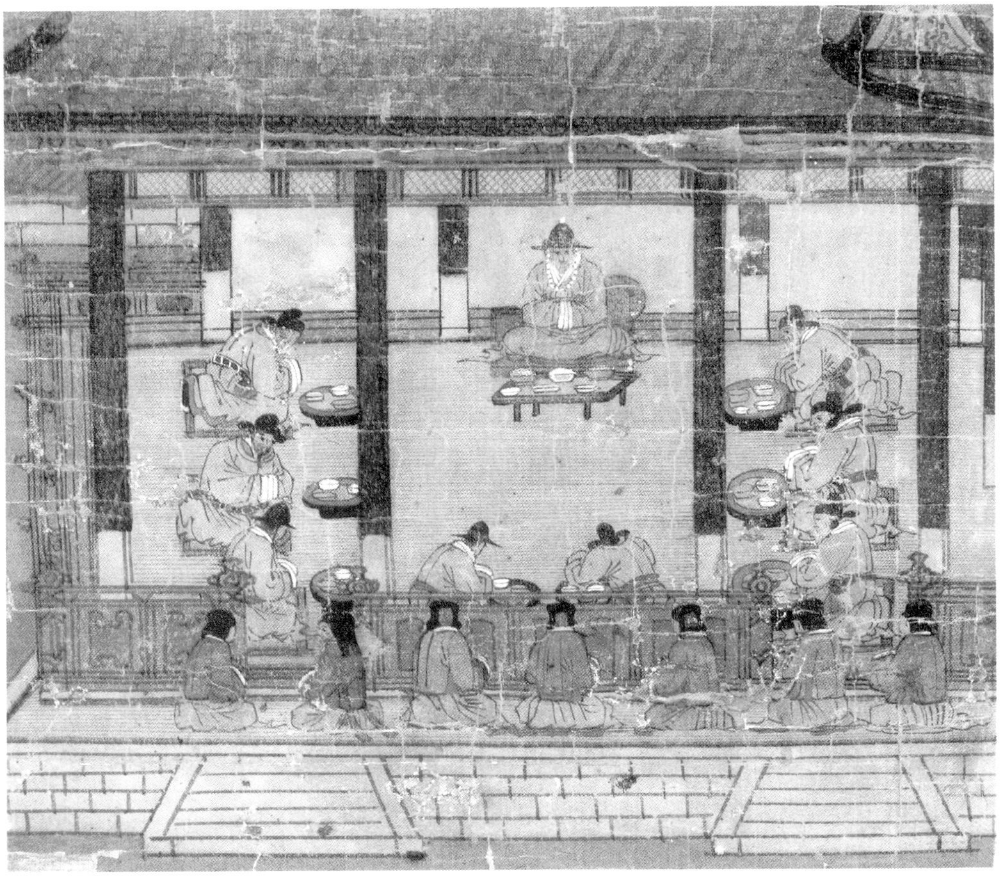

FIGURE 2
Anonymous, Gathering of Government Officials *(detail), Chosŏn dynasty, ca. 1550. Light colors on silk; 121 x 59 cm. Seoul, National Museum of Korea.*

for the writing of spoken Korean and thus ideally suited for the edification of those who, unlike male *yangban*, had no opportunity to become literate in Chinese. Similarly, the unprecedented expansion in the production of plain white porcelain vessels (Kor. *paekcha*) during the Chosŏn period is directly related to the austere aesthetic tastes and ritual requirements of the Neo-Confucian elite (figs. 2 and 3).

The pronounced Neo-Confucian orientation of the Korean court and bureaucracy also had important repercussions for the dynasty's foreign relations. On the one hand, this orientation prompted Korean officials to regard loyal subservience as a positive virtue in their relations with the Ming regime in China. On the other hand, the greater prestige accorded civil officials over military officials within Korea engendered a chronic decline in the state's ability to defend itself against aggression from without or insurrection from within. Japanese pirates, long feared and hated for their incursions along Korea's coasts, became a scourge of sufficient magnitude during the early decades of the fifteenth century to elicit a determined and highly successful response from the still youthfully vital Chosŏn regime. The pirates' lairs on Tsushima Island were overwhelmed and strict limitations were placed on Japanese interactions with Korea.

FIGURE 3

Yangban *family holding a* chesa *(memorial service honoring its ancestors). Photograph, late nineteenth/early twentieth century. From Yi Kyu-tae,* The Modern Transformation of Korea, *trans. Sung Tong-mahn et al. (Seoul, 1970), p. 64.*

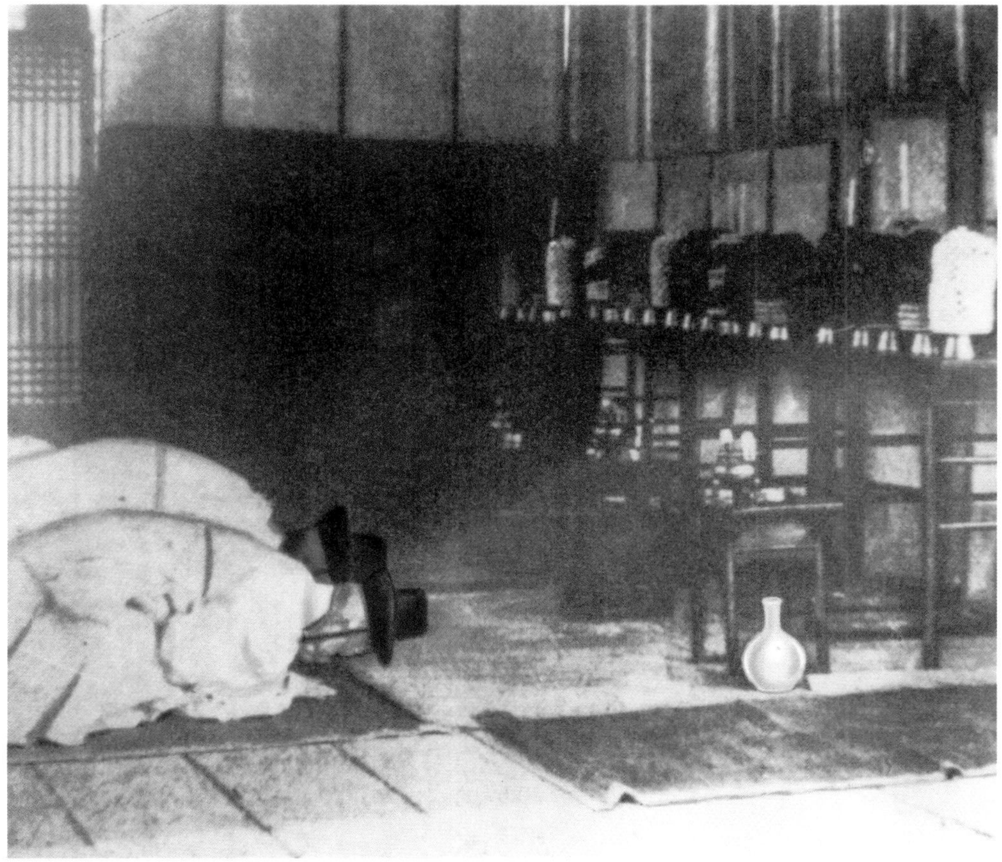

By the end of the sixteenth century, however, Korea's military forces had been allowed to deteriorate, while in Japan Toyotomi Hideyoshi (1536–1598) had just brought centuries of internecine war to a halt and had assumed supreme command of the archipelago's many battle-hardened armed forces. Confronted by the internal threats posed by such a vast standing army, Hideyoshi conceived the remarkable notion of conquering Ming China and forthwith requested that Chosŏn's rulers allow his troops free passage through the peninsula. Both common sense and Neo-Confucian loyalty to the Ming argued against acquiescence to Hideyoshi's wishes, and, accordingly, in 1592 and again in 1597 devastating Japanese assaults were loosed against Korea. The first swept north as far as P'yŏngyang, but the second was stopped less than halfway up the peninsula. The Chosŏn court's loyalty to the Ming was rewarded to the extent that Chinese armies were sent to Korea, where they lived off the land and frequently joined in the fight against the Japanese. Between their initial onslaught in 1592 and their final withdrawal from the peninsula in 1598, the Japanese forces maintained themselves in massive fortifications erected along the southern coast while they, too, lived off the backs of the Korean peasantry. During the long years of the Japanese occupation, large areas of southern Korea were plundered and laid waste, and as part of the booty borne off to Japan by the tea-ceremony aficionados among Hideyoshi's field commanders were numbers of Korean potters. Southern Korean *punch'ŏng* wares, in particular, had become much esteemed by Japanese tea masters, and the securing, entirely by force, of the services of skilled Korean potters for Japanese kilns both immediately enriched the production of tea-ceremony stonewares and also hastened the development of porcelain production in Japan.

One effect of Ming participation in the war against Hideyoshi's troops was the further weakening of the already declining Chinese dynasty which, during the early decades of the seventeenth century, was to show itself incapable of containing the rising military power of the seminomadic Manchu people, whose homeland, Manchuria, still bears their name. Initially, the Chosŏn government prudently managed to avoid seriously antagonizing either the Ming or the Manchus, but then a group of pro-Ming officials gained the upper hand at court and their overt support of the ailing Chinese regime provoked major Manchu invasions of the northern portion of the peninsula in 1627 and 1636. The chastened Korean court was thereupon compelled to swear fealty to the Manchus, who completed their conquest of China soon after and, in 1644, replaced the Ming with their own Qing dynasty.

The shock of the fall of the respected Ming, the only Chinese sovereign power with which the Chosŏn court had ever dealt, coupled with the great sufferings visited upon the peninsula during the Japanese and Manchu invasions, caused Korea's *yangban* elite to reassess the premises that had guided them in the past and to formulate responses to the many immediate problems besetting their country. One reaction to the dynastic change in China was the development of an enhanced interest in, and appreciation of, that which was Korean. Neo-Confucian scholars began to debate less about philosophical abstractions and ceremonial minutia and to attend more to researches concerning Korean history, geography, language, medicine, and agriculture. Scholarship in the latter vein, now termed Sirhak, or "Practical Learning," began to appear in the early seventeenth century and continued in vogue until the start of the nineteenth century. Bureaucratic factionalism, which from early in the dynasty had compromised the effectiveness of the central government, was largely brought under control in the late eighteenth century. Tax regulations were altered in such a way that the burden on the peasant was somewhat lightened, the revenues of the government were increased, and commercial activity throughout the nation was stimulated. Legal changes also encouraged the private production of the handcrafted articles needed by the citizenry and the government, and with the passage of time the numbers of unfree artisans under state control gradually decreased. In 1801, the central government formally freed all the slaves remaining in its possession.

The rise in attention to specifically Korean concerns that had been manifest initially in Sirhak scholarship also began, in the eighteenth century, to find expression in paintings by both professional artists and gifted *Yangban* amateurs. Increasingly, the paintings produced for Korea's cultural elite depicted unmistakably Korean landscape and genre scenes in place of the stereotypical Chinese vistas and personalities that had previously pervaded the art. A comparable transformation is apparent in the brushwork of the iron and cobalt (blue) underglaze decorations on white porcelains of the late Chosŏn period. On vessels of pre-seventeenth-century date, the brushwork tends to be staid and fairly representational in its definition of the painted designs—commonly such stock Confucian motifs as pine and bamboo. After the start of the seventeenth century, however, the brushwork became increasingly bold and expressive and the repertoire of frequently encountered designs expanded to include such dynamic forms as dragons and the Korean tiger. It is not too much to say that the painted decorations of the later porcelains are much more distinctively "Korean" than those appearing on the earlier ones. At the present time, any explanation for this change is necessarily speculative, but its coincidental occurrence with—and its underlying complementarity to—contemporary developments in *yangban* culture, such as the emergence of Sirhak and the thematic changes in painting, should not pass unnoticed. Also likely relevant in this regard is the relaxation of governmental control over white porcelain potteries in the eighteenth and nineteenth centuries, a development that resulted in the widely expanded use of their once-restricted products (fig. 4).

In spite of the positive institutional and cultural developments of the seventeenth and eighteenth centuries, many causes of social unrest and political instability remained unaddressed within Korea. The debilitating effect of these problems was exacerbated in the nineteenth century, first by the protracted domination of the central government by rapacious royal in-law families and second by the disruptive

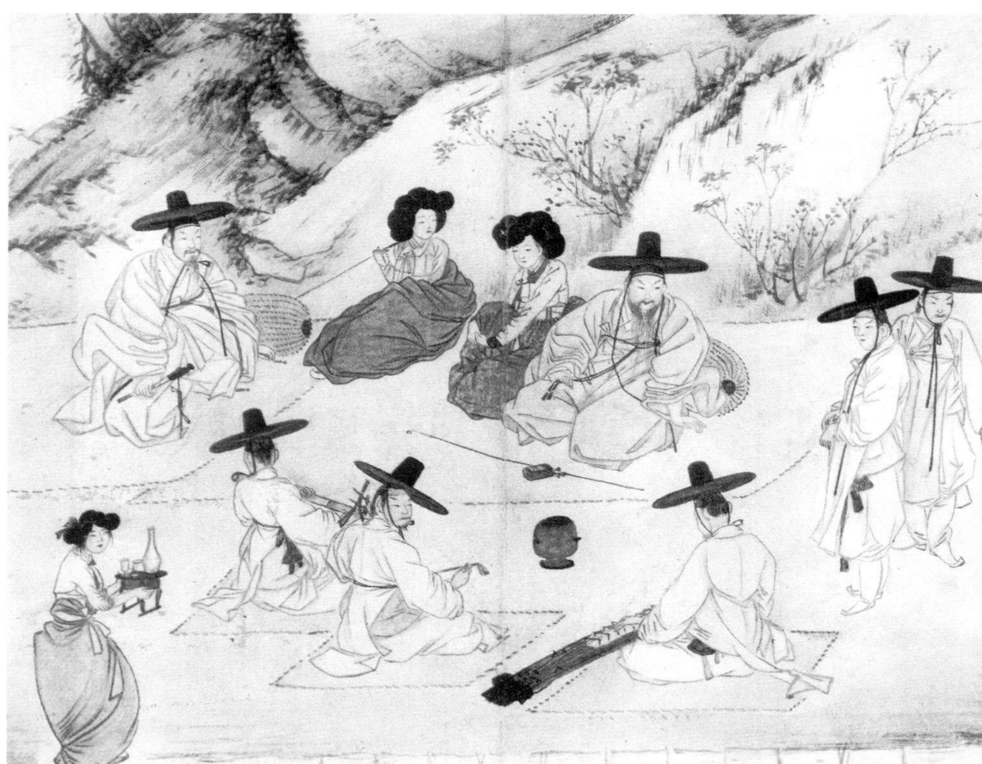

FIGURE 4
Sin Yun-bok. Outdoor Drinking Party *(album leaf), Chosŏn dynasty, late eighteenth/early nineteenth century. Colors on paper; 28.3 x 35.2 cm. Seoul, Kansong Art Museum.*

pressures exerted on the government by foreign imperialist powers. At the start of the twentieth century, political and economic conditions on the peninsula had deteriorated to such an extent that the Korean people found themselves lacking in all but the will to resist annexation by Japan and, in 1910, the long reign of the Chosŏn dynasty came to a bitter and ignominious end with the abdication of the last Korean monarch and the establishment of the Japanese military government.

SELECTED BIBLIOGRAPHY

Buswell, R. E. *The Korean Approach to Zen: The Collected Works of Chinul.* Honolulu: University of Honolulu Press, 1983.

Chun Hae-jong. "Sino-Korean Tributary Relations in the Ch'ing Period." In J. K. Fairbank, ed., *The Chinese World Order.* Harvard East Asian Series, no. 32. Cambridge: Harvard University Press, 1968, pp. 90–111.

Cleary, J. C. *A Buddha from Korea: The Zen Teachings of T'aego.* Boston: Shambala Press, 1988.

De Bary, W. T., and Haboush, J. K., eds. *The Rise of Neo-Confucianism in Korea.* New York: Columbia University Press, 1985.

Gompertz, G. St. G. M. "Hsü Ching's Visit to Korea in 1123." *Transactions of the Oriental Ceramic Society* 33 (1960–62): 1–21.

Joe, W. J. *Traditional Korea: A Cultural History.* Seoul: Chungang University Press, 1972.

Lee Ki-baik [Yi Ki-baek]. *A New History of Korea.* Translated by E. W. Wagner. Cambridge: Harvard University Press, 1984.

Lee, P. H., ed. *Anthology of Korean Literature from Early Times to the Nineteenth Century.* Honolulu: University of Hawaii Press, 1981.

Rogers, M. A. "Cultural Relics Found off the Sinan Coast." *Orientations* 14, no. 12 (December 1983): 32–40.

Rogers, M. C. "Sung-Koryŏ Relations: Some Inhibiting Factors." *Oriens* 11 (1958): 194–203.

Koryŏ and Chosŏn dynasty ceramics

YUTAKA MINO

WHEN ONE THINKS of the arts of Korea, it is often ceramics that come first to mind. Strikingly decorated celadon-glazed wares of the Koryŏ period and the more freely stamped, incised, and slip-painted vessels of the Chosŏn—these uniquely styled, individual examples of ceramic art are among the most widely appreciated of Korean art forms. Understandably, it is in this area of ceramic manufacture that Korea can claim a significant share of its most important artistic contributions.

This exhibition focuses on the two major periods during which Korean ceramic art flourished, the Koryŏ and the Chosŏn. During the Unified or Great Silla period that directly preceded the Koryŏ, Korean potters mastered fundamental techniques of glazed ceramic manufacture, laying the foundation for the innovative styles characteristic of the ceramic pieces of later periods. Typical of the Unified Silla is a gray stoneware with a yellowish-brown or olive-green glaze. Pots of this type are usually covered with small-scale, stamped designs that repeat over the entire surface. Funerary urns ornamented with stamped designs, such as the one now in the collection of the National Museum of Korea (fig. 1), came into use as receptacles for the ashes of the dead as a result of the introduction of the Buddhist practice of cremation and reflect the growing influence of Buddhism in this period. The use of stamped decoration recurs in later centuries. In addition, the manufacture of stoneware from a native clay body and the application of a feldspathic, iron-bearing glaze to its surface point to significant technological advances in ceramic manufacture. Indeed, the very existence of stoneware demonstrates that improvements had been made in kiln design and in high-temperature firing that were to prove vital to the later development of Korean ceramics.

The Koryŏ dynasty was a golden age for ceramic making. Two major centers of production from this period are known, one located at Kangjin, near the southernmost point of the Korean peninsula, and the other at Puan, on the southeastern coast (map, p. 14). Outstanding among the wares of this period are the celadons; the covered jar in the Ataka Collection (no. 13), with its delicately incised ornament of lotuses under a blue-green glaze, is an excellent example of this type. Its lovely color derives from iron present in the glaze during firing in a strongly reductive kiln atmosphere with little oxygen present. In the early Koryŏ dynasty, potters made wares with a grayish-green glaze that imitated the glaze and overall appearance of the well-known southern Chinese Yue ware of the ninth and tenth centuries. By the eleventh century, however, Korean celadon had been perfected to the point where even the Chinese had to admire it as one of the "first under heaven," that is, as equal to the Song imperial Ru ware manufactured in northern China in the eleventh and early twelfth centuries. The more general influence of Song ceramics can also be seen in the harmonious, integrated shapes and subdued ornament of Koryŏ celadons. Some pieces are even undecorated, allowing the luminous beauty of the glaze to be fully appreciated. It is this glaze that was described by the Chinese envoy Xu Jing (1091–1153) as possessing "the radiance of jade and the crystal clarity of water."[1]

The simplicity of celadon is also emphasized in Koryŏ ceramic sculpture.

Xu Jing, who was dispatched by the Song court to visit Korea in 1123, commented upon the great skill and charm with which Koryŏ potters rendered figures of birds, animals, and people.[2] The mandarin duck seated on the cover of a three-legged incense burner (fig. 2) is a fine example of such ceramic sculpture. Another appealing figure is that of a girl designed as a water dropper (no. 1); the dark pupils of her eyes are painted iron brown under the celadon glaze.

In analyzing the evolution of Korean celadon, the influence of northern Chinese ceramic wares is usually mentioned. However, the impact of the southern wares, particularly export wares such as Yingqing (Chingbai) that could easily have been transported to Korea by sea, should not be overlooked. For instance, a Koryŏ celadon ewer with a lobed warming bowl and a carved lion on its cover (fig. 3) is strikingly similar in form to a Yingqing ewer found in a tomb, dated to 1087, in Susong in the southern Chinese province of Anhui (fig. 4). A vase in the shape of a Chinese bronze *zun* (no. 6), with its long neck, widely everted mouth, and flared foot with carved fluting, reflects a type derived from Song vessels such as the Yingqing vase illustrated here (fig. 5). The Koryŏ example varies from its Song model only in its foliated mouth rim and fine celadon glaze. Another Koryŏ bowl incised with a large peony on the interior (fig. 6) is quite similar in both shape and technique of decoration to Chinese Yingqing bowls.

Incision and inlay are two of the main forms of surface decoration utilized by the Koryŏ potter. Inlay on celadon, which is unique to Korea, is one of the Korean potter's important inventions; in this technique, designs carved or incised on the body of the pot are filled in with black-and-white slips prior to glazing and firing. One such example, a long-necked bottle with an eight-sided body decorated with inlaid floral sprays (no. 17), derives its shape from a tenth-century Yue ware bottle (fig. 7). Pots with inlaid decoration are a major category of Koryŏ celadons and are thought to have been made for the first time in the twelfth century. Inlaid celadons have also been excavated from datable tombs; the earliest known example is a bowl (fig. 8) found in a tomb believed to be dated 1159. Another inlaid bowl was discovered in the tomb of King Myŏngjong, who died in 1197.[3] Both these examples have reserved floral decoration on an inlaid white ground, in addition to inlaid black-and-white floral sprays. Some of the floral designs used on celadons are virtually the same as those seen on inlaid Korean lacquer ware; the two types of media, inlaid ceramic and lacquer, were probably worked with during the same period. Some celadons with carved openwork and inlaid decoration (see fig. 9) borrowed their ornamental technique directly from metalwork. Examples of such metalwork were discovered in a Koryŏ tomb in Yusan-ri, which yielded as well a bronze mirror, a silver-plated needle case, and a small oil bottle.

One of the most impressive and better known of the inlaid celadons is a tall *maebyŏng* ("plum vase") in the Kansong Art Museum (fig. 10). Though large in size, it is very finely worked, with the main part of its black-and-white inlaid decoration arranged in and around forty-six evenly spaced circles. Each of the circles is drawn with a black line within a white line and contains a crane flying upward and to the right among clouds. Another twenty-three cranes in the background glide downward

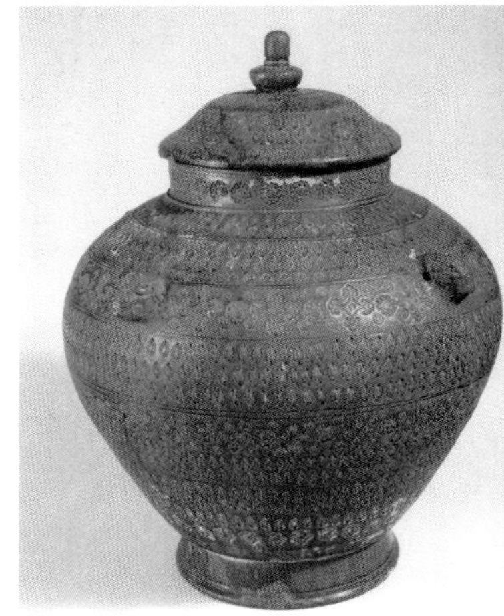

FIGURE 1
Funerary urn and cover with stamped designs, *United Silla period, eighth/ninth century. Stoneware with olive-green glaze; h. 39 cm. Seoul, National Museum of Korea. Photo: Seoul, National Museum of Korea, and San Francisco, Asian Art Museum,* 5000 Years of Korean Art, *exh. cat. (1979), pl. 19.*

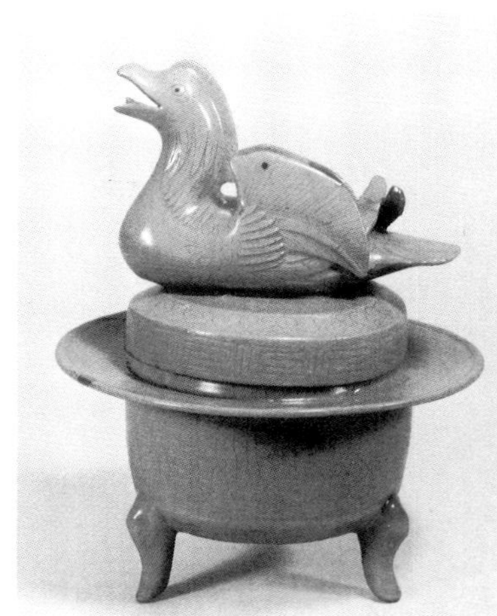

FIGURE 2
Three-legged incense burner and cover with incised design, *Koryŏ dynasty, twelfth century. Celadon; h. 23.7 cm. Osaka, Museum of Oriental Ceramics. Photo: Ikutarō Itoh,* Koryŏ Dynasty, *vol. 1 of* Masterpieces of Chinese and Korean Ceramics in the Ataka Collection *(Tokyo, 1980), pl. 133.*

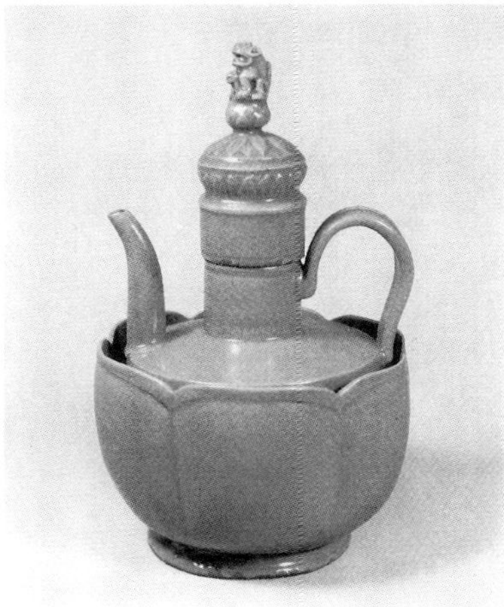

FIGURE 3
Ewer with lobed warming bowl, *Koryŏ dynasty, twelfth century. Celadon; h. 30 cm. Seoul, National Museum of Korea. Photo:* Sekai tōji zenshū *(Catalogue of world ceramics), vol. 18 (Tokyo, 1978), pl. 28.*

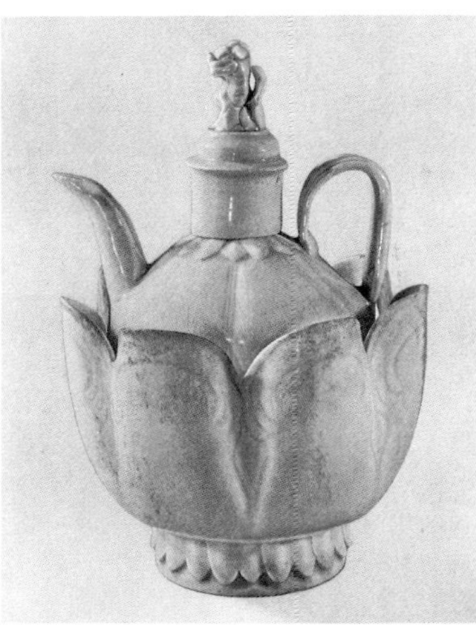

FIGURE 4
Ewer with warming bowl, *Chinese, Northern Song dynasty, late eleventh century (excavated at Susong, Anhui province, tomb dated 1087). Yingqing ware, porcelain with bluish-white glaze; h. 26.5 cm. Photo:* Wen wu *(Cultural relics) 5 (1973): pl. 1.*

to the right, also through drifting clouds. The uniform spacing of the circles and the orientation downward of all the cranes in the background suggest that a textile pattern may have served as the model for this ornamental design.[4] Koryŏ paintings provide some remarkably detailed illustrations of the textile patterns of Buddhist robes, as seen in an exhibition of Korean Buddhist paintings of that period at the Yamato Bunkakan in Nara.[5] In one painting, dated 1286 (fig. 11), the Buddha is depicted wearing a robe with a pattern of large circles filled with floral designs. On the skirt of his lower garment is a design of clouds with long, curving trails similar to the clouds on the *maebyŏng*. A detail of another Koryŏ Buddhist painting in the Shohō-ji in Kyoto (fig. 12) provides a clearer picture of this circular pattern.

Dating of these inlaid celadons is problematic. Although currently all inlaid celadons are still assigned to the twelfth and early thirteenth centuries, this now seems highly questionable. The shape of the Kansong *maebyŏng* alone suggests that it was not made in the twelfth century. The Chinese *meiping*, or "plum vase," comparable in form, is not known to have existed until much later. In fact, a blue-and-white porcelain vase of very similar shape can be securely dated to the Yuan dynasty.[6] Furthermore, recent archeological evidence indicates that the manufacture of inlaid celadons may have continued well into the fourteenth century. A bowl decorated with an inlaid design of cranes and clouds was recovered from a shipwreck found in the coastal waters near Sinan-gun.[7] The ship, whose cargo contained several pieces of Korean ceramics, as well as hundreds of Chinese ceramics, coins, and other artifacts of the Mongol period, is believed to have sunk late in that century.

The rise of the Mongols to power in Korea in the thirteenth century is generally regarded as having brought about a decline in the arts of the Koryŏ dynasty. The Koryŏ court was exiled to Kanghwa Island, off the west coast of the peninsula, for nearly three decades, until it finally surrendered to the Mongols in 1258. From this time on, Mongol influence dominated the Korean kingdom, but would not necessarily have discouraged artistic endeavor; certainly Mongol rule in China did not. It is therefore quite possible that many fine Koryŏ celadons were made throughout the thirteenth and fourteenth centuries. A thorough reexamination of the dating of these inlaid wares will eventually be necessary on the basis of the growing body of archeological material now being brought to light.

In addition to inlay work, Korean potters apparently also evolved a technique for underglaze decoration in copper red on celadon. The process involved painting in copper-oxide pigment on the body of the pot before the glaze was applied. A magnificent ewer excavated from the tomb of Choi Hang, who died in 1257, is one of the finest examples of this rare type (fig. 13). This technique seems to be a Korean invention, because its appearance in Korea predates its use on Chinese porcelain of the Yuan dynasty.

Although celadons are by far the largest group of Koryŏ ceramics, other wares were also produced; one small group consists of white porcelaneous ware and a second of transparently glazed stonewares with inlaid and painted decoration. The second group of non-celadon wares is of a less refined type, with a grayish stoneware body decorated in either black or white inlay or painted in black under an olive-tinged transparent glaze. This type of inlaid ornament achieves a variety of effects quite distinct from those of inlays used on fine celadon. In a pear-shaped bottle with carved black-slip designs on an inlaid white-slip ground (no. 28), the overall effect of the slip decoration recalls that of wares with cut-glaze decoration from Quanzhou in Fujian province in China.[8] In Korea, this type of slip-decorated pottery is known to have been made at the Sadang-ri kilns in Kangjin.[9]

In Koryŏ ceramics, the effect of painted decoration under a transparent glaze is not unlike that of some Chinese *meiping* of the Cizhou type with sgraffito decoration in black slip over white slip. Indeed, this entire group of transparently glazed Koryŏ wares is frequently compared to ware of the Cizhou type from northern China. There is little evidence, however, of any direct influence from the Cizhou kilns. Many painted Koryŏ pieces actually show a greater degree of similarity in style to wares from Xicun, in the vicinity of Canton in the far south of China.[10] Like the patterns on the Koryŏ pieces, ornament on the wares from Xicun is painted directly on the clay body. Decoration on the surface of Cizhou pieces, on the other hand, is applied over the coat of white slip covering the body of the pot.

The Koryŏ period came to an end soon after the collapse of Mongol power in China. One of the Koryŏ generals, Yi Sŏng-gye (1335–1408), seized power and established a new dynasty formally named Chosŏn, but sometimes referred to as Yi after

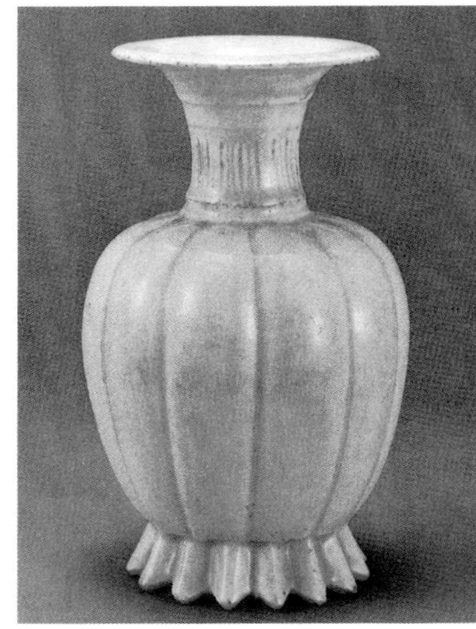

FIGURE 5
Lobed vase, *Chinese, Northern Song dynasty, late eleventh century. Yingqing ware, porcelain with bluish-white glaze; h. 17.6 cm. Seoul, National Museum of Korea. Photo:* Sekai tōji zenshū *(Catalogue of world ceramics), vol. 12 (Tokyo, 1977), pl. 31.*

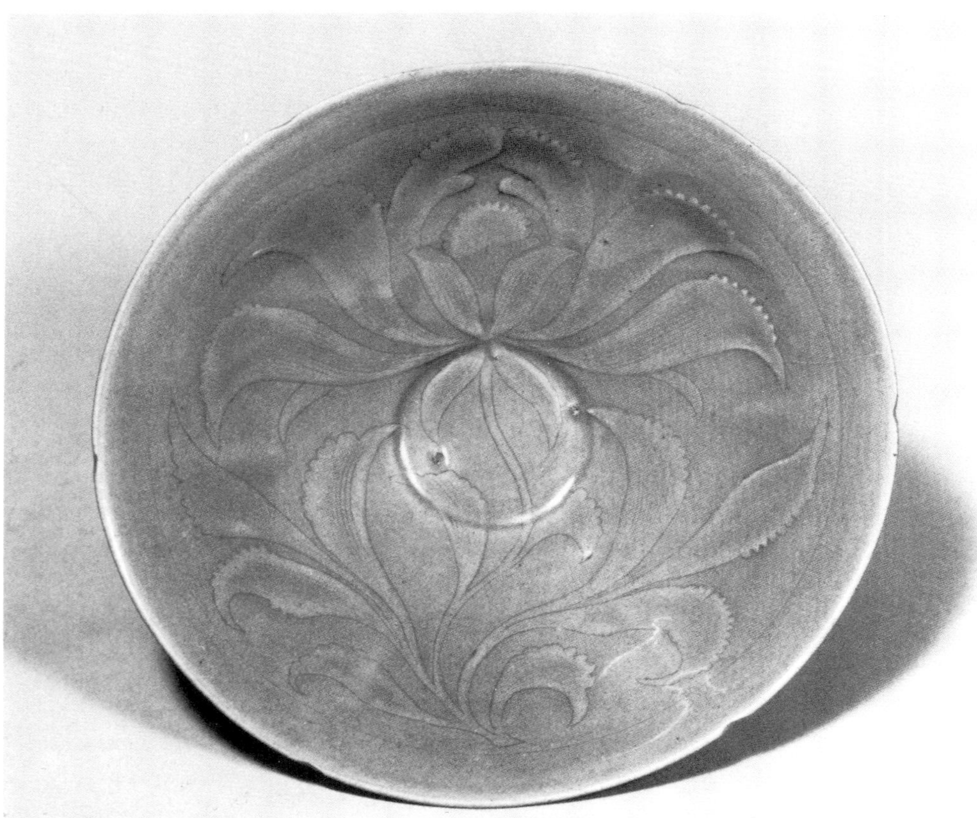

FIGURE 6
Bowl with peony-scroll design, *Koryŏ dynasty, twelfth century. Celadon; h. 6.6 cm. Osaka, Museum of Oriental Ceramics. Photo: Ikutarō Itoh,* Koryŏ Dynasty, *vol. 1 of* Masterpieces of Chinese and Korean Ceramics in the Ataka Collection *(Tokyo, 1980), pl. 199.*

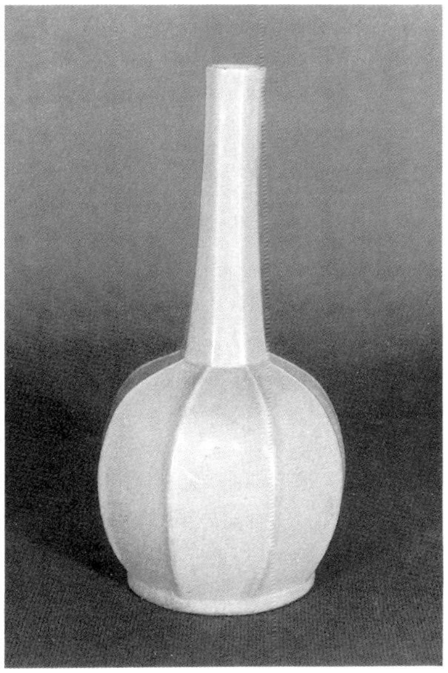

FIGURE 7
Long-necked bottle, *Chinese, Five Dynasties, tenth century. Yue ware, celadon; h. 21.5 cm. Photo:* Nisennen no bi *(Two thousand years of Chinese art) (Tokyo, 1965), pl. 12.*

his family name. The very long-lived Chosŏn dynasty, which lasted from 1392 to 1910, made Neo-Confucianism the official state philosophy and banned Buddhism in the capital and court. Up to the seventeenth century, this state maintained a close relationship with China, which was under Ming rule. But the Japanese and Manchu invasions of the early seventeenth century disrupted the economic and political life of Korea, forcing it to turn inward and draw on its own resources.

During the Chosŏn period, two basic types of ceramics were manufactured, one a robust stoneware called Punch'ŏng ware and the other a fine white porcelain. Punch'ŏng ware carried on the techniques of inlay and underglaze painting in iron pigment on a gray stoneware body that were first used in Koryŏ ceramics. Inlay work on Punch'ŏng ware was frequently simplified through the introduction of stamped decoration. The inlaying was usually done in white slip only, and the general appearance of these vessels was less refined than that of the Koryŏ wares. Small dots, circles, and florets were stamped close together, producing an effect reminiscent of the stamped wares of the Unified Silla period. Sometimes the stamped ornament was applied in dense rows over broad areas of the vessel, which created an effect rather like printing. In fact, the Japanese name for Punch'ŏng ware, which is Mishima, seems to derive from the resemblance between this type of stamped ornament and the finely printed almanacs compiled at the Mishima Shrine in Shizuoka Prefecture. Occasionally, a dark-slip inlay was used on a white-slip ground. The same style of incised and

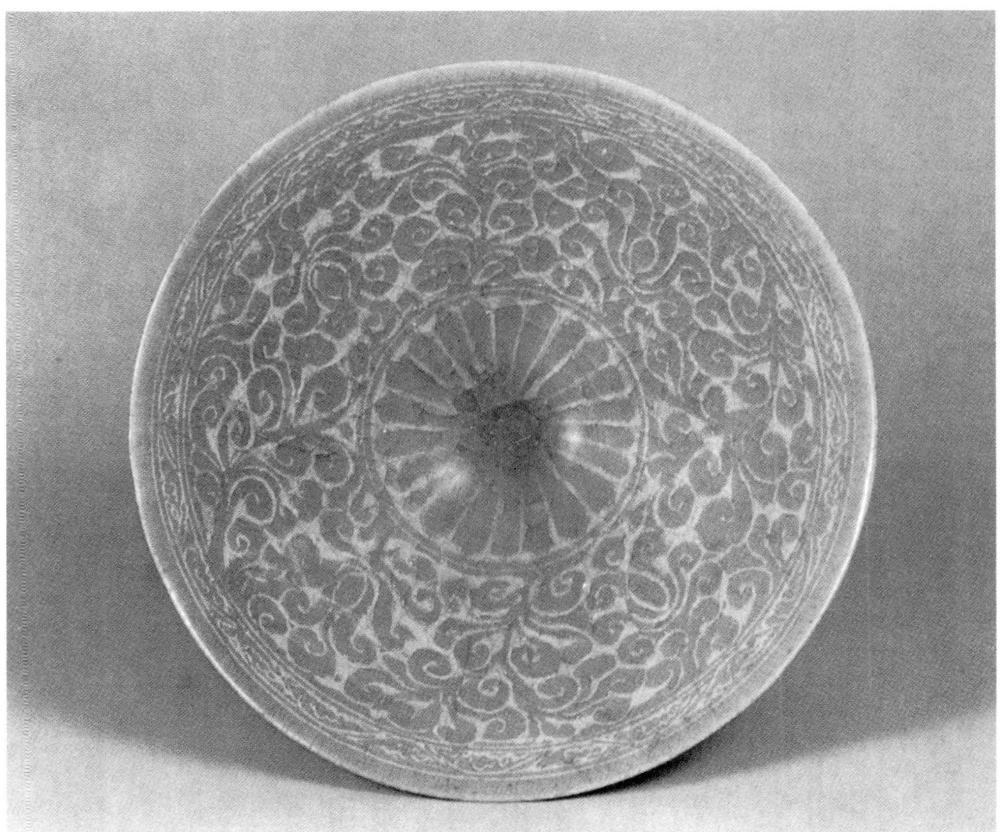

FIGURE 8
Bowl with inlaid peony-scroll design, *Koryŏ dynasty, twelfth century. Celadon. Seoul, National Museum of Korea. Photo:* Sekai tōji zenshū *(Catalogue of world ceramics), vol. 18 (Tokyo, 1978), pl. 50.*

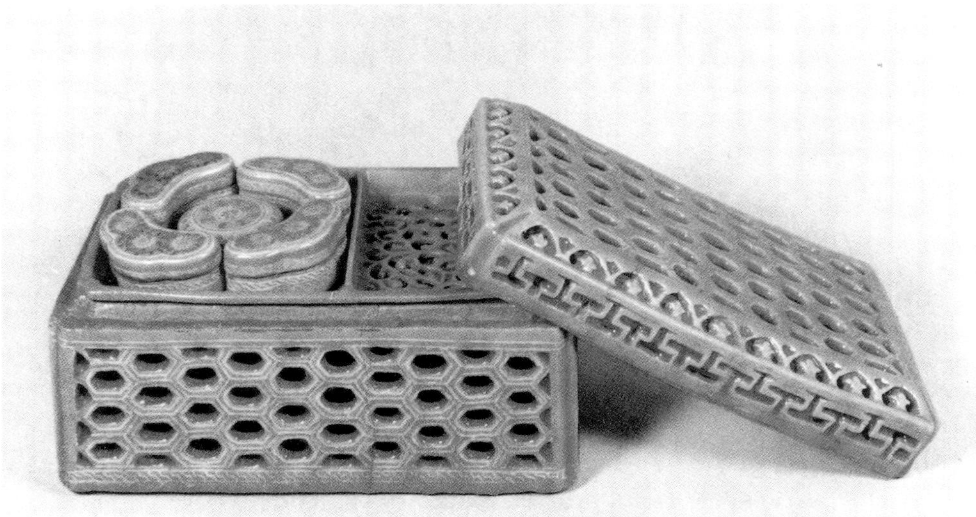

FIGURE 9
Rectangular toilet box with openwork and inlaid design, *Koryŏ dynasty, twelfth century. Celadon; h. 12.1 cm. Seoul, National Museum of Korea. Photo:* Sekai tōji zenshū *(Catalogue of world ceramics), vol. 18 (Tokyo, 1978), pl. 67.*

inlaid ornament can be seen on a mortuary tablet from the tomb of a man named Chung, who died in 1466.[11] Other examples of Punch'ŏng ware were not inlaid but simply incised.

The development of Japanese ceramics was almost as strongly influenced by Chosŏn dynasty wares as Koryŏ ceramics were by their Chinese models. The Japanese admired Chosŏn dynasty ceramics so much that they imported Korean potters as well as Korean pots. Punch'ŏng, or Mishima, ware and a group of rough, utilitarian bowls known in Japanese as Ido were especially prized because they were regarded as suitable for the tea ceremony. By the late sixteenth century, Korean ceramics were so desirable in Japan that Toyotomi Hideyoshi (1536–1598), its de facto military ruler, dispatched two military expeditions to Korea in 1592 and 1597. Part of the purpose of the expeditions was to raid the Korean kilns and capture potters, who were then taken back to Japan to build up the native ceramic industry.

The other group of ceramic wares made in the Chosŏn period consists of porcelains that were either plain white or were decorated with underglaze painting. The earliest white porcelains imitate the rather rigid forms of bronzes, while those of the later periods have more graceful, flowing shapes. The peak of refinement in the manufacture of plain white porcelain occurred in the seventeenth and early eighteenth centuries. At the highest point of development of these vessels, both their form and the luminous quality of their white surfaces evoke the radiant beauty of the moon. Slight imperfections in the glaze and body and some unevenness in shape, which are characteristic, do not detract from this beauty but endow it with warmth and spontaneity.

The development of plain white porcelain is paralleled by that of porcelain decorated with underglaze painting. The earliest blue-and-white porcelain made in Korea dates from the fifteenth century and reflects the influence of Ming China. The motif of bamboo, pine, and plum, which was known in China as "the three friends of winter," expresses Confucian ideals of scholarly fortitude and purity. Frequently seen on

FIGURE 10
Maebyŏng with inlaid design, *Koryŏ dynasty, thirteenth century. Celadon; h. 42 cm. Seoul, Kansong Art Museum. Photo:* Sekai tōji zenshū *(Catalogue of world ceramics), vol. 18 (Tokyo, 1978), pl. 61.*

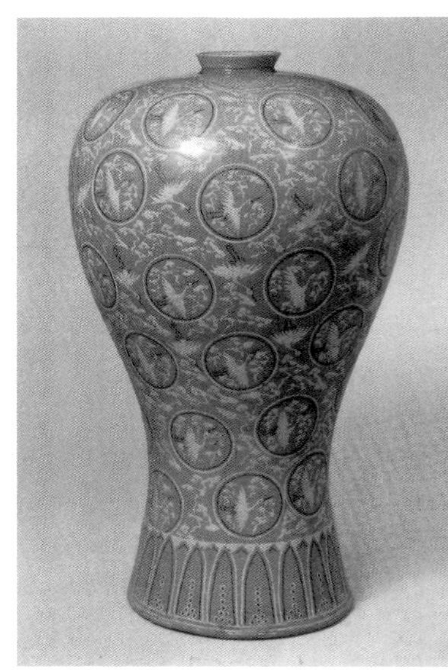

FIGURE 11
Standing Buddha, *Koryŏ dynasty, dated 1286. Ink and colors on silk; 203.5 x 105 cm. Photo: Yutaka Mino, "A Re-examination of an Inlaid Celadon* Maebyŏng *Decorated with Cranes and Clouds,"* Archives of Asian Art *39 (1986), p. 72, fig. 2.*

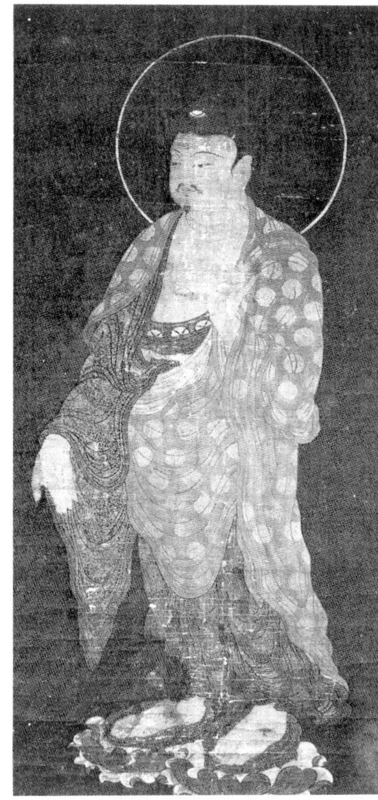

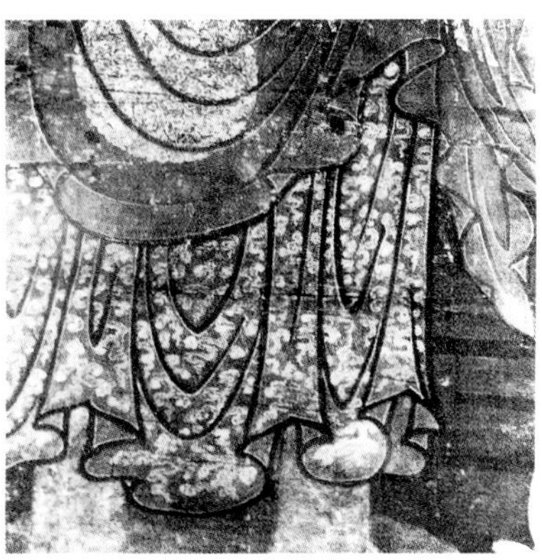

FIGURE 12
Standing Buddha, *Koryŏ dynasty, thirteenth century (detail). Ink and colors on silk; 184.5 x 86.5 cm. Kyoto, Shobō-ji. Photo: Nara, Yamato Bunkakan,* Korai Butsuga *(Korean Buddhist paintings of the Koryŏ dynasty), exh. cat. (1978), pl. 26.*

Chinese porcelain from the Xuande period (1426–35), the motif was imitated by Korean potters on their blue-and-white ceramics (see no. 85). In the Chosŏn dynasty, it also became common to use Chinese cyclical dates and imperial reign marks.

The Japanese invasions of 1592 and 1597 damaged the official kilns and disrupted their activity. When the kilns were restored, a new period of activity began that lasted from 1625 to 1751. During this time, some of the finest and most distinctly Korean porcelains were made. Among the loveliest of these are jars decorated with delicately painted flowering plants, a motif called "autumn grasses" (Kor. *chucho*) especially prized in Japan. Perhaps the most famous example is a piece from the Ataka Collection, an eight-sided blue-and-white jar with floral decoration arranged within four lobed panels (no. 86). The cobalt pigment of this period is typically blurred and of a pale grayish-blue hue tinged with lavender, and the glaze on the warm-toned body is thick and slightly bluish in color. Underglaze iron brown was also used in the decoration of porcelain during the seventeenth through nineteenth centuries, as seen on the large jar ornamented with a painted tiger (no. 87). A popular subject in painting, tigers were depicted in numerous Korean scrolls of the period. Finally, underglaze copper-red painting reappeared in the eighteenth century after an absence of about five centuries. This technique is represented here by a jar ornamented with a large lotus flower (no. 98).

From 1752 on, in the late Chosŏn period, the official kilns were in Punwon-ri, to which they had been relocated from Kwangju. Ceramics of this epoch are somewhat less strong in their designs and less refined in their decoration. For example, on the eight-sided bottle included here (no. 78), the painted decoration has been applied with broader strokes of the brush; the blue is also not as subtle, but darker and slightly blackish in tone. Another wide-mouthed jar (no. 88), painted with tigers and magpies in a landscape, has a large ovoid body; here, too, the surface design and the tone of the cobalt-blue pigment are not as understated as they were in ceramics of the earlier Chosŏn period.

Korean ceramics play an important role in the history and development of ceramics in East Asia. From Chinese origins, Korean ceramicists developed their own techniques and styles. By drawing upon past ceramic methods and combining them with novel techniques introduced from China, Korean potters were eventually able to develop styles whose aesthetics were unique to Korea. Native Korean techniques may at points have influenced the development of ceramic art in China. In Japan, where the aesthetic preference for certain Korean ceramic techniques and styles was quite pronounced, the impact of Korean ceramics is clear. The formation of the Ataka Collection itself stands as a tribute to the Japanese admiration for Korean ceramic art.

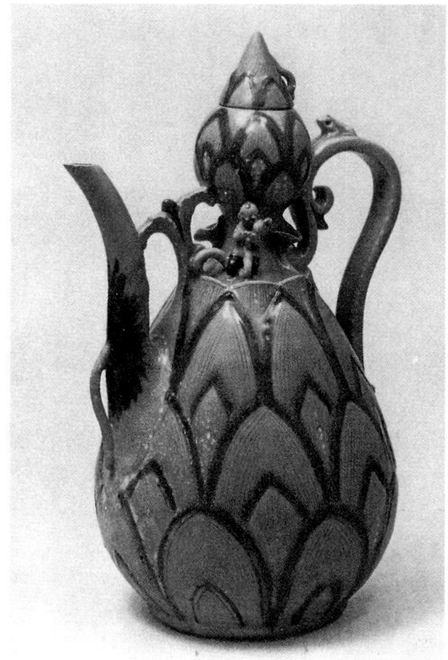

FIGURE 13
Ewer in the shape of a lotus bud with inlaid design, *Koryŏ dynasty, mid-thirteenth century. Celadon with copper-red underglaze; h. 32.7 cm. Kyŏnggi province, Yongin-gun, Hoam Art Museum. Photo:* Sekai tōji zenshū *(Catalogue of world ceramics), vol. 18 (Tokyo, 1978), pl. 85.*

NOTES

1. Xu Jing, *Xuanhe fengshi Gaoli tujing* (Illustrated description of the Chinese embassy to Korea during the Xuanhe period; 1124), in *Congshu jicheng chupien* (Primary edition of collected works of ancient texts; 1937), chap. 32, p. 109.

2. Ibid., p. 110.

3. Nomori Takeshi, *Korai tōji no kenkyū* (Study of Korean ceramics) (Tokyo, 1972), pp. 28–29.

4. Yutaka Mino, "A Re-examination of an Inlaid Celadon Maebyŏng Decorated with Cranes and Clouds," *Archives of Asian Art* 39 (1986): 71–74.

5. Nara, Yamato Bunkakan, *Korai Butsuga* (Korean Buddhist paintings of the Koryŏ dynasty), exh. cat. (1978).

6. Yabe Yoshiaki, *Gen no Sometsuke* (Blue-and-white wares in the Yuan dynasty), *Tōji taikei*, vol. 41 (Tokyo, 1974), pl. 39.

7. Seoul, National Museum of Korea, *Special Exhibition of Cultural Relics Found off the Sinan Coast*, exh. cat. (1977), pl. 2.

8. *Wenwu* (Cultural relics) 1 (1973): 63.

9. Nomori (note 3), pp. 56–57.

10. *Guangzhou Xicun gu yao yizhi* (The ancient kiln site of Xicun in Guangzhou) (Peking, 1958).

11. *Sekai tōji zenshū* (Catalogue of world ceramics), vol. 19 (Tokyo, 1980), pl. 152 (Kyŏnggi province, Yongin-gun, Hoam Art Museum).

The ceramics

Chronology

KOREA

Three Kingdoms:	
Koguryŏ dynasty	37 BC–AD 668
Paekche dynasty	18 BC–AD 663
Old Silla dynasty	57 BC–AD 668
Unified Silla dynasty	668–935
Koryŏ dynasty	918–1392
Chosŏn (Yi) dynasty	1392–1910

CHINA

Shang dynasty	c. 1700–1050 BC
Zhou dynasty	c. 1050–221 BC
Qin dynasty	221–207 BC
Han dynasty	206 BC–AD 220
Six Dynasties	221–589
Sui dynasty	589–618
Tang dynasty	618–906
Five Dynasties	907–960
Song dynasty	960–1279
Yuan dynasty	1260–1368
Ming dynasty	1368–1644
Qing dynasty	1644–1912

Editor's note

Unless indicated otherwise, when only a single measurement is given for an object, that dimension is its height. If two measurements are cited, it is the height and width or width and depth that are being provided.

The Japanese authors whose work is cited here are listed, in western practice, with surnames last, if the publication is in English; and, in eastern practice, with surnames first, if the publication is in Japanese.

Many Koryŏ celadon wares modeled on human figures, animals, and plants are known, particularly among those pieces dated to the first half of the twelfth century, a period during which Koryŏ celadon glaze was at its most beautiful. In Korea, this kind of celadon ware is called 'celadon that imitates life"; a highly representative example of this type is the duck-shaped water dropper in the Kansong Art Museum, Seoul.

The water dropper illustrated here is made in the shape of a young girl sitting with knee bent, holding a water bottle in her arms. Her topknot has been made into a lid, which can be removed to fill the dropper with water. The water bottle cradled in her arms then acts as the spout. With the exception of touches of iron dust to indicate the girl's eyes, her entire costume is decorated with finely incised designs. She wears a shirt with a three-petaled floral design, over which is an embroidered vest with a cloud and classical scroll design. The deliberately artless manner in which the floral pattern of the shirt naturally continues onto the front of the water bottle is a stylistic feature seen from time to time during the Koryŏ period. The glaze is of a translucent grayish-green color, except for a slightly burnt area on the lower part of the girl's body. The vessel's flat base is glazed as well and bears five spur marks.

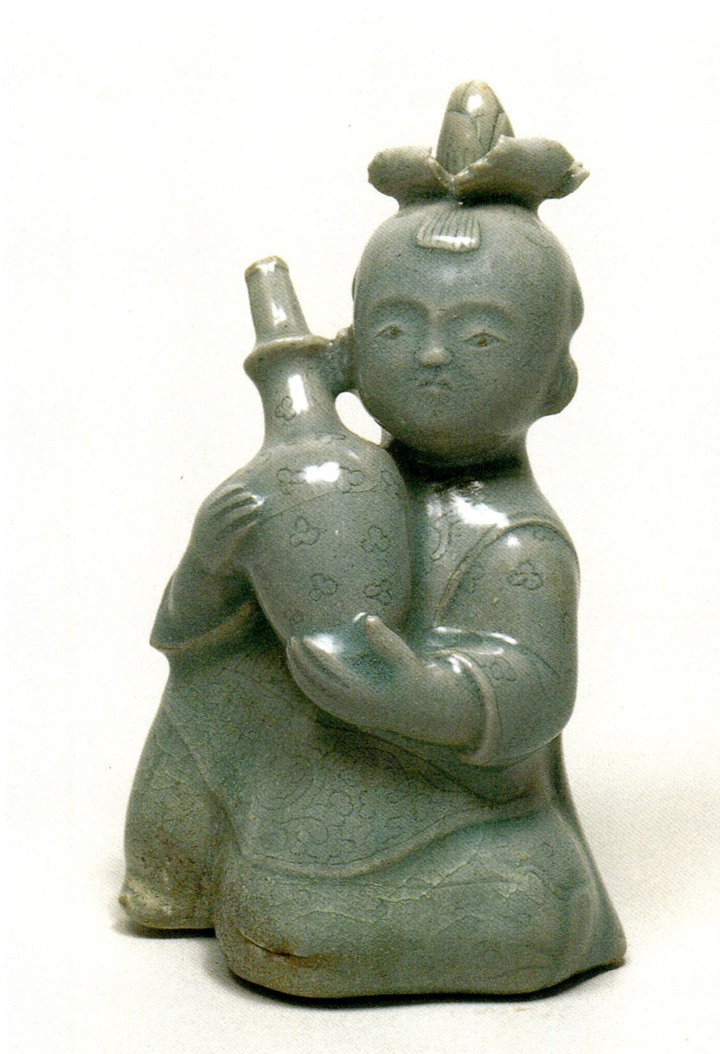

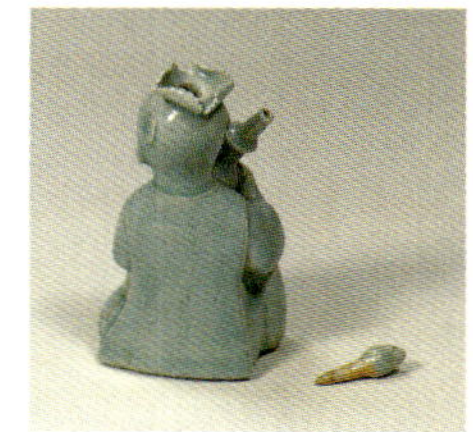

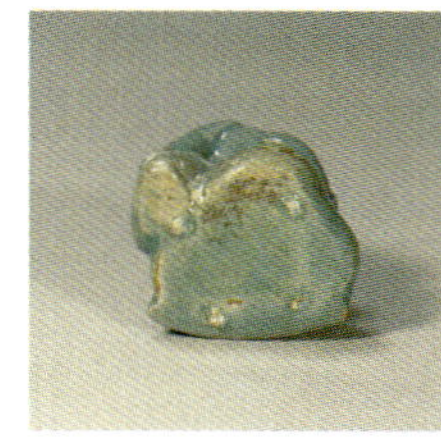

1

Water dropper in the shape of a girl

Celadon

Koryŏ dynasty, first half of the twelfth century

11.1 cm

Acc. no. 20171

Important art object

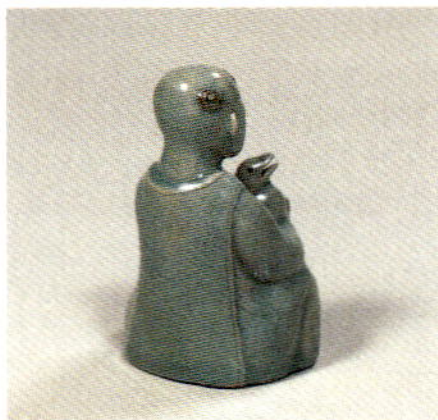

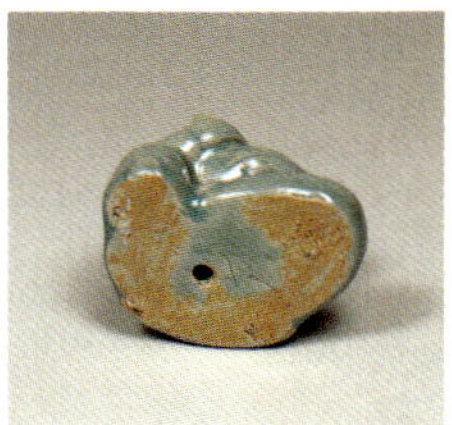

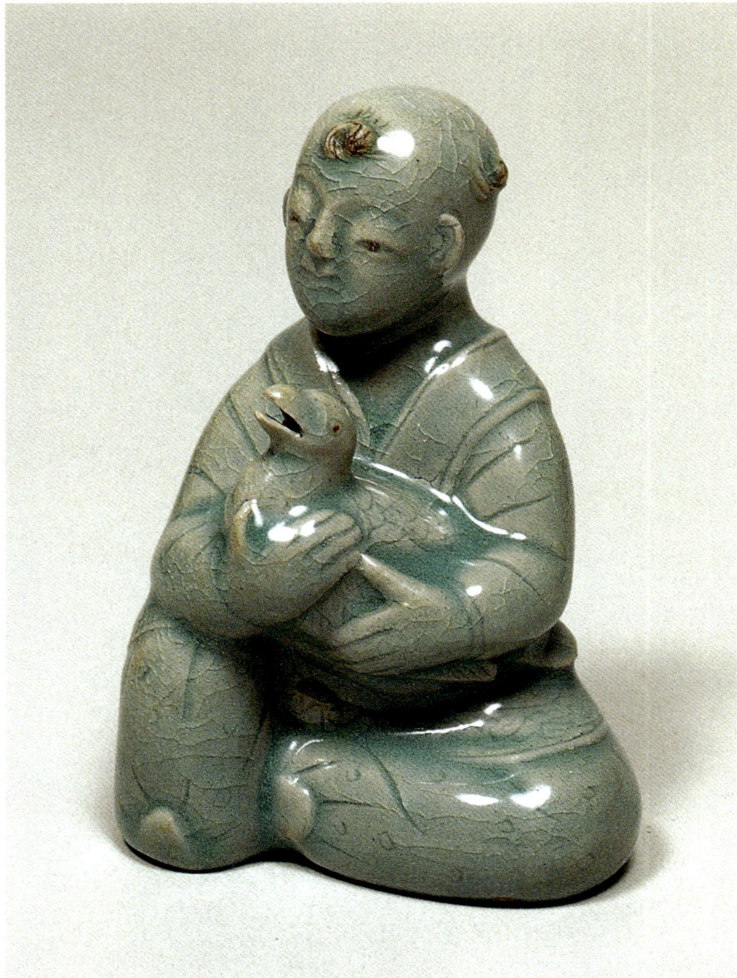

2

Water dropper in the shape of a boy

Celadon

Koryŏ dynasty, first half of the twelfth century

11 cm

Acc. no. 21131

This water dropper is in the shape of a boy holding a duck. It is thought to have been made at about the same time and in the same studio as the water dropper in the shape of a young girl in no. 1. Although the two pieces are similar in size and workmanship, the glaze on this example is thicker and the greenish color somewhat darker. Similar celadon figures have been excavated at the Yaozhou kiln site in China, but those pieces were mass-manufactured from molds. This Koryŏ figure was not molded but was instead carefully potted and is a unique work. Its base has a suction hole; water rises through a vertical tube inside the body of the vessel and the overflow remains in the vessel's cavity. Water is then poured from the mouth of the duck held in the boy's arms. This system of construction for the intake and outflow of water continued into the first half of the Chosŏn period in Korea. The same design is seen, for example, in the Chosŏn water dropper in no. 45.

Historical evidence clearly indicates that celadons of the first half of the twelfth century, with their beautiful glazes and superb workmanship, were extremely popular. The water droppers illustrated here and in no. 1 support this. The glaze applied to the base of this vessel has been in large part scraped off; four spur marks can be discerned.

Many Koryŏ celadon incense burners assume the form of Chinese bronze vessels. This incense burner, for example, is modeled after a Chinese bronze *ding* (Kor. *chong*). It stands on three tall legs and has two loop handles on the rim of its body. The body is separated into upper and lower sections by a deep horizontal groove, and these areas are further divided by six raised ridges. The surface is decorated overall with a key-fret pattern; on the upper register is a raised dragon design, and on the lower half a *taotie* (Kor. *hochan*) mask is visible. Both impressed designs, however, have been simplified and are not as accomplished as the Chinese bronze decoration from which they were derived. The legs of the vessel and the two handles are also covered with more simplified incised designs. The body is somewhat roughly formed as well, and the tall legs are attached to it in an unbalanced manner. But what compensates for the incense burner's deficiencies and serves to make it appealing is the beauty of its clear, grayish-green glaze. Only in Koryŏ celadon ware are such incense burners with tall legs known; examples are not even found among Chinese celadons. On the underside of each leg are small holes, suggesting that a special tool was used to permit hot air to escape from the hollows of the legs during the firing process.

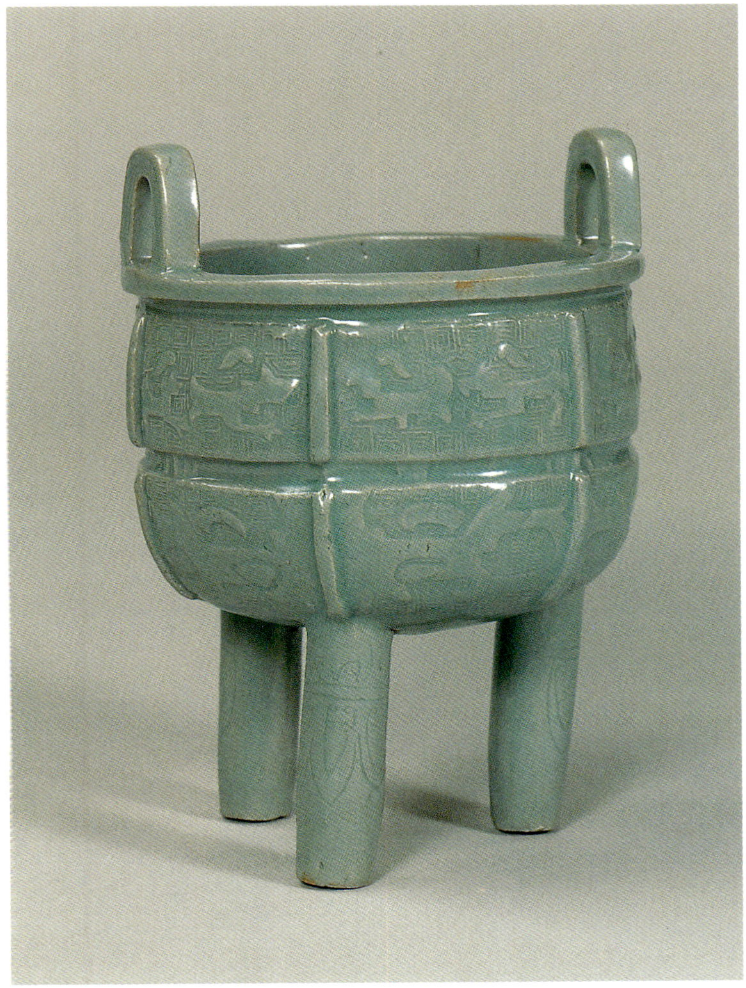

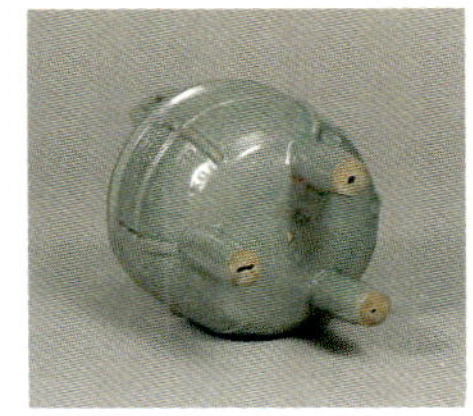

3

Incense burner with an impressed taotie *design*

Celadon

Koryŏ dynasty, first half of the twelfth century

17 cm

Acc. no. 20358

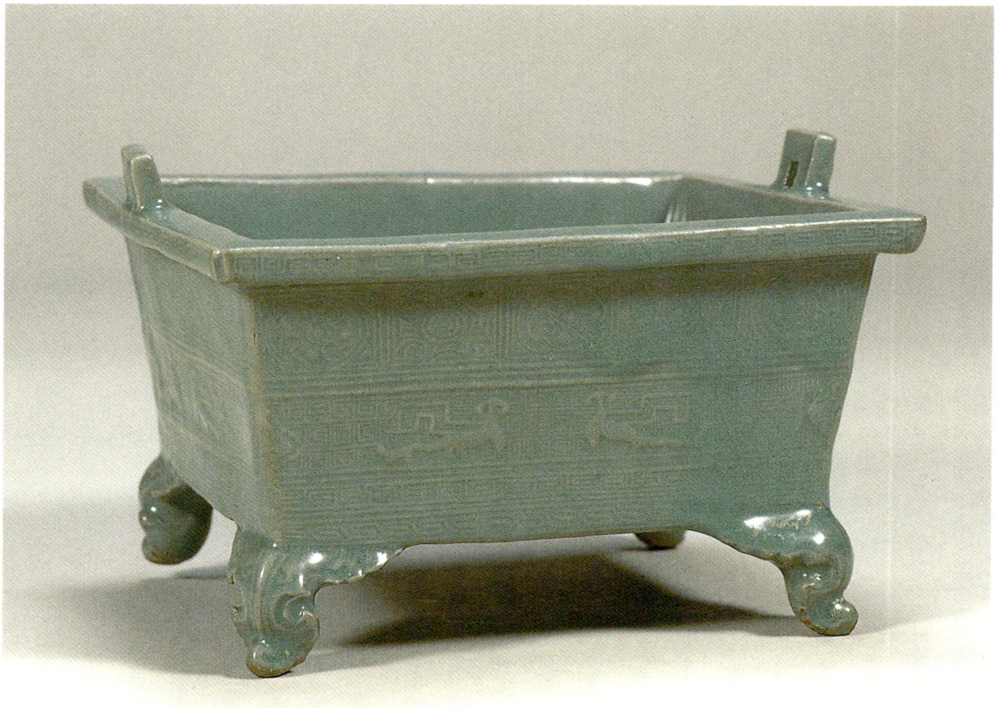

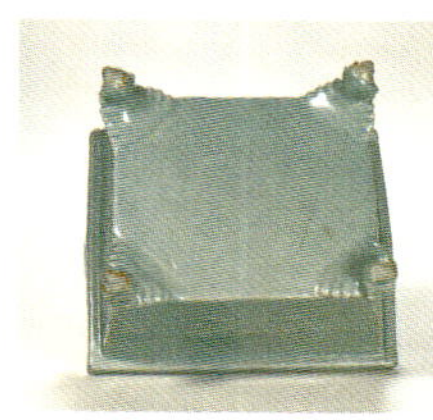

4

Incense burner with an impressed dragon design

Celadon

Koryŏ dynasty, first half of the twelfth century

11.9 cm

Acc. no. 20176

Important art object

Like the incense burner in no. 3, this piece is also modeled after a Chinese bronze; its date of production is thought to be the first half of the twelfth century. Shards of a similar type were discovered at the no. 7 kiln site at Sadang-ri, Taegu-myŏn, Kangjin-gun, South Chŏlla province. The impressed designs found on the four sides of the body are organized into three horizontal bands. The uppermost band is divided into five areas of geometric designs; the middle band has a mythical bird design on a key-fret-pattern ground; and the lowest band is ornamented with two kinds of interlocking key-fret patterns. All of these designs are borrowed from Chinese bronzes but have been modified through adaptation to Koryŏ style. The mouth of the body has an everted, flat rim, the top of which is delicately incised with a classical scroll design. The sides of the rim and the exteriors of the two loop handles are decorated with impressed key-fret patterns. In comparison with the designs on the incense burner in no. 3, these are somewhat more carefully rendered. The thickness of the glaze produces a strong green cast characteristic of its so-called jade color. Only a few square incense burners are known to exist; two such pieces are found in the Museum of Fine Arts, Boston, and the Kansong Art Museum, Seoul. The bottoms of the legs on this piece are unglazed.

With its trumpet-shaped mouth, slender neck, ovoid body, and large, low foot, this undecorated bottle displays a well-balanced shape. Examples of undecorated bottles of this sort are found in Chinese imperial Ru ware, a type that probably influenced this bottle. The influence of Ru on Koryŏ celadon was, in fact, noted by Xu Jing, a member of a Chinese delegation that visited the Koryŏ capital in 1123, in his *Xuanhe fengshi Gaoli tujing*. Even the treatment of the base of the bottle verifies this observation by reflecting the characteristics of Ru ware. The entire base is carefully finished, and small spur marks are discernible in three places. On the basis of this evidence, it seems clear that this piece can serve as a typical example illustrating the connection between Koryŏ celadons and Chinese Ru ware. However, the glaze on this bottle is somewhat thin when compared to that of Ru ware. Consequently, the celadon is a highly transparent grayish blue a hue that, in Koryŏ, according to Xu Jing, was called "jade-colored."

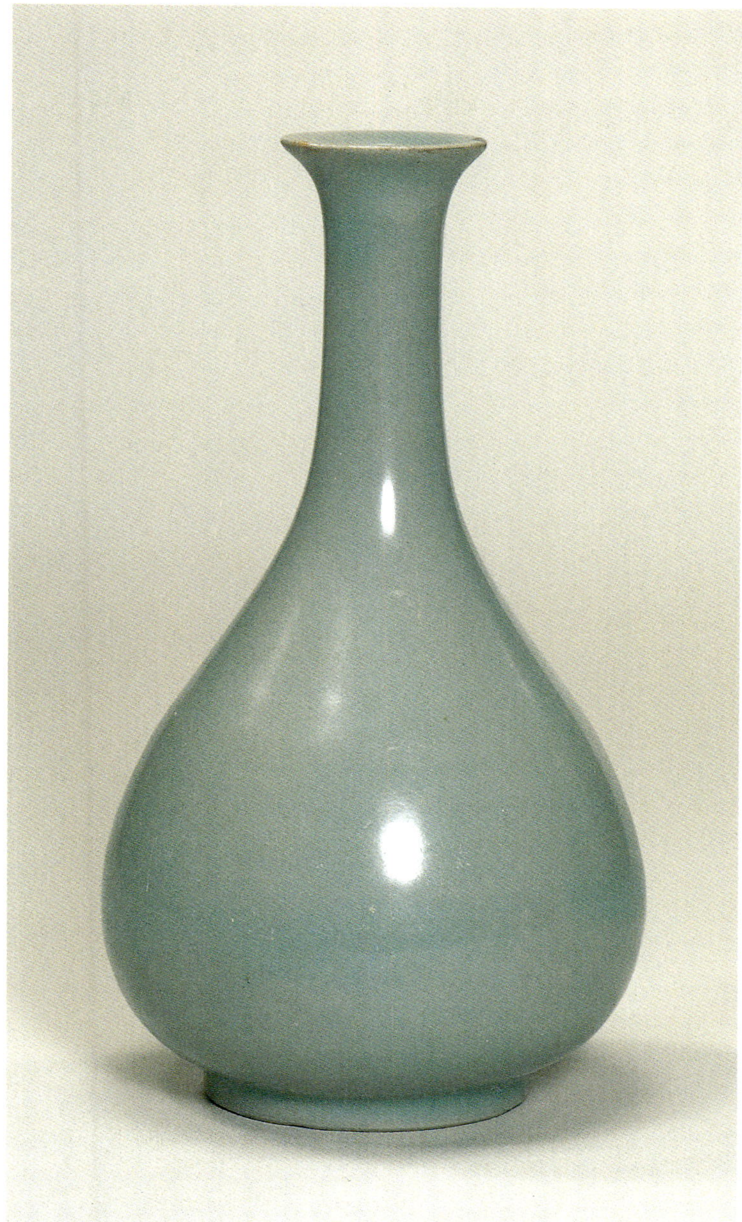

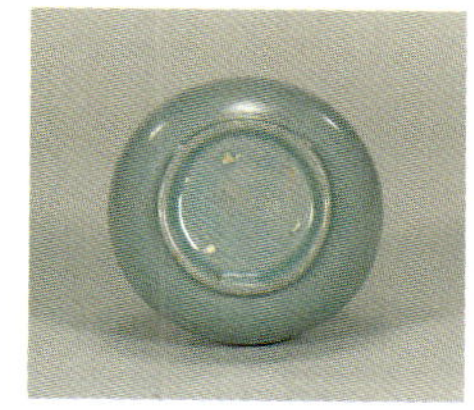

5

Bottle

Celadon

Koryŏ dynasty, eleventh/twelfth century

20.4 cm

Acc. no. 20244

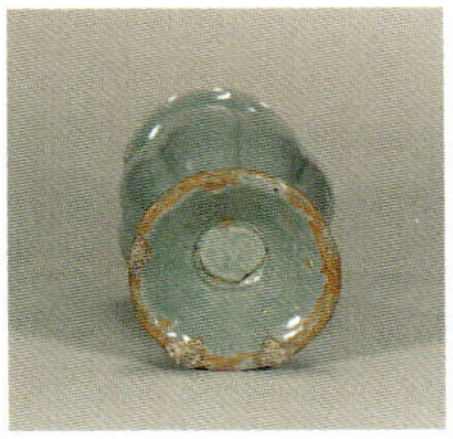

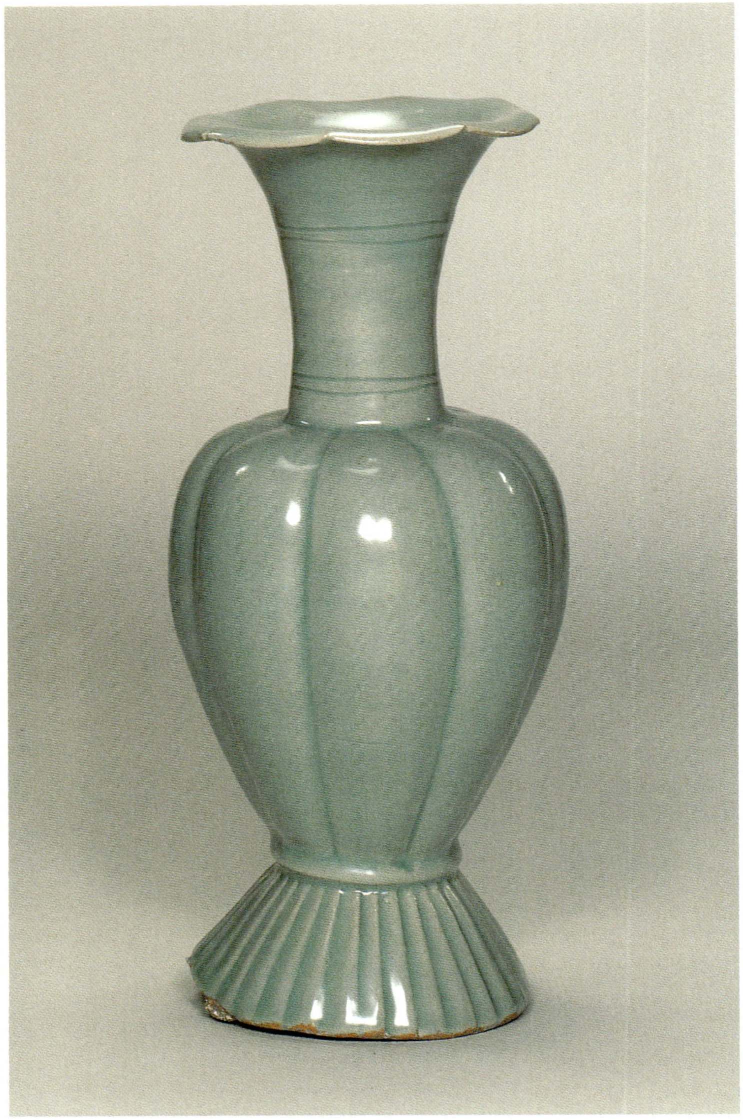

6

Lobed vase

Celadon

Koryŏ dynasty, first half of the twelfth century

25.7 cm

Acc. no. 20483

This vessel has a trumpet-shaped mouth and a lip in the form of an eight-petaled flower. The body, an elongated melon shape divided into eight vertical lobes, is exquisitely curved. The lip, body, and fluted foot are all thought to have been made separately and to have been assembled into one piece later.

This bottle is very similar to some other, rare examples of Koryŏ celadon ware dated on the basis of documents. The dated pieces, now in the collection of the National Museum of Korea, Seoul, were excavated from the tomb of King Injong (r. 1122–46), the seventeenth king of the Koryŏ dynasty, along with documents dated 1146. Therefore, this vessel too is important and datable material, and, as such, shares certain characteristics with other examples of Koryŏ celadon ware of the first half of the twelfth century. These include no decoration; a beautiful grayish-blue glaze of the so-called jade color; no crackle; a beautiful shape; and meticulous attention to the creation of an individual piece. On the bottom of the vessel, six spur marks can be seen.

In Japan, vessels with a flat mouth, long neck, and vertical, tubular body are called fulling-block-shaped after the wooden tool (Jap. *kinuta*) used to pound cloth in washing. Celadons of a similar shape, at least the glaze of celadon wares of that shape, are called *kinuta*-type celadons. In China, such bottles appeared in the Northern Song period and were made extensively until the Yuan. In Korea, numerous examples of such Koryŏ celadon wares date from the late eleventh through the first half of the twelfth century.

This bottle has no decoration, its shape is simple and elegant, and its glaze is exquisite—all characteristics that indicate its similarity to Chinese Ru ware. Even the treatment of the base of the vessel, which is flat, glazed, and bears four small spur marks, is similar to examples of Ru ware. The completely uniform and perfect green- and grayish-blue hue of the glaze is typical of wares of the jade color. At least ten other similar pieces are known, including one in the Fitzwilliam Museum in Cambridge, England, but the piece illustrated here is of a larger size than the other examples in this group.

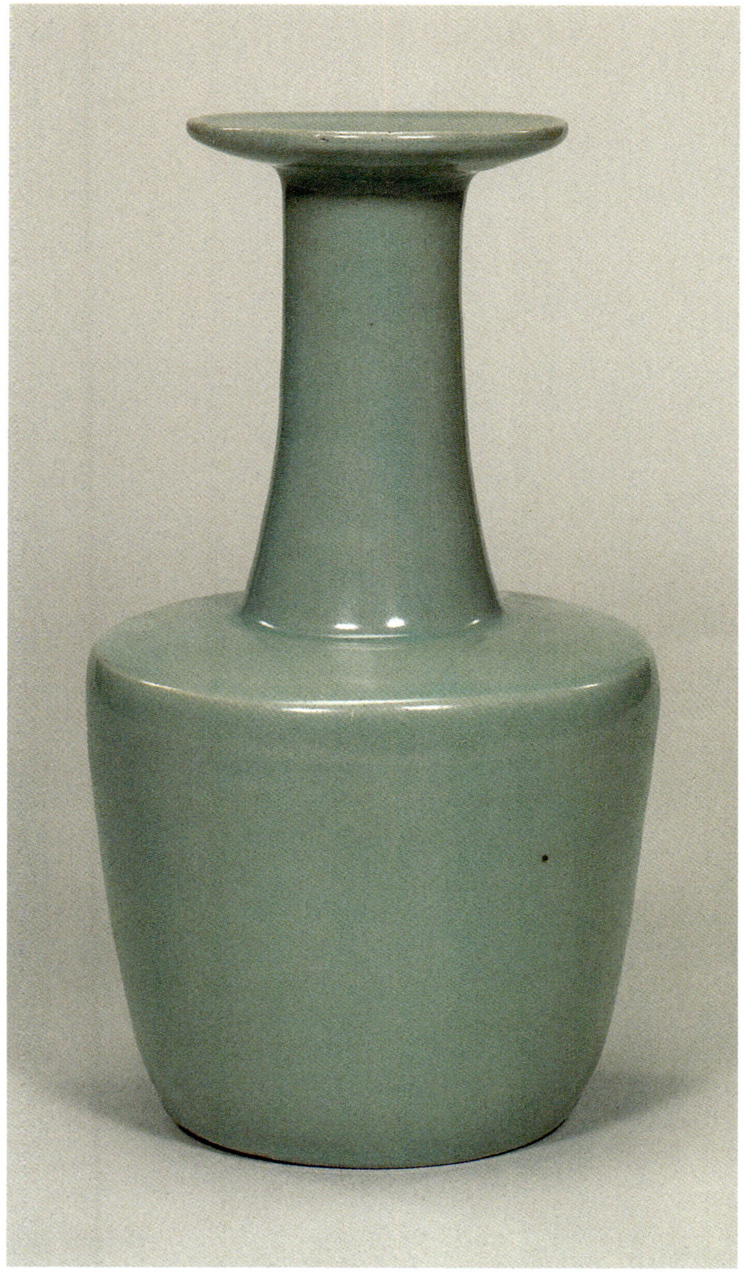

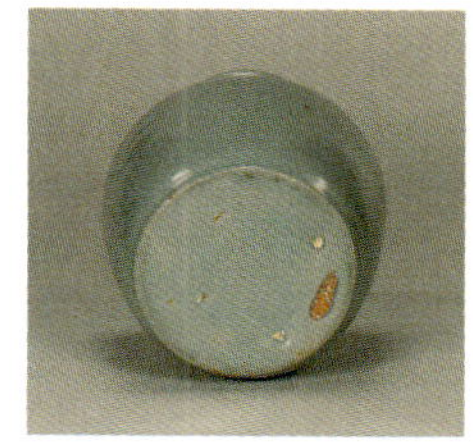

7

Bottle in the shape of a fulling block

Celadon

Koryŏ dynasty, late eleventh/first half of the twelfth century

21.6 cm

Acc. no. 20273

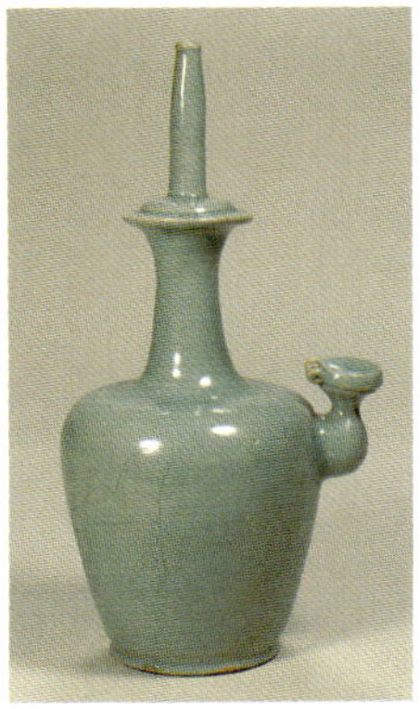

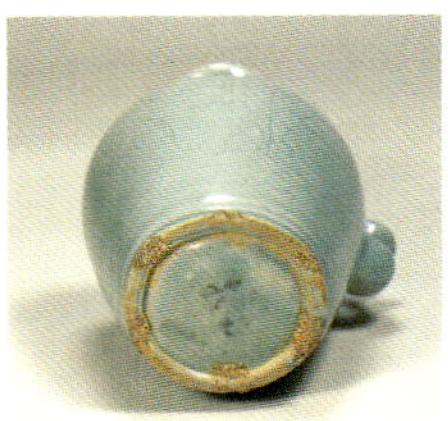

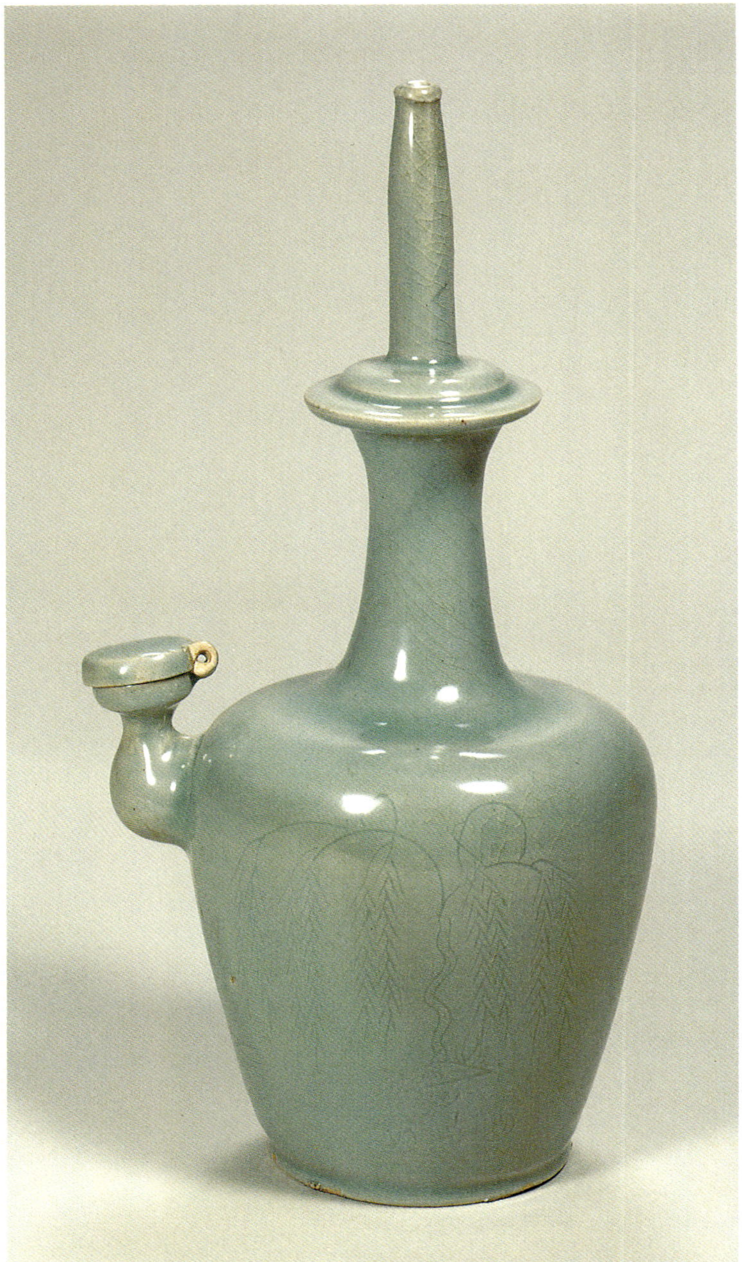

8

Kuṇḍikā *bottle with an incised willow and bird design*

Celadon

Koryŏ dynasty, first half of the twelfth century

33.1 cm

Acc. no. 20343

Originally, the *kuṇḍikā* (Kor. *chŏngbyŏng*) bottle was a vessel type used in Buddhist ritual. But the *Gaoli tujing*, a work of the first half of the twelfth century, states that "nobility, officials, temples, and common people alike used these [i.e., *kuṇḍikā*] to store water," thus indicating their extensive use in daily life. During the Koryŏ period, there were not only numerous ceramic *kuṇḍikā* but also many of metal. In addition to ceramic bottles made in celadon, examples in white wares and black-glazed wares exist. The particular characteristics of the *kuṇḍikā* vessel type are a long neck with a rim in the shape of a sword guard, topped by an even narrower tubelike mouth extending above that. This opening is the one through which water is poured. On the shoulder is a spout in the shape of a tobacco pipe to which a removable lid is attached. It is this opening that is used for water intake. The *Gaoli tujing* likens the shape to an official's appearance when he sticks a writing brush in his hair.

The principal decoration on the surface of the pot consists of simplified drawings of ducks among willows and ducks among reeds shallowly incised on its sides. A translucent glaze tinged with green was evenly applied over the body, and the appearance of the glaze imparts to the vessel a gemlike beauty. The shape, glaze color, and design of this piece are typical of Koryŏ celadons of the first half of the twelfth century. There are six spur marks on the bottom of the vessel.

Ewers of the Koryŏ period were sometimes made in the shape of plants, such as melons or gourds. This ewer in the shape of a bamboo shoot is one such example. The body and lid are one continuous form that skillfully depicts a shoot that has just begun to sprout through the soil. The edges and tips of the sheath are done in relief with the body of the vessel made in three layers, or four layers if one includes the lid. Most ewers in the shape of a bamboo shoot are three-layered; four layers are rare. Because of the additional detail, this ewer is a most impressive example of the type.

There is extensive detail in the lines of the bamboo sheath carved onto the surface of the pot. The execution is freehand; and, in the flow of the lines, done with a delicacy similar to the drawing of the lines for a woman's hair, one can see the possible range of expression that sometimes manifests itself in Koryŏ celadon ware. Both the spout and the handle are represented as bamboo nodes; the handle is shaped by two parallel nodes. The rings on the lid and on the handle and even on the spout are rendered appropriately as bamboo sprouts. The glaze is jade-colored with a strong green tone, and its thick pooling adds to the refinement of the ewer. Even the foot of the vessel is glazed, and on it are three spur marks.

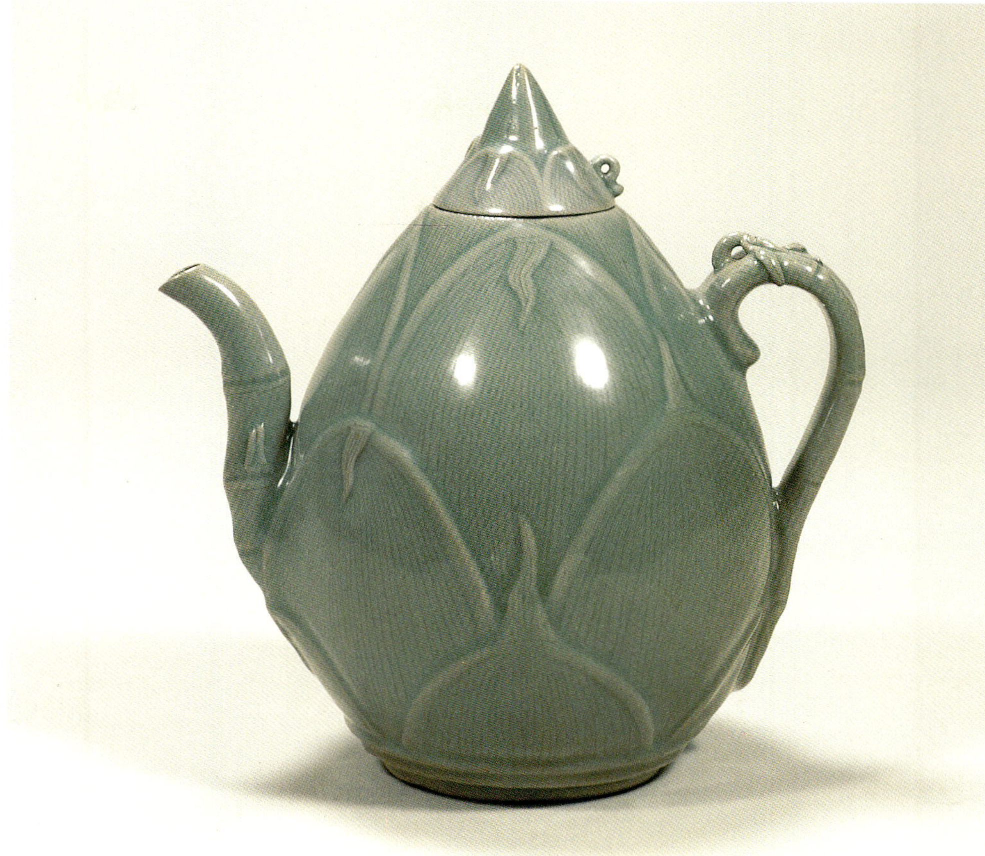

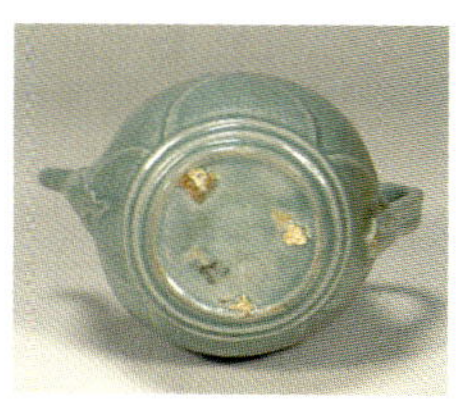

9

Ewer in the shape of a bamboo shoot with a carved and incised bamboo design

Celadon

Koryŏ dynasty, first half of the twelfth century

22.5 cm

Acc. no. 20401

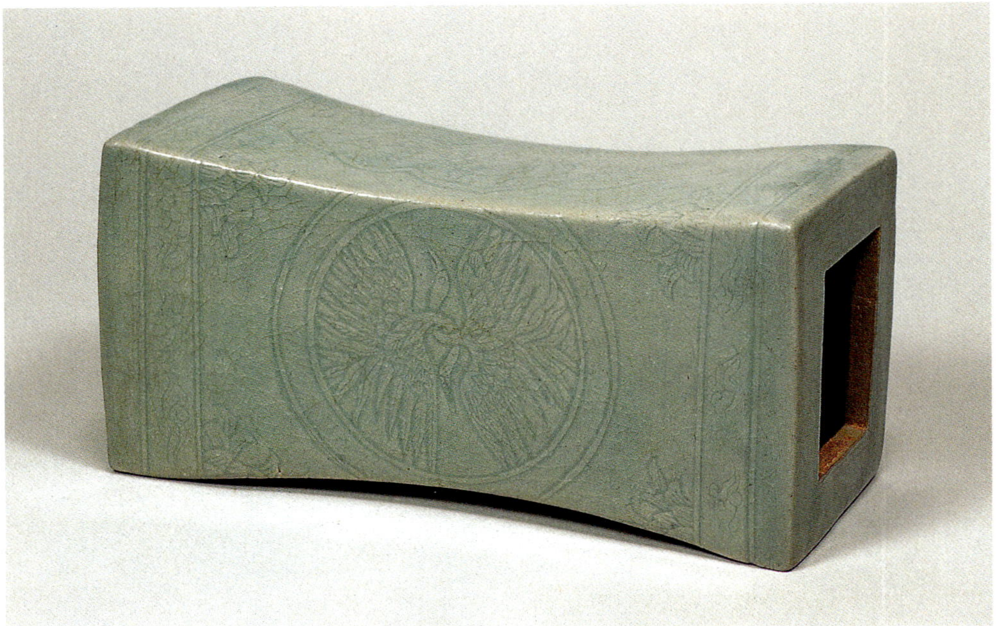

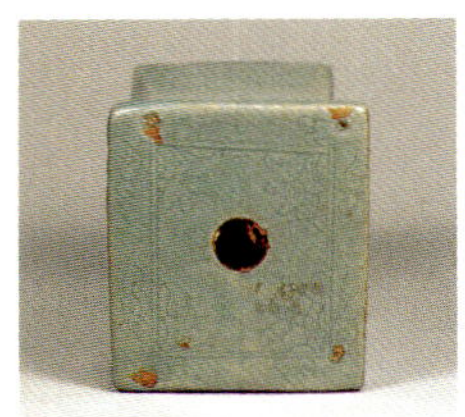

10

Rectangular pillow with an incised crane and cloud design

Celadon

Koryŏ dynasty, mid-twelfth century

9.1 x 21.5 cm

Acc. no. 20467

In Koryŏ celadon ware, two types of pillow shapes are known. One is box-shaped; the other has a boardlike headrest mounted on an animal-shaped base. Both types trace their origins to ceramic pillows made in China from the late Tang-Five Dynasties period through the Northern Song.

The oldest example of the box-shaped pillow in Koryŏ celadon ware, done in iron-ground celadon with a white-spot design, is thought to be of the second half of the eleventh century. In comparison to that piece, the shape, glaze color, and design of this pillow are somewhat more refined and thus can be considered to date from the mid-twelfth century. The moderately concave form of the pillow's midsection, which resembles a thread spool, derives from the original Chinese shape; however, the greater length and gentle curve of its concavity can be seen as Koryŏ stylistic adaptations. Its translucent glaze is jade-colored. The midsections of the front and back are covered with designs of pairs of cranes, and on the top and bottom of the pillow are lotus scrolls. Both designs are carefully incised with a carving technique. This object has been used as a flower vase since its arrival in Japan sometime during the second half of the seventeenth century. For that purpose, one of the two holes on the surface of this piece was filled in, and the other was made larger and carved into a square. Originally, one end of the pillow served as the base; that the pillow was fired in this position is made clear from the spur marks found at the four corners on that end.

The second type of Koryŏ celadon pillow incorporates animal figures into its base (cf. no. 10). In the present example, two crouching lions, back to back, make up the base of the pillow, on top of which is a ceramic headrest. The concavity of the headrest's curved surface is designed to accommodate one's head. It appears that these pillows were actually produced to be used.

Other known examples of Koryŏ celadon sculpture include an incense burner with the knob of the lid in the shape of a lion. One entry of the *Gaoli tujing* mentions such an incense burner, describing it as having a beautiful jade color and a carving technique of unbounded delicacy. An extant example of this type is the famous incense burner in the National Museum of Korea, Seoul. The lion on that piece and those on the pillow illustrated here are somewhat roughly carved, and it would be difficult to consider them delicate. However, strong modeling with less attention to detail is definitely indicative of the height of Koryŏ sculptural achievement. Thus, this work is thought to date to the first half of the twelfth century. The transparent glaze is well dissolved but is so thin that the ground seems to penetrate through its grayish-blue hue. In one spot, the glaze also ran onto the flat base.

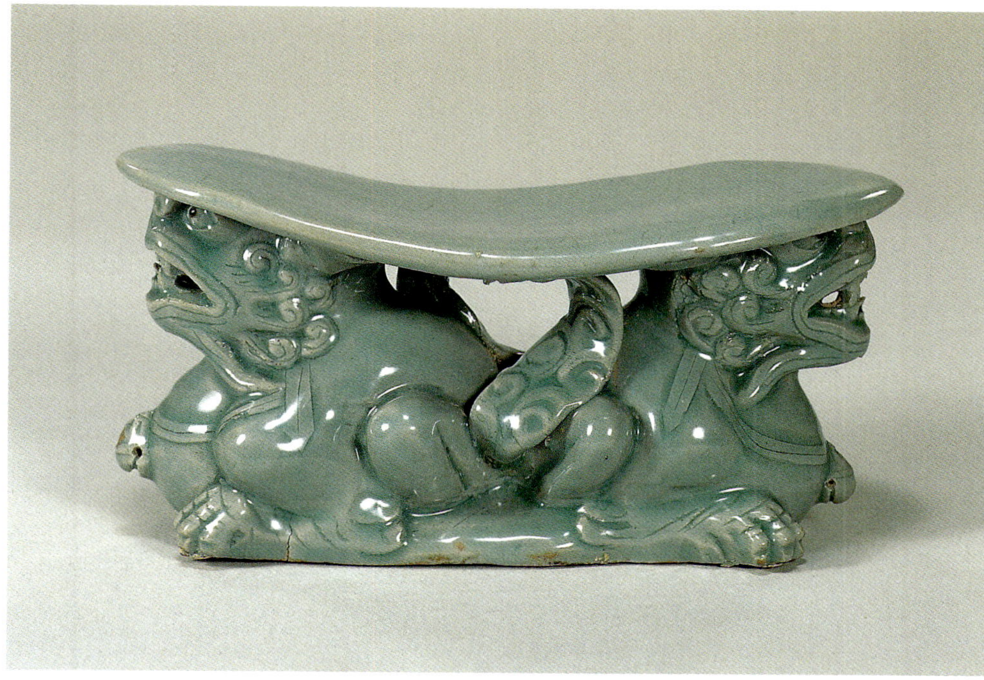

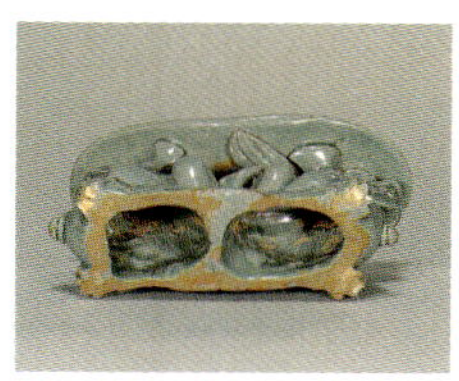

11

Pillow with a stand in the shape of two lions

Celadon

Koryŏ dynasty, first half of the twelfth century

9.1 x 18.2 cm

Acc. no. 20805

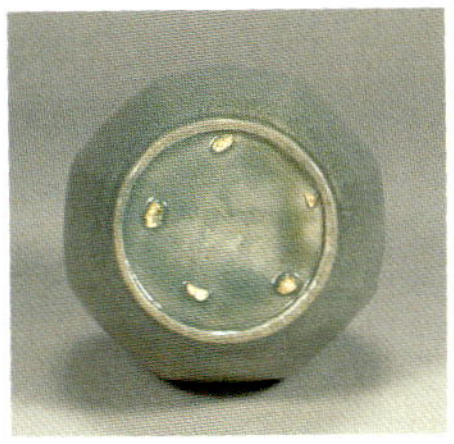

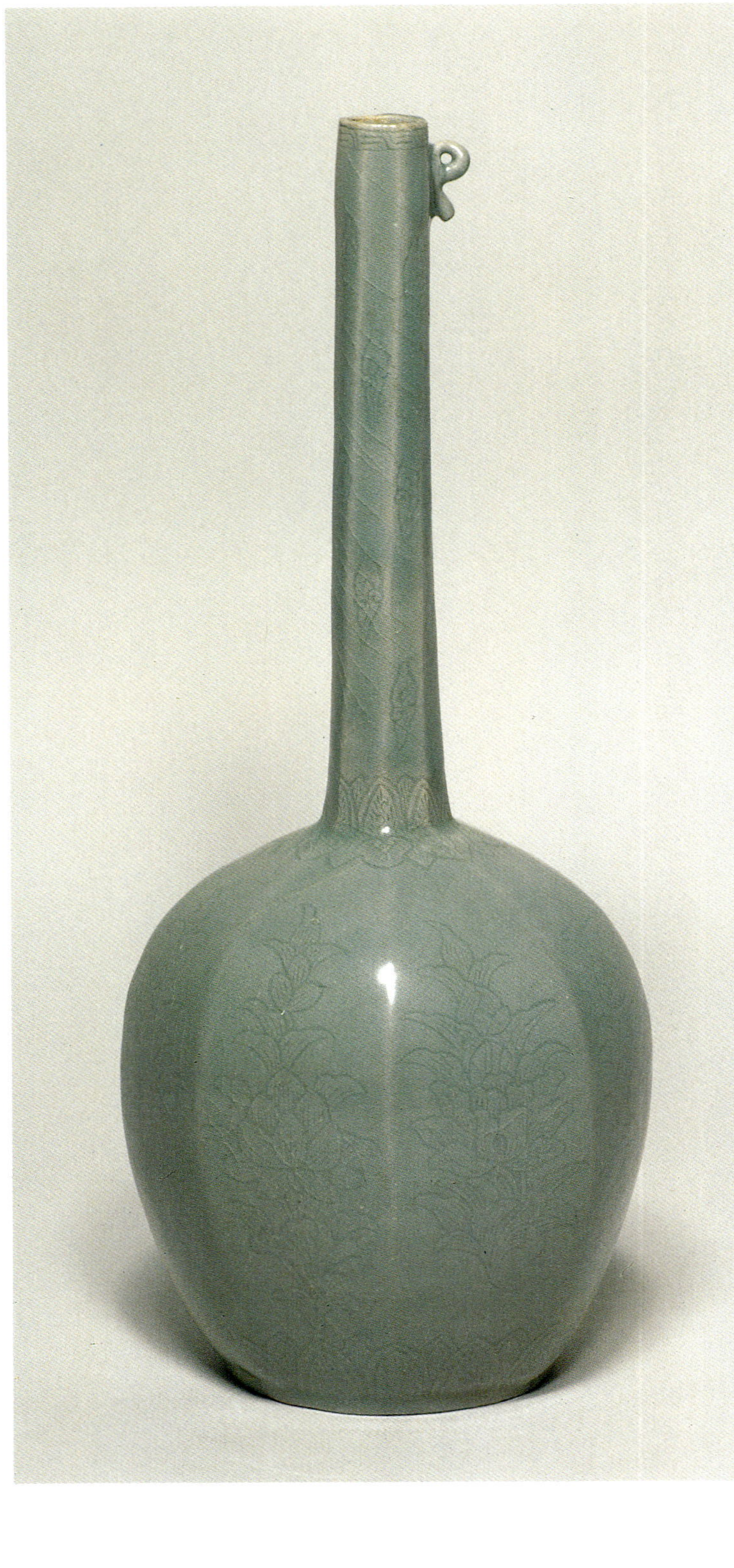

12
Beveled long-necked vase with an incised peony and lotus design
Celadon
Koryŏ dynasty, mid-twelfth century
36.7 cm
Acc. no. 20845

The characteristically long, cylindrical neck of this vessel, described in Japan as "crane-necked" or "gun-barrel-shaped," first appeared in Chinese ceramics of the ninth century. Excavated examples are known from a burial mound in Shaanxi province dated to 871, and shards similar to those were excavated at a celadon kiln site in Yuyaoxian in Zhejiang province. The neck of this bottle is a little longer and narrower than the typical examples of crane-necked bottles of Koryŏ celadon or of Chinese manufacture. With its gently curved shoulder and slender body, this vessel shape, whose origins are Chinese, is now rendered in purely Koryŏ style. The modification of its shape illustrates the perfection of the potter's original artistic statement.

The body was beveled into eight sides and decorated with lotus scrolls alternated with scrolls of peony sprays. The outlines were carved fairly thickly, and the flower petals and leaves are finely incised. In addition, incised designs are found on the lip, neck, base of the neck, shoulders, and outer rim of the base. The grayish blue-green glaze has been evenly applied to the entire body. The shape of the vessel, its glaze color, and its design suggest a mid-twelfth-century date. The underside is entirely glazed and bears five spur marks.

Small jars with relatively large mouths, four loop handles mounted on the shoulders, and tapering bases are not particularly rare worldwide; however, a jar as small as this one, with only three loop handles on the shoulder, is unique. The classic shape of this vessel, the beauty of its glaze, the exquisite rendering of its surface design, and the skill with which its shoulder loop handles are made all indicate that it is an extraordinary and superlative example of its kind. The top of the lid is decorated with a design of flowers and grasses, around which is a key-fret pattern. On the shoulder of the body is a cloud-collar pattern, on the three sections are sprays of lotus blossoms, and at the base is a key-fret pattern, all of which are carefully incised. These designs are so subtly carved that they are not visible from a distance. In addition to the vessel's fine workmanship, the cross-shaped loop handle on the lid and the three shoulder loop handles, each of which assumes the shape of an animal's foot, suggest that this container was inspired by metalwork.

This vessel form is first seen in southern Chinese celadon jars of the tenth century. Many other examples of Koryŏ celadon dated to the beginning of the twelfth century were modeled on the shapes of Chinese celadon vessels of about the tenth century. The refined workmanship of such pieces reached new heights on the basis of borrowings from ancient forms and adaptations to Koryŏ style. The underside of the lid and the bottom of the jar each have four spur marks.

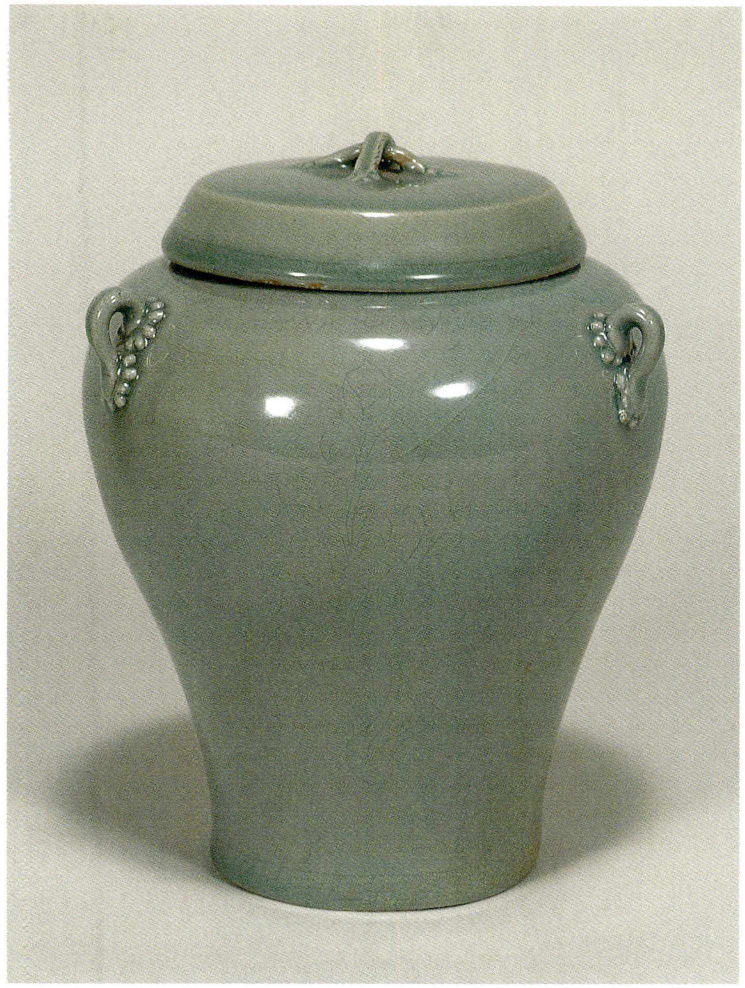

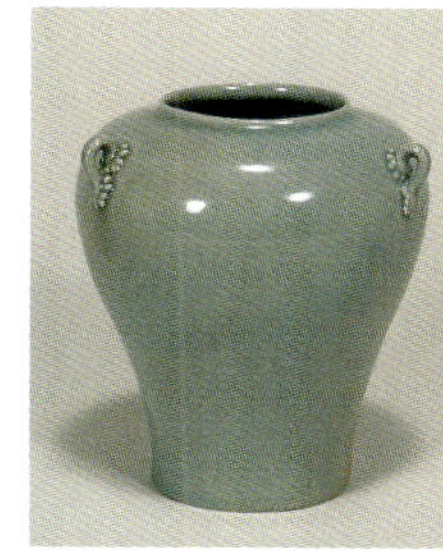

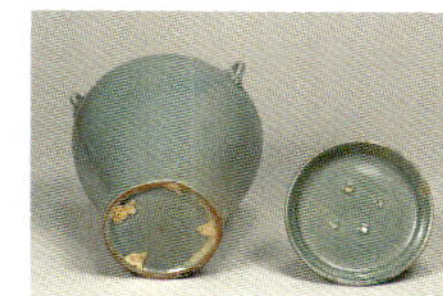

13

Three-handled jar with an incised lotus design

Celadon

Koryŏ dynasty, early twelfth century

18.9 cm

Acc. no. 20205

Important art object

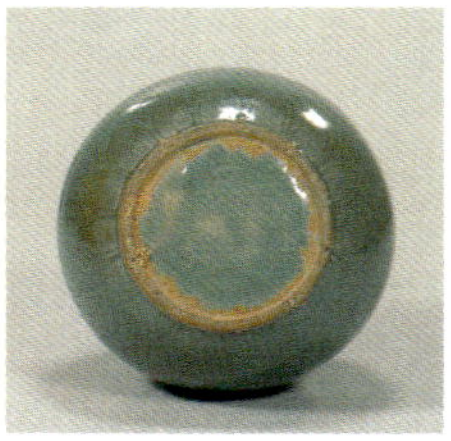

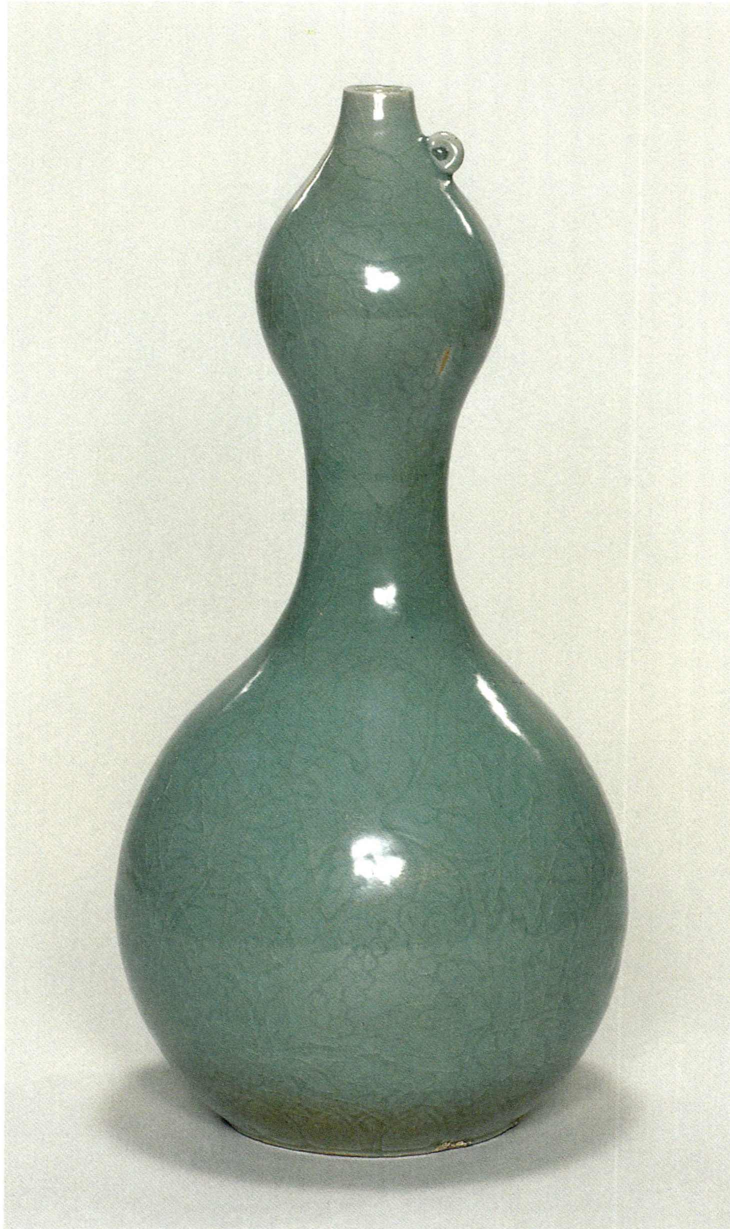

14
Bottle in the shape of a gourd with an incised grapevine-scroll design
Celadon
Koryŏ dynasty, mid-twelfth century
36.4 cm
Acc. no. 20340

Koryŏ celadon vessel shapes include relatively few round jars and gourd-shaped bottles. This gourd-shaped bottle is similar in size and form to one inscribed with a poem in black inlay in the National Museum of Korea, Seoul. In general, a comparison of the gourd-shaped bottle of the Koryŏ period with its ancestor, the Chinese gourd-shaped bottle, shows that the swelling of the upper and lower parts of the Koryŏ form is somewhat less full, rendering its shape slenderer and the curves over its entire surface less pronounced.

The body of this piece is decorated overall with the thick contours of the fruit, leaves, and vines of the grape motif, with details executed in finely incised lines. Around the base is another incised design of lotus petals in a band. Since the surface of the vessel is heavily glazed, the thick coat of glaze tinged with dark green obscures the sharply incised design, which cannot be clearly seen. A ring for securing a lid has been attached close to the mouth, but again, the glaze was so thickly applied that the hole is blocked. After a coat of glaze was applied to the entire base of the vessel, the glaze on the foot was wiped off and as many as ten spur marks are identifiable.

A vase with a small mouth, shallow neck, and broad shoulder that tapers toward the base is called a *maebyŏng* ("plum vase"). The name suggests a vase decorated with a single plum branch, but the origin of the term is unclear. In China, the shape is utilized in Ding, Cizhou, Yaozhou, Yue, Longquan, and Jingdezhen wares of the eleventh through the thirteenth centuries. The Koryŏ *maebyŏng* originated under the influence of Chinese *meiping*, and its distinct characteristics appeared beginning in about the first half of the twelfth century. While almost all Chinese *meiping* have mouths shaped either as apricots or sword guards, an overwhelming number of Koryŏ *maebyŏng* have mouths in the form of shallow cups. Moreover, the line from the shoulder to the bottom of the vessel takes the form of an *S* curve, a shape almost never seen in Chinese *meiping*. These distinguishing traits of the *maebyŏng* together suggest a feeling of elegance rather than of strength.

The main decoration on this *maebyŏng* consists of an inlaid pattern of bamboo, cranes, and clouds around the body. The peacefulness of the depiction is an expression of the refined and elegant Koryŏ sensibility. The presence of the four pendantlike shapes on the shoulder reflects the then-current practice of covering the mouth of the vessel with a cloth. The band of lotus petals and key-fret patterns at the base of the vase skillfully heightens the contrast between the black-and-white inlays and emphasizes the tautness of the vessel's form. The glaze color, an elegant, dark grayish blue, is unusual for inlaid celadon ware. Only the foot of the vessel is unglazed, and around its interior are six spur marks.

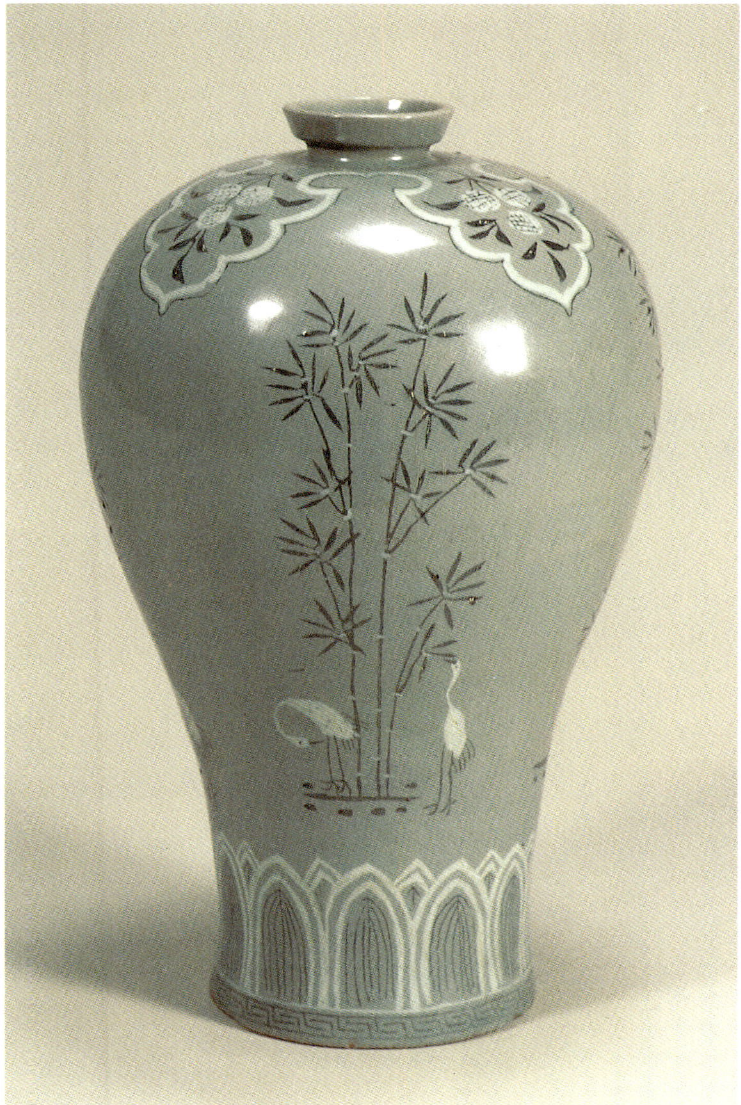

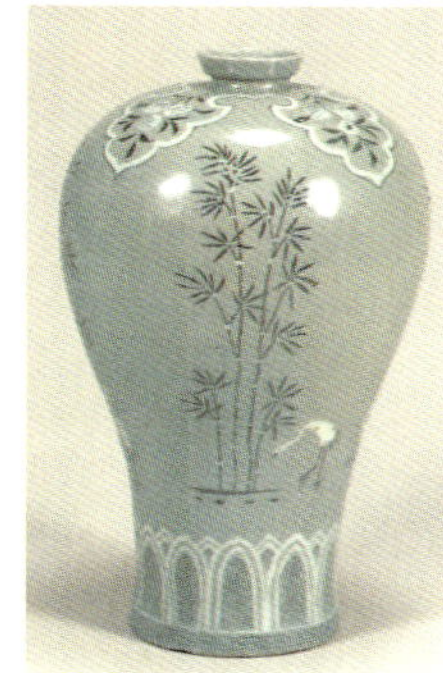

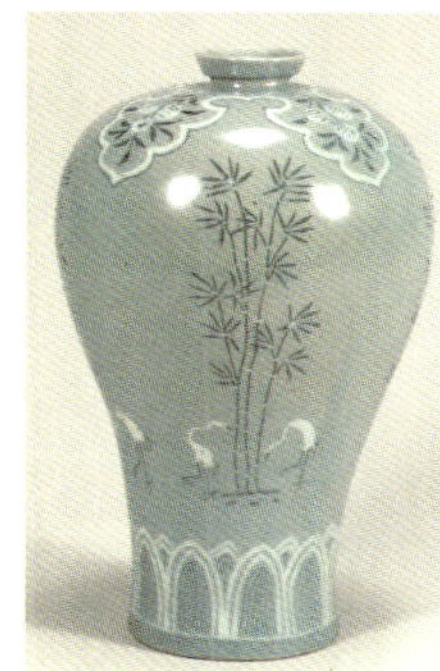

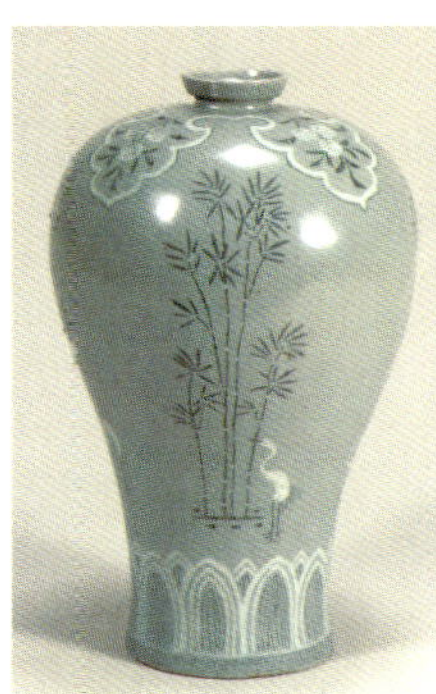

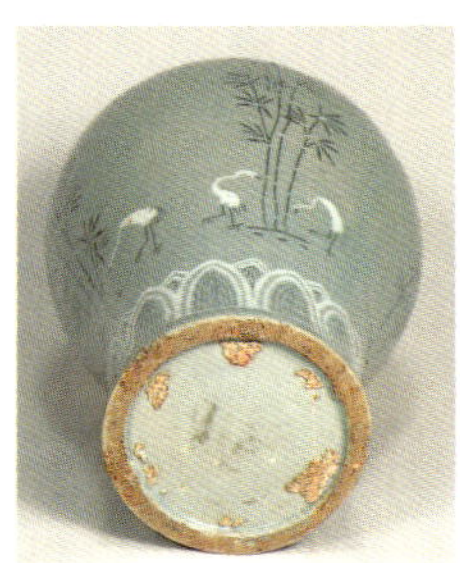

15

Maebyŏng *with an inlaid crane, bamboo, and cloud design*

Celadon

Koryŏ dynasty, twelfth century

29.2 cm

Acc. no. 20206

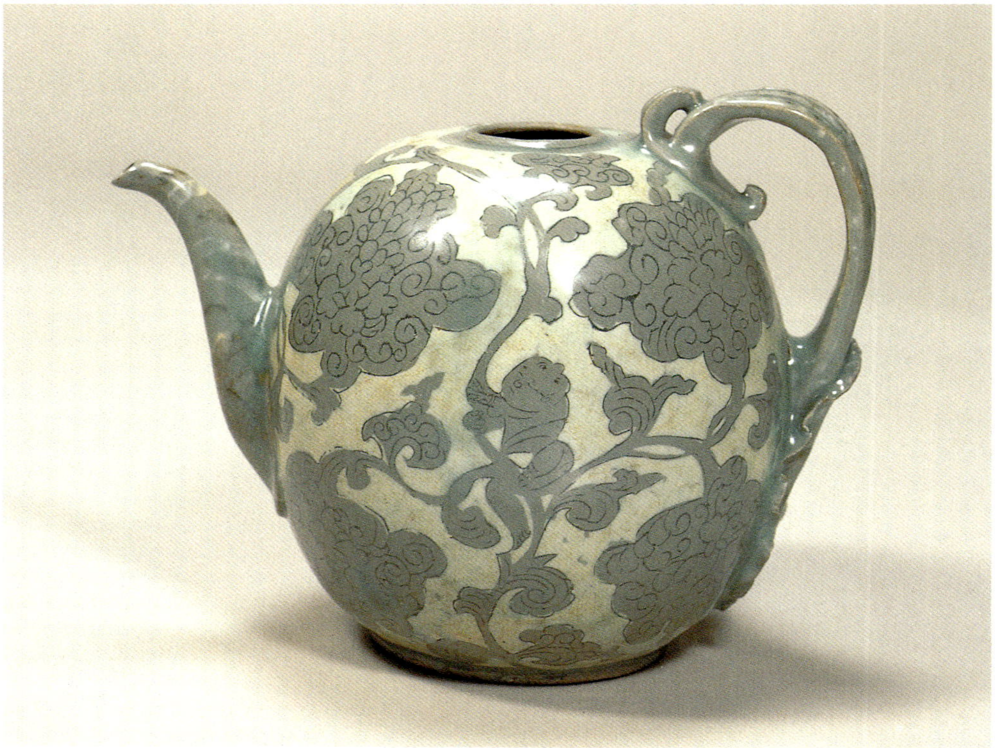

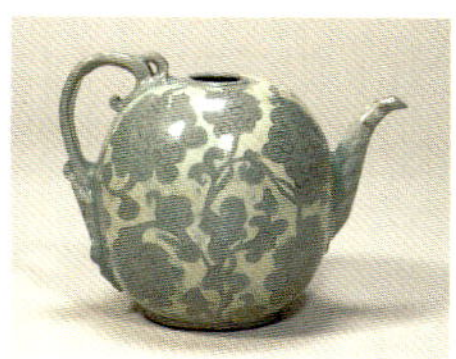

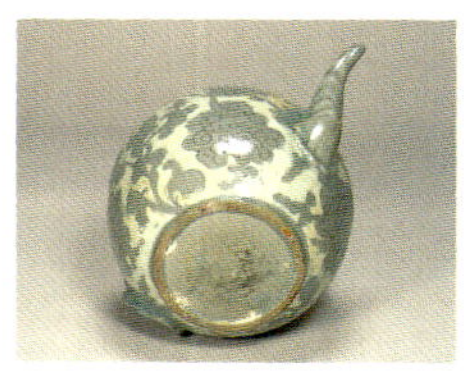

16

Ewer with a reverse inlaid design of children and floral sprays

Celadon

Koryŏ dynasty, twelfth century

19 cm

Acc. no. 20286

Important Cultural Property

The fully rounded body of this ewer is inspired by the shape of a melon. Melon and gourd forms are seen from time to time in ewers of the Koryŏ period. The spout on this piece is somewhat short in relation to the body, but the body is short as well; its stoutness is skillfully complemented by the simple, strong contour of the handle. Applied to the base of the spout and the handle are large lotus leaves, a Buddhist decoration that serves to strengthen the structural connection between these appendages and the main part of the vessel.

Distinct Buddhist overtones can be seen in the design appearing on the central section of the ewer. Young boys, Chinese in appearance, are represented climbing plant vines; the motif undoubtedly derives from Buddhist folklore, but its actual source is not clear. The flowers (Kor. *bosanghwa*; Chin. *baoxianghua*) encircling the boys are not based on nature but belong to the realm of Buddhism; what is depicted here is the Buddhist world of peace and contemplation. The cream-colored ground is done in white inlay, a reverse of the usual inlay technique, in which the surface design is executed in inlay. The details of the boys' eyes and noses and the details of the flower designs are done in black inlay to stand out against the overall pattern of figures and flowers glazed in celadon. This ewer is an excellent example of inlay technique at its peak. Only the foot of the vessel is unglazed, and inside it are six spur marks.

There are two types of long-necked vases, one of which is beveled and the other round. This piece, which was thrown on a potter's wheel, is of the second type; another example of about the same size, design, and style is in the Victoria and Albert Museum, London.

The main decoration on this vase consists of four double circles enclosing alternating patterns of peony sprays and sprays of three chrysanthemums. Willows and flowers appear alternately between the circles. These designs, which cover the central area of the body, are all done in white inlay. Only the willow, executed in black inlay, brings out the definitive character of the black-and-white contrast. This interplay is emphasized even more by the band of classical scroll patterns in reverse inlay encircling the shoulder and the second band of lotus petals in white inlay encircling the lower part of the body. The band of classical scroll patterns on the shoulder, in particular, is similar to the inlaid design on an excavated bowl bearing an inscription dated 1159. The glazes of inlaid celadon wares were kept thin so that their designs would show through clearly. In order to meet that requirement, there are many instances where the glaze color of a piece was sacrificed, resulting in an inferior hue. This vase is a successful example of that rare combination: refined color and clearly revealed designs. The glaze on the foot and its surrounding area was scraped off; the spur marks are unclear but can be discerned in six spots.

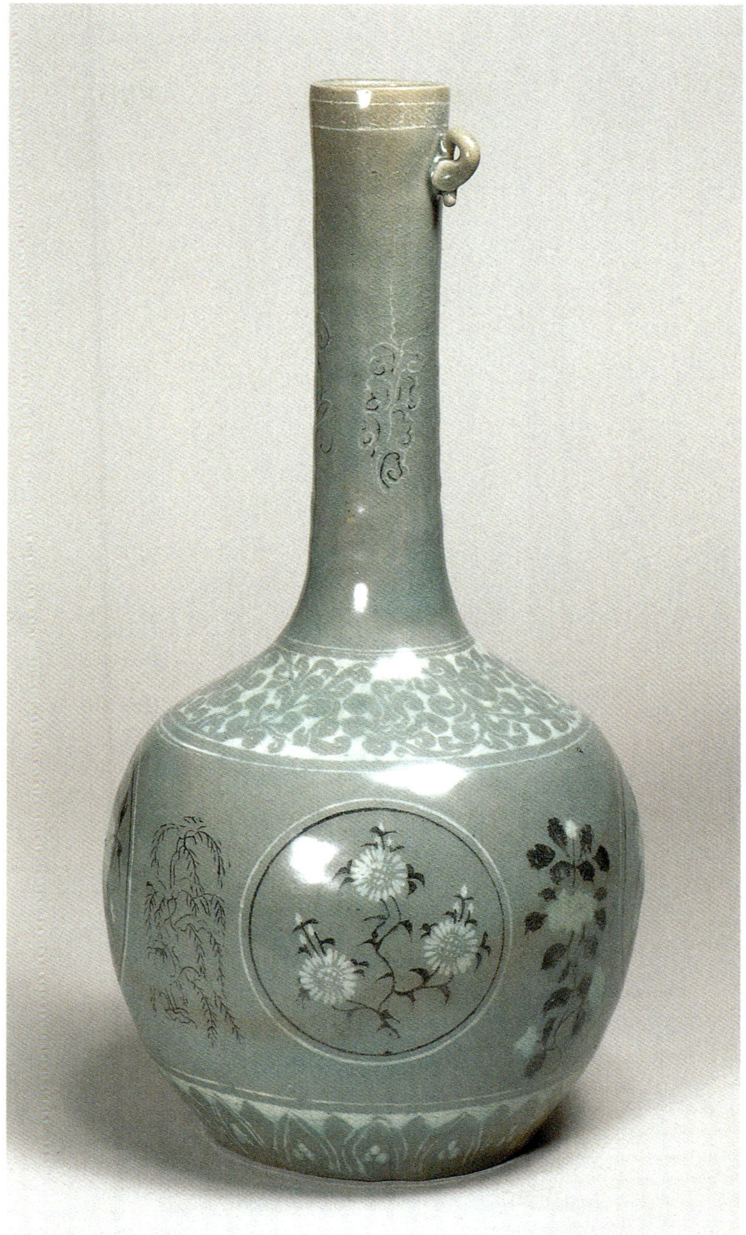

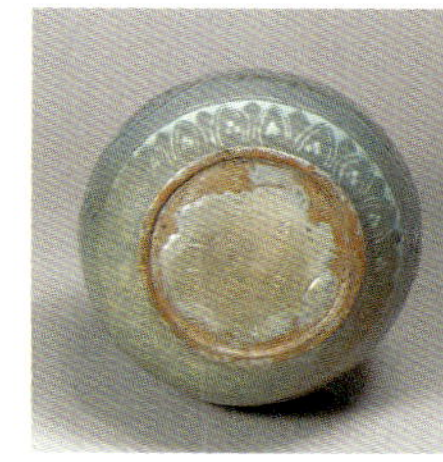

17

Long-necked vase with an inlaid chrysanthemum and peony design

Celadon

Koryŏ dynasty, mid-twelfth century

35.5 cm

Acc. no. 20474

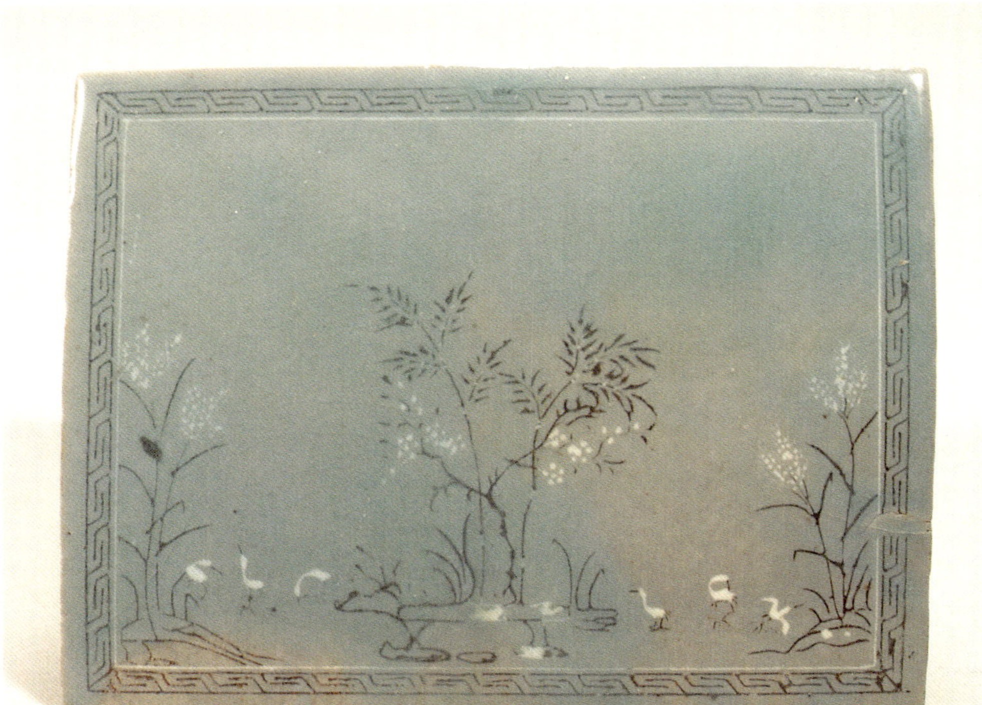

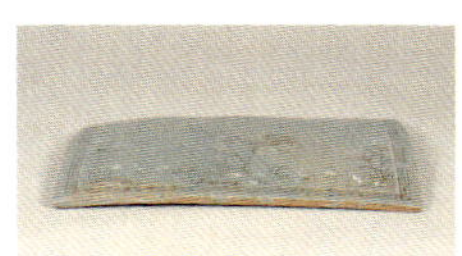

18

Plaque with an inlaid design of reeds and birds

Celadon

Koryŏ dynasty, twelfth century

15.9 x 20.5 cm

Acc. no. 20054

The purpose of ceramic plaques of this type has yet to be determined. One large ceramic plaque and another many-sided example were discovered at a kiln site in Yuch'ŏn-ri, Poan-myŏn, Puan-gun, North Chŏlla province. Judging from the shapes and designs of those two samples, it is possible to speculate that plaques of this type were perhaps used as architectural ornaments.

Of the rectangular ceramic plaques presently known, this one is the smallest, and there are no other examples similar to it. Moreover, the decoration, which is presented like a painting, is unique. The four borders of the piece are ornamented with key-fret patterns in black inlay. These bands frame a distinct pictorial area, which displays a paintinglike scene of ducks frolicking happily on a shore where dense clumps of reeds and bamboo grow. Some ducks cry out with their necks extended, some search about for food or preen their feathers, and others swim in a pond or take to the air. While the movement of the ducks and bamboo swaying in the breeze is lively, the tone of the entire scene is peaceful. And there is poetic lyricism in the arrangement of space largely left in white. In areas where the glaze has been omitted or is thin, the light red of the clay body is revealed; but luckily these faults give the impression of mist or fog. This piece exhibits the exquisite refinement and near-spiritual calm and sense of beauty that Koryŏ celadon wares can possess. The glaze color is a rich grayish blue-green; its backside is unglazed.

Koryŏ celadons include many melon-shaped ewers, of which there are two types: a smooth kind without subdivisions, and a second with eight vertical ridges that section the body. This ewer belongs to the latter type but is quite different from the typical example in that category. First, while the usual melon-shaped ewer has a low mouth with a drop lid, this one has a tall, straight mouth with a cover. (The existing lid, it should be noted, is not original to the ewer.) Second, there is a clear separation between the shoulder and body of the vessel, and the melon shape of the body does not continue into the shoulder. Third, the design on the shoulder of the vessel is incised, but on the body the technique of inlay is used. Such an elaborate example is rare.

The body is covered with a pattern of willowy flowers, and on the shoulder are lotus petals. The design is carefully applied and harmonizes well with the simple shape of the vessel, which recalls a metal ewer. This piece can be dated to the mid-twelfth century, a period when inlaid celadon was at its peak. The glaze is of a rich, glossy grayish blue; and, particularly where it pooled, it appears thick and refined. The glaze on the foot of the ewer was scraped off, and there are seven spur marks.

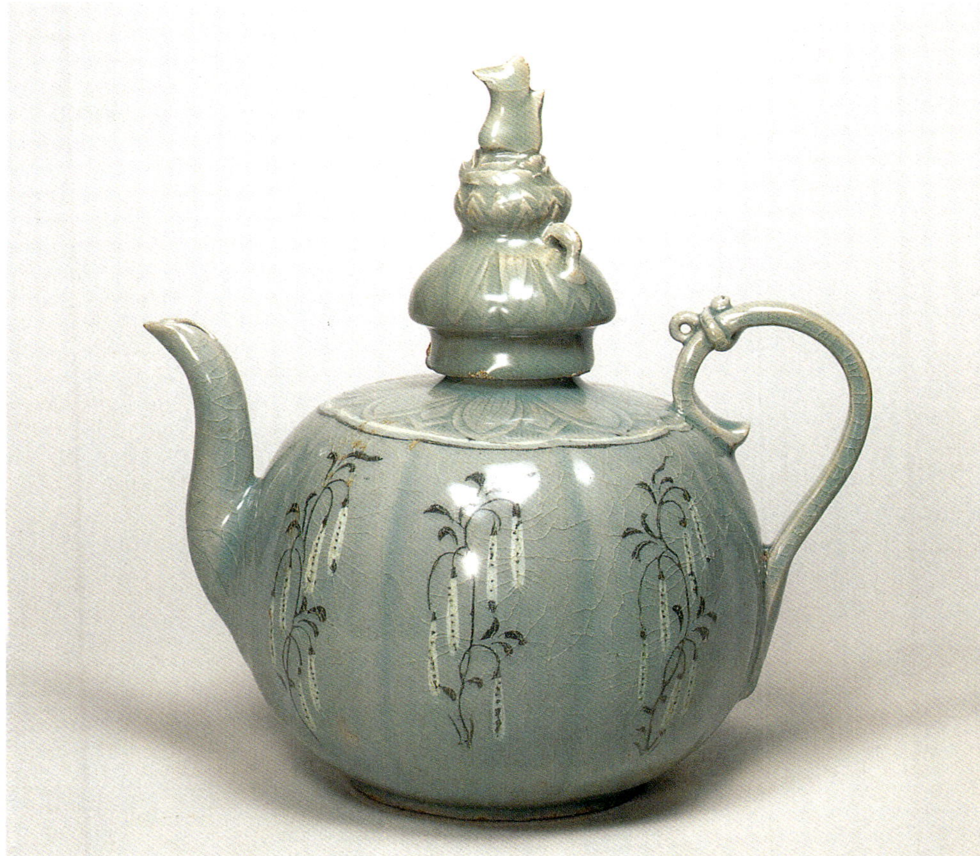

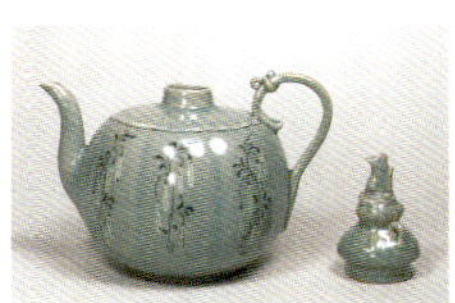

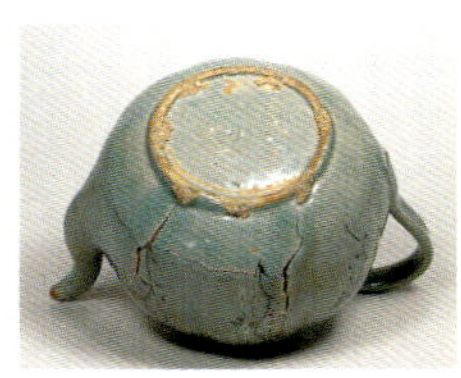

19

Ewer with an inlaid floral-spray design

Celadon

Koryŏ dynasty, mid-twelfth century

24.2 cm

Acc. no. 20345

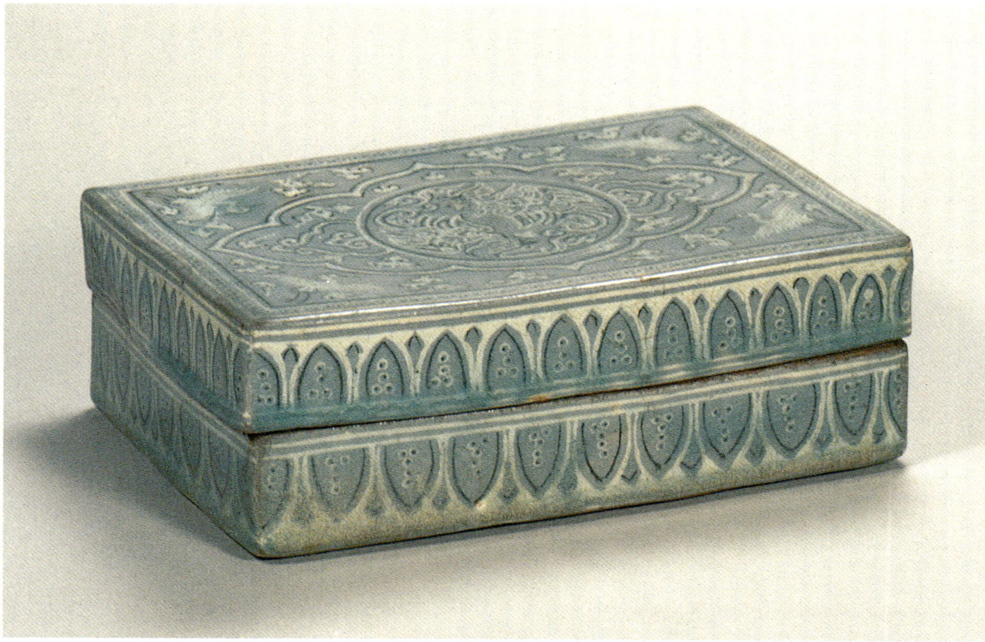

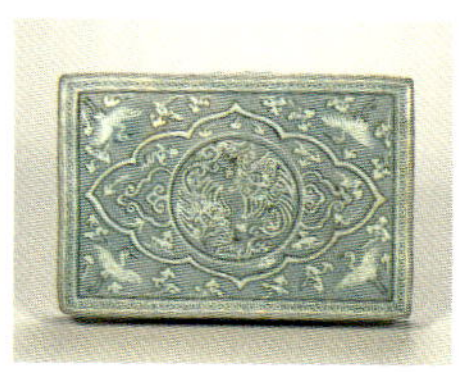

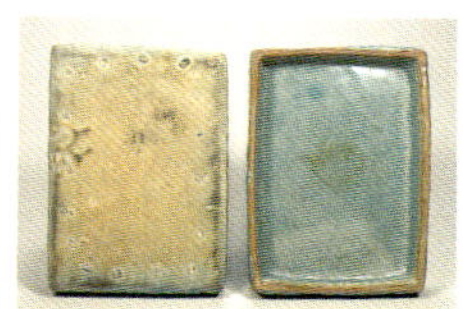

This type of square box is believed to have been used for toilet articles. Surviving examples are rare. There are two types, a large one which holds an inner tray, such as the inlaid box in the National Museum of Korea, Seoul, or the openwork box in the Tokyo National Museum; and a second type which is shallow, such as the example illustrated here.

The sides of this box and cover are decorated with lotus petals, while the top is ornamented with a dense pattern focusing around a pair of phoenixes. More precisely, a medallion with a pair of phoenixes is surrounded by a large diamond-shaped flower design, outside of which, in the four corners, are flying crane and cloud designs. The filling in of most of the surface with black-and-white inlay, leaving little open space, indicates a stylistic date of the thirteenth century for this piece. A fine crackle pattern appears over its entire surface, which is of a luminous grayish blue-green. The glaze has pooled along the bottom rim of the cover, where it is particularly beautiful and appears jadelike. Both the box and its cover are thick-walled. Since the clay body is heavily potted, any warping that occurred in firing can hardly be detected. The bottom of the box has spur marks in seventeen places, indicating that great care was taken in its firing.

20

Covered cosmetic box with an inlaid phoenix design

Celadon

Koryŏ dynasty, first half of the thirteenth century

22 x 15.5 cm

Acc. no. 20324

This covered box, which has an unusually large diameter of 23.5 centimeters, is almost unique among surviving examples of inlaid celadon ware. A fair number of such boxes were made, however, in iron-painted and incised celadon ware. Pieces of the same shape, produced in metal, are believed to have been used in Buddhist ritual.

At the center of the box's cover is a spray of peonies in black-and-white inlay in a circular frame; surrounding this medallion are alternating designs of flying cranes and clouds. The side of the lid was beveled, and below that, along the rim, are six cloud designs in white inlay. The main pattern on the body consists of six sprays of peonies in white inlay. Narrow bands of key-fret patterns in white inlay encircle the edges of both body and cover. The well-spaced arrangement of designs over the entire surface indicates that this piece dates to the latter half of the twelfth century. The color of its glaze is grayish green tinged with light brown. Since the glaze is thick, the designs are submerged by it and appear dull. The undecorated interior surfaces of the body and lid are also covered with a thick glaze, as is the base; marks on the foot and surrounding area indicate that those surfaces were roughly scraped. Spur marks are visible in thirteen places.

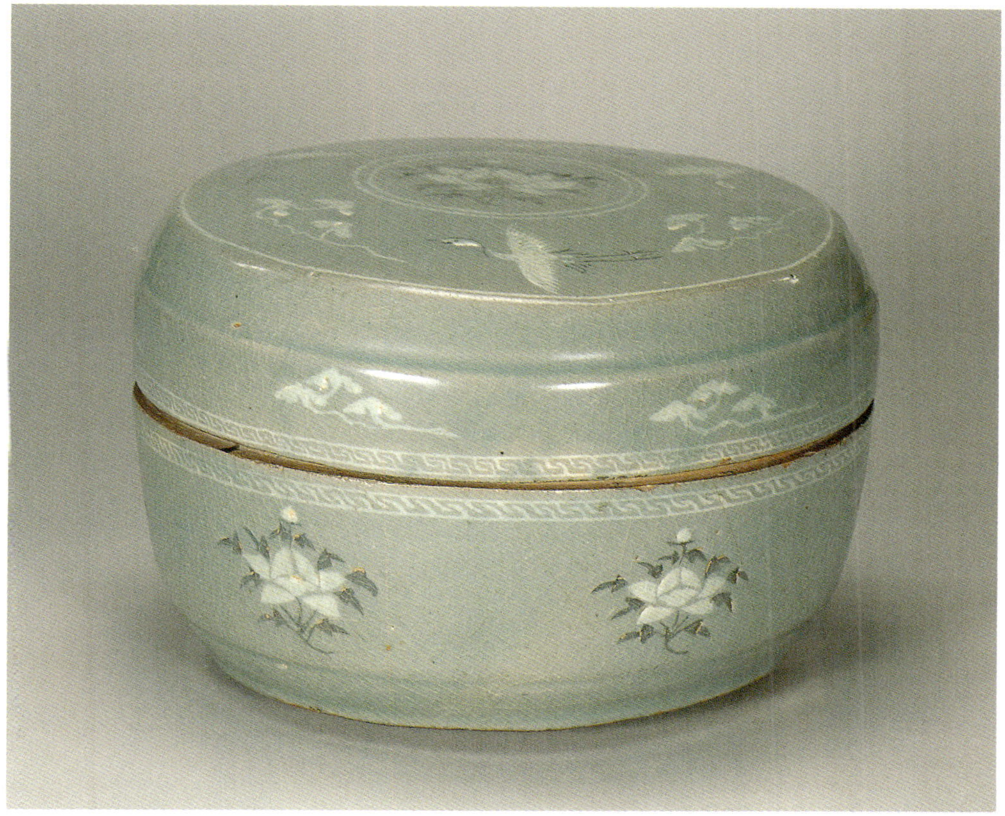

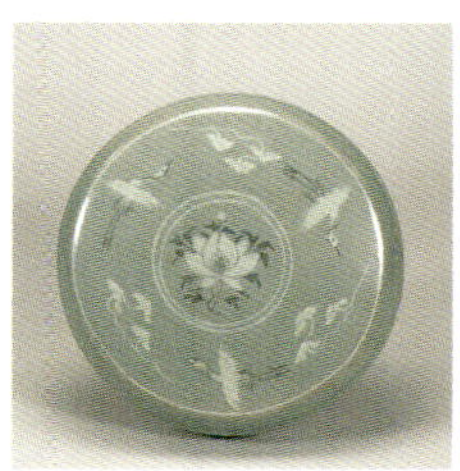

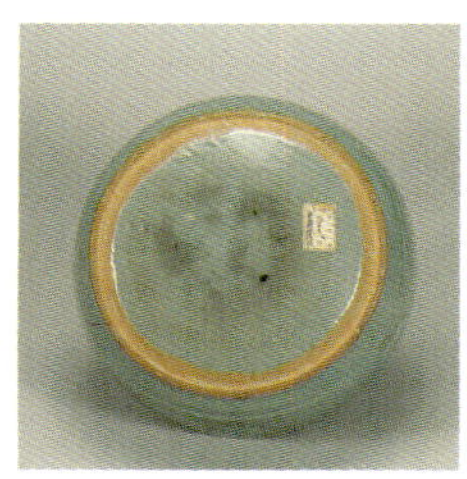

21

Covered round box with inlaid peony, crane, and cloud designs

Celadon

Koryŏ dynasty, second half of the twelfth century

16.2 cm

Acc. no. 20333

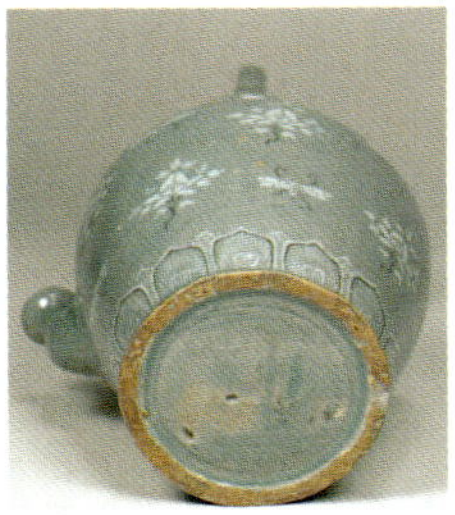

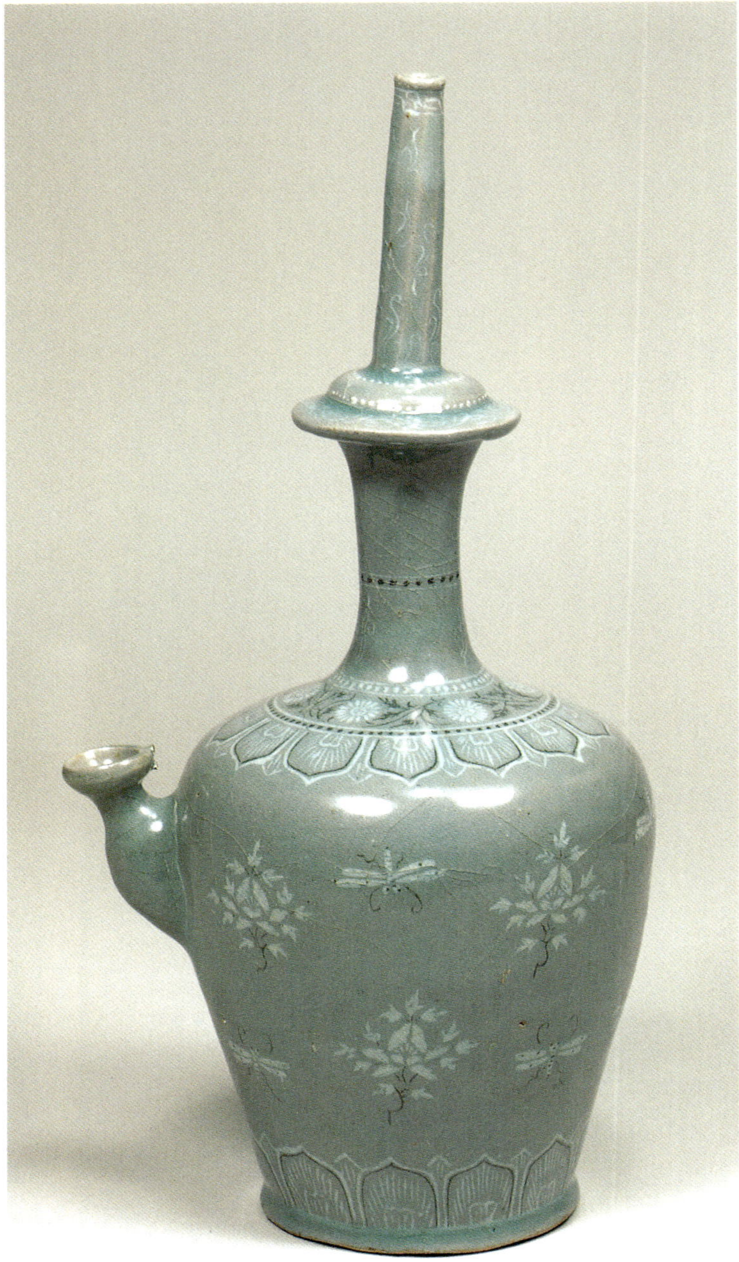

22

Kuṇḍikā *bottle with an inlaid peony and butterfly design*

Celadon

Koryŏ dynasty, end of the twelfth century

33.5 cm

Acc. no. 20029

This *kuṇḍikā* (Kor. *chŏngbyŏng*) is simpler and more slender than the *kuṇḍikā* vessel of no. 8. The principal surface design consists of eight sprays of peonies and seven butterflies that alternate with each other over the upper and lower halves of the body. These patterns are fairly stylized, indicating a date of production a little after the peak period for manufacture of inlaid celadon. Even in the bands of lotus petals on the shoulder and around the base of the body, the expressive power of the design is less forceful. Inlaid celadon is believed to have reached the pinnacle of its artistic achievement in about the mid-twelfth century; this bottle demonstrates the stylization that became widespread in about the first half of the thirteenth century. As this tendency grew even more pronounced in the latter half of the thirteenth century, glaze color for inlaid pieces also became duller. The trend toward stylization is apparent in this design, but the color of the glaze on this vessel is a beautiful grayish blue with a strong blue tinge. Moreover, the bottle can also be dated to about the end of the twelfth century on the basis of its simple shape. Shards similar to this piece have been discovered at Yuch'ŏn-ri, Poan-myŏn, Puan-gun, North Chŏlla province. The glaze, beautiful in color, was heavily applied over all surfaces except the foot. Within the foot ring are eight spur marks, but they are not clearly discernible. The lid to the spout has been lost.

There are few extant examples of rounded jars in Koryŏ celadon ware. This jar has a wide mouth with a very low rim; the contour of the vessel curves gently downward from the shoulder and tapers sharply toward the base. The lower part of the body is tightly proportioned, creating a perfect and well-balanced form.

The inlaid design consists of four peony medallions encircling the body. The flowers and stems of the peonies are depicted in white inlay; only the leaves are done in black. Surrounding the two-ringed medallion of the main design are cloud-collar patterns, and the rim of the shoulder is graced with a string of jewels. These are rendered in a stamping technique that was apparently first introduced in white-inlay decoration beginning in the early thirteenth century. Moreover, its thick walls, heavy body, and dim, grayish glaze color are characteristic of the period. The glaze on this jar is thin, and as a result of that, its grayish-blue color is tinged with light brown. The vessel's refined shape and the expressiveness of its design do not reflect any trace of stylistic decline and preserve the jar's quietly serene mood. This piece was damaged in a kiln accident and, before its main firing, was heavily patched with clay to prevent cracking. The base is unglazed, and no spur marks are discernible.

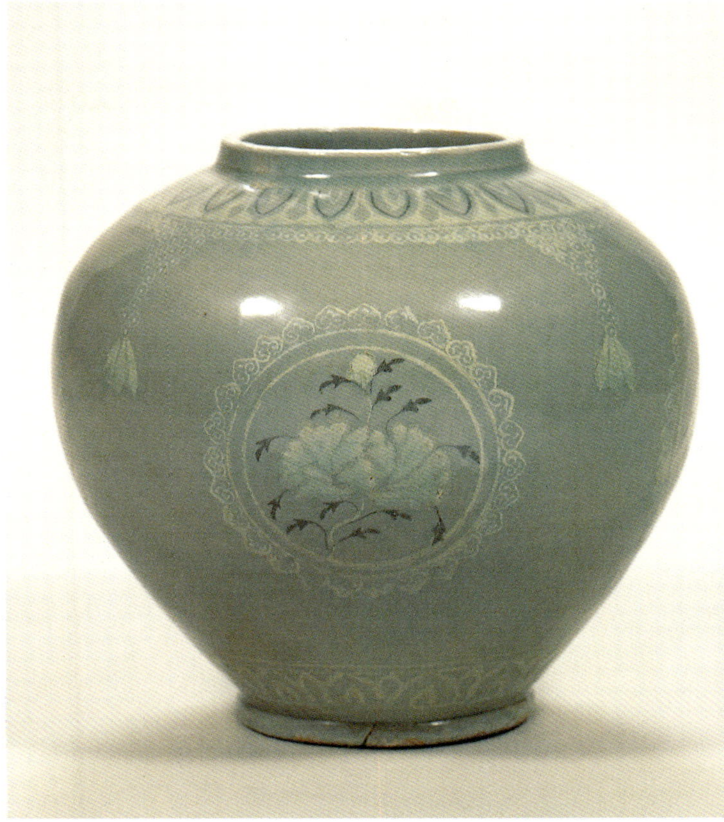

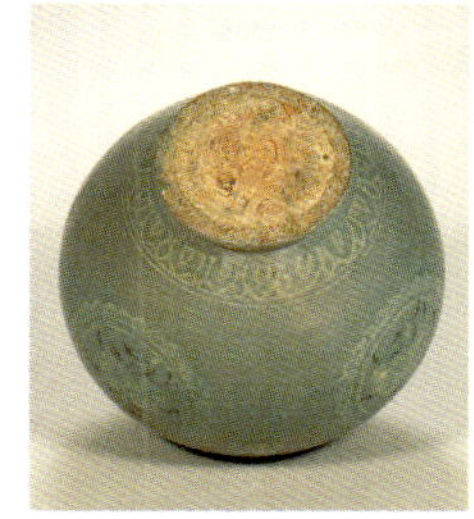

23

Jar with an inlaid peony design

Celadon

Koryŏ dynasty, first half of the thirteenth century

21.5 cm

Acc. no. 20279

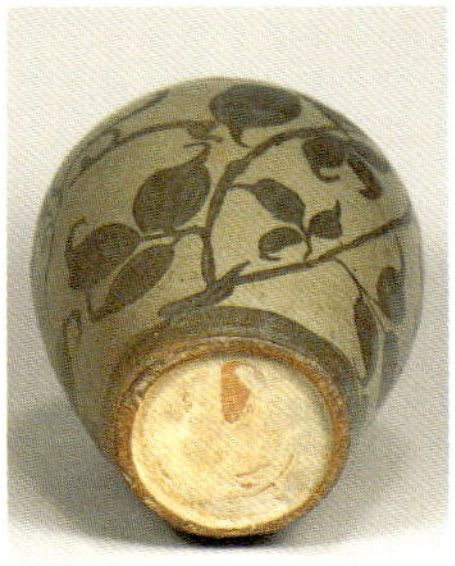

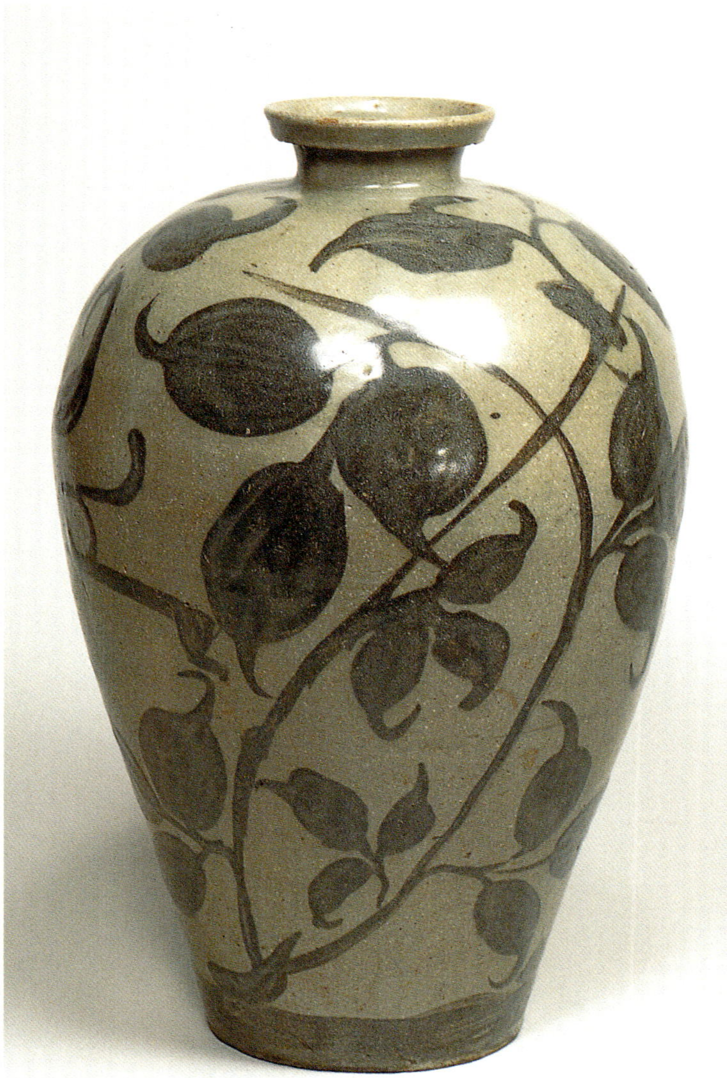

24

Maebyŏng *with an underglaze iron-brown vine-scroll design*

Celadon

Koryŏ dynasty, first half of the twelfth century

27.2 cm

Acc. no. 20438

This *maebyŏng* is decorated with a design drawn in iron paint under its celadon glaze. It seems that iron-painted celadon ware was produced mainly at kilns whose lineage differed from that of the kilns producing pure celadon and inlaid celadon wares. Several factors support this conclusion: There are many examples of glazed vessels that are brown in color, their glaze having oxidized during firing. The shapes of these vessels are also slightly different; there are cases where, for example, the shape of such a vessel perpetuates an ancient Chinese form of the late Tang-Five Dynasties period that is not reproduced in other Koryŏ wares. Finally, in many instances, the type of surface ornament utilized on such iron-painted celadons is unique to that category of ware. Moreover, this hypothesis seems to have been confirmed by the discovery in 1980 of a large-scale kiln site at Jinsan-ri, Sani-myŏn, Haenam-gun, South Chŏlla province, where iron-painted celadon was made.

Even among examples of iron-painted celadon ware, this *maebyŏng* is unique. A creeper plant with large leaves emerges from two stems that intersect and encircle the decorated surface of the vase. The bold and carefree expression of the design is characteristic of iron-painted celadon ware. The glaze is of a dark greenish-gray color, and there are places where the iron pigment pooled. The shape of the vessel is simple, and in the creation of the cup-shaped mouth there is a feeling of restraint. A strong sense of orderliness pervades the entire surface. Consequently, it is probably a work of the first half of the twelfth century, when celadon reached its peak. On the underside of the vessel, a shallow base has been hollowed out. Glaze was applied to the entire vase; marks appear only on the foot, where the glaze was scraped off. No spur marks are found.

Rounded jars are rare in Koryŏ celadon ware, and iron-painted celadon jars of this size, in particular, are not found in other collections. The wide mouth of this piece has a low rim, and the body swells into a full, ample curve. In the glaze are spots that are brown from oxidation, but the degree to which the glaze retains its grayish-blue color is exceptional. In iron-painted celadon, the glaze covering the entire surface often oxidizes to a brown color; from that point of view, this example is unusual in preserving much of its celadon color.

Three designs of floral sprays (Kor. *bosanghwa*; Chin. *baoxianghua*) are drawn around the jar, with the intertwining leaves of the classical scroll pattern depicted over the entire surface. The application of glaze to the interior of the jar also differentiates it from vessels of pure celadon or inlaid celadon ware. The flat bottom of this piece is slightly concave at the center, demonstrating a construction peculiar to the jar. Its base is unglazed, and fine, white sand has adhered to its outside rim. In many ways, this iron-glazed celadon jar is unusual; nevertheless, in terms of shape, glaze, and design, it is a representative example of the category in general.

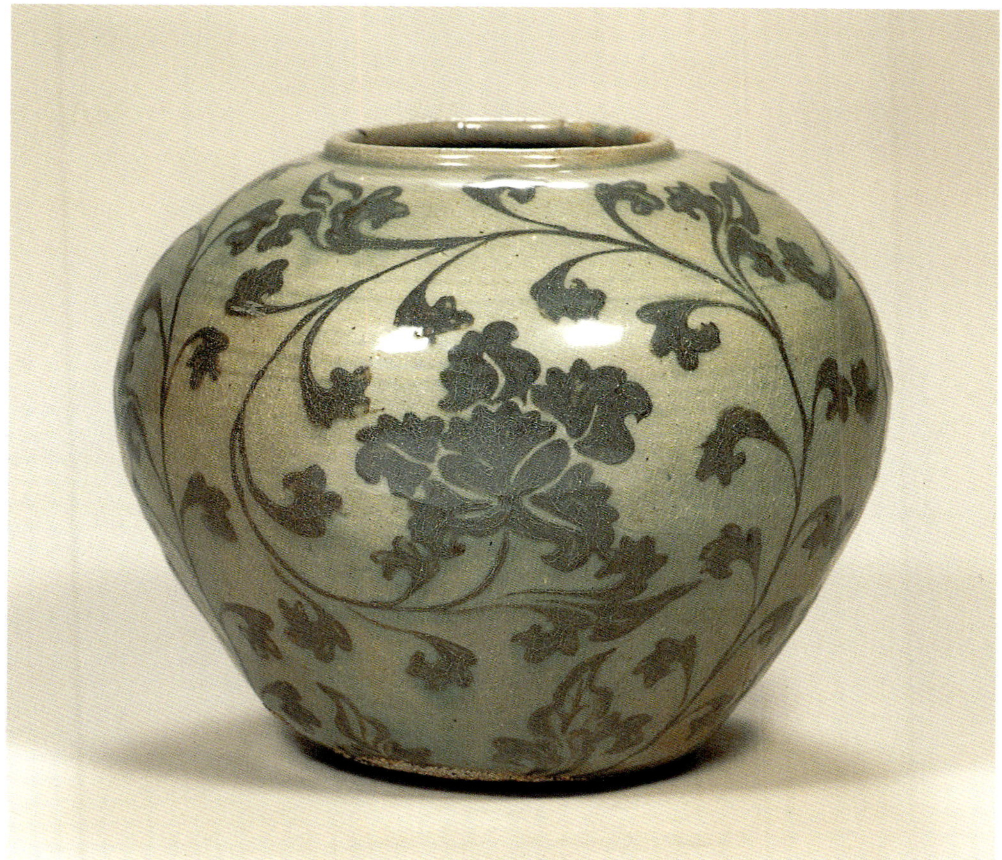

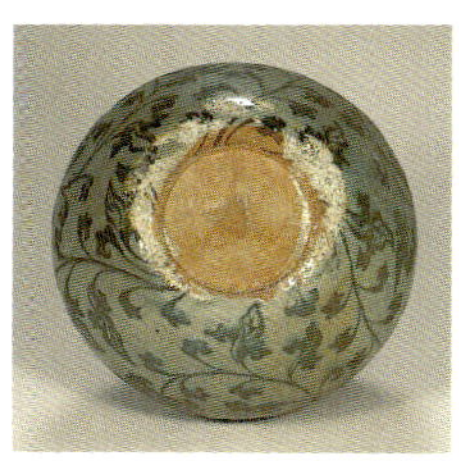

25

Jar with an underglaze iron-brown floral-scroll design

Celadon

Koryŏ dynasty, twelfth century

22.7 cm

Acc. no. 20342

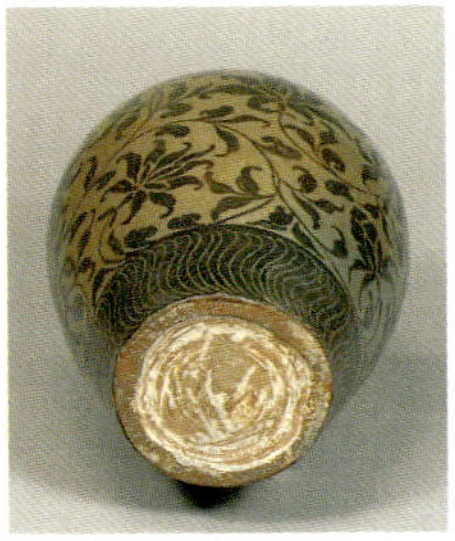

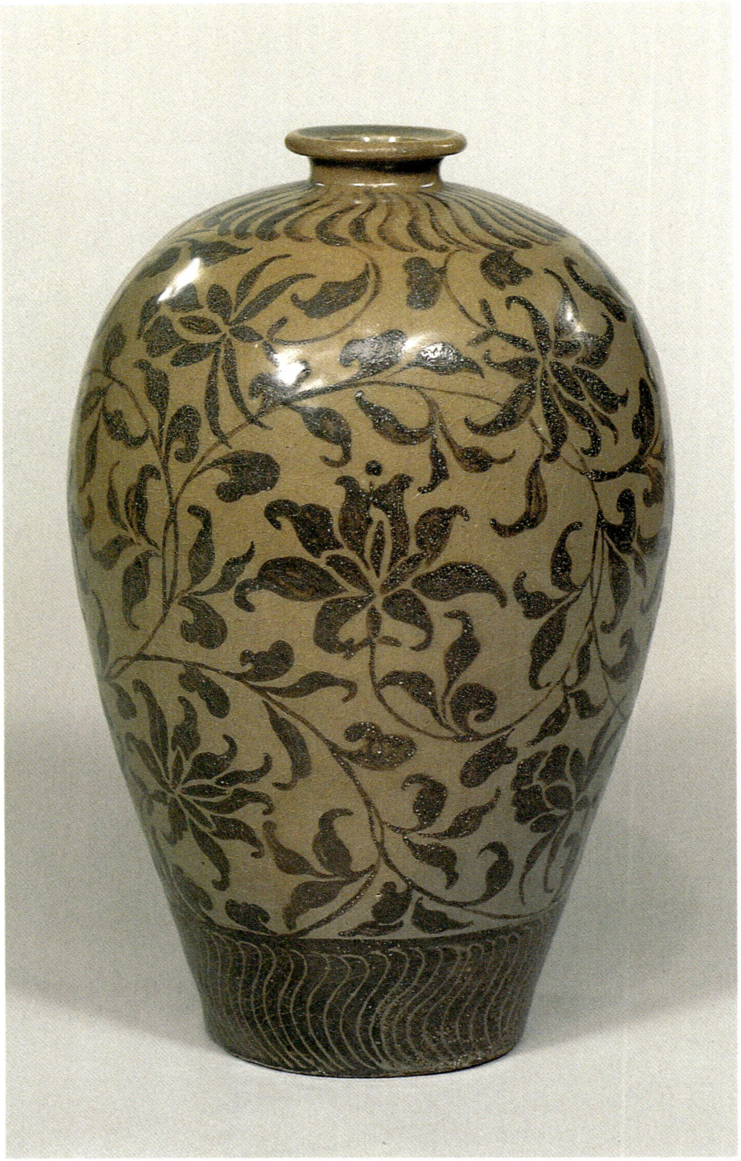

26

Maebyŏng *with an underglaze iron-brown floral-scroll design*

Celadon

Koryŏ dynasty, first half of the twelfth century

28.2 cm

Acc. no. 20862

From an early date, it was known that iron-painted celadon wares were produced in Sadang-ri and Yongun-ri in Taegu-myŏn, Kangjin-gun, South Chŏlla province, both of which were also centers of production for Koryŏ celadon wares. With the discovery of a large-scale kiln site at Jinsan-ri, Sani-myŏn, Haenam-gun, South Chŏlla province, in 1980, however, it became clear that iron-painted and iron-ground celadon wares were produced in great quantities. Since the investigation of the kiln site has not been completed, its period of activity has not yet been determined; but scholars speculate that it extended from the middle of the eleventh through the beginning of the twelfth century.

The design of this *maebyŏng*, which consists of sprays of flowers (Kor. *bosanghwa*; Chin. *baoxianghua*) encircled by classical scrolls, is elegantly organized. On the shoulder, the *S*-shaped lines of the pattern were painted with iron pigment, while on the lower part of the body they were executed in the sgraffito technique. This *S*-shaped design is clearly related to Chinese Cizhou wares. While development of the technique of iron-painted celadon is thought to have been influenced by Yue wares or the ceramics of the Xicun kiln near Canton, the impact of Cizhou and Yaozhou designs is also evident. But in this vase, which is more typical, the impact of those influences has ceased, revealing a style that has undergone a complete transformation, having been modified by Koryŏ adaptations. Evidence of oxidation can be discerned over most of the surface. The glaze, which is light brown in color, was applied to the base of the vessel, after which the foot and inner area were scraped off.

Iron pigment can be used in two ways on celadon ware: the design can be painted in iron pigment; or the entire vessel, with the exception of the underside, can be coated with the pigment. In both instances, a celadon glaze is then applied and the vessel is fired. Iron-coated vessels may be categorized into four types according to decoration: the first are those without decoration; the second, those with a white inlaid design; the third, those with a white-slip-painted design; and the fourth, those with an incised design. Most numerous among extant examples are those with white inlaid designs.

Most iron-oxidized-slip vessels with white inlaid designs, such as the example here, are made in the shape of *maebyŏng*, but other forms decorated in this manner include bottles, ewers, and bowls. The types of designs on these vessels are seldom found on either pure celadon or inlaid celadon ware. The willow design of this *maebyŏng* is unique; its rough, unpretentious expression suggests that it is a product of a South Chŏlla province kiln. Although iron-oxidized-slip celadons, like iron-brown celadons, had been found at Sadang-ri, Taegu-myŏn, Kangjin-gun, South Chŏlla province, it was not clear until the discovery at Jinsan-ri, Sani-myŏn, Haenam-gun, South Chŏlla province in 1980 that they were produced in great numbers. The treatment of the base is unusual; it was painted with iron slip but is unglazed.

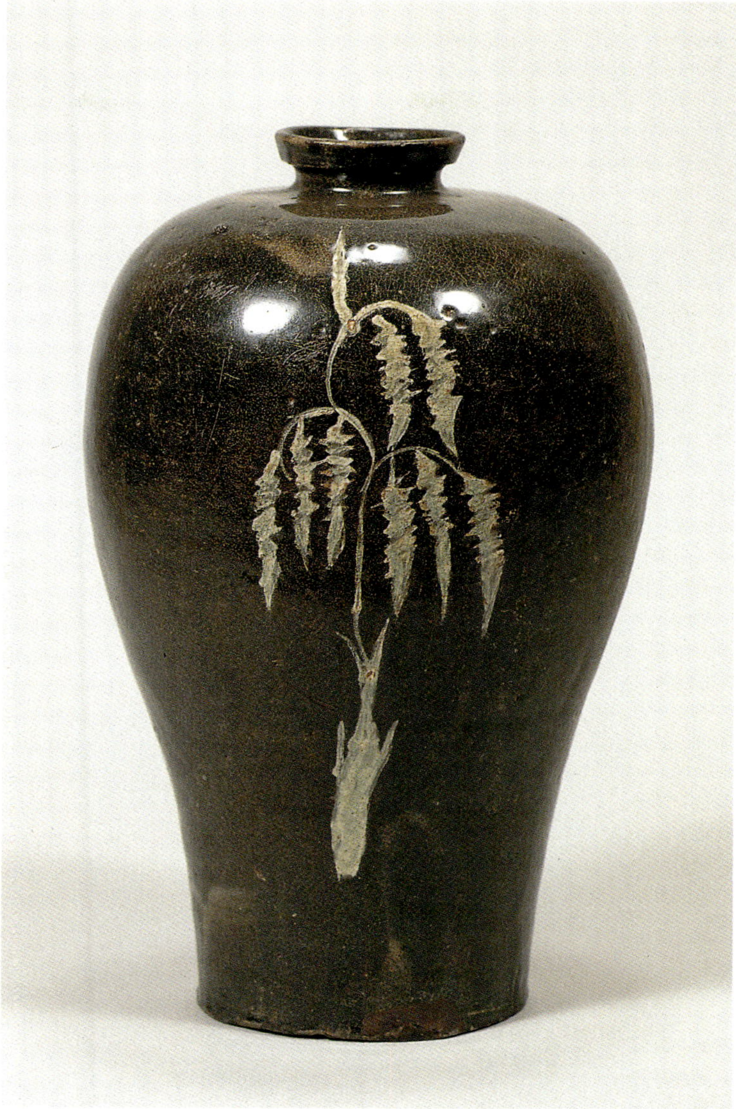

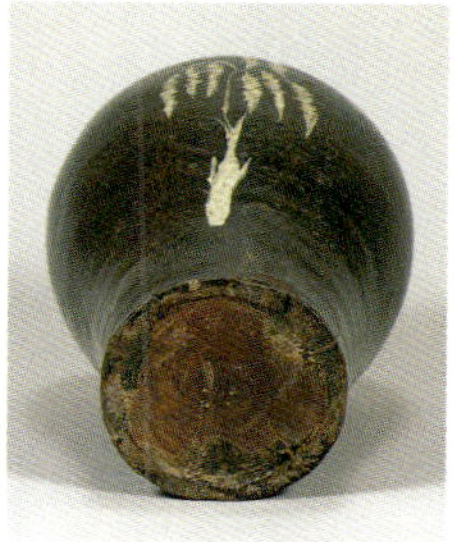

27

Maebyŏng *with an iron-oxidized slip and an inlaid willow design*

Celadon

Koryŏ dynasty, twelfth century

25.7 cm

Acc. no. 20523

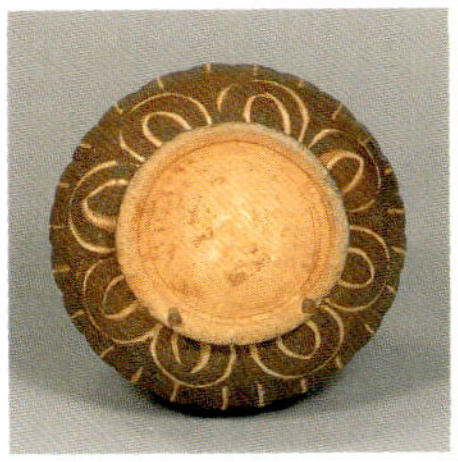

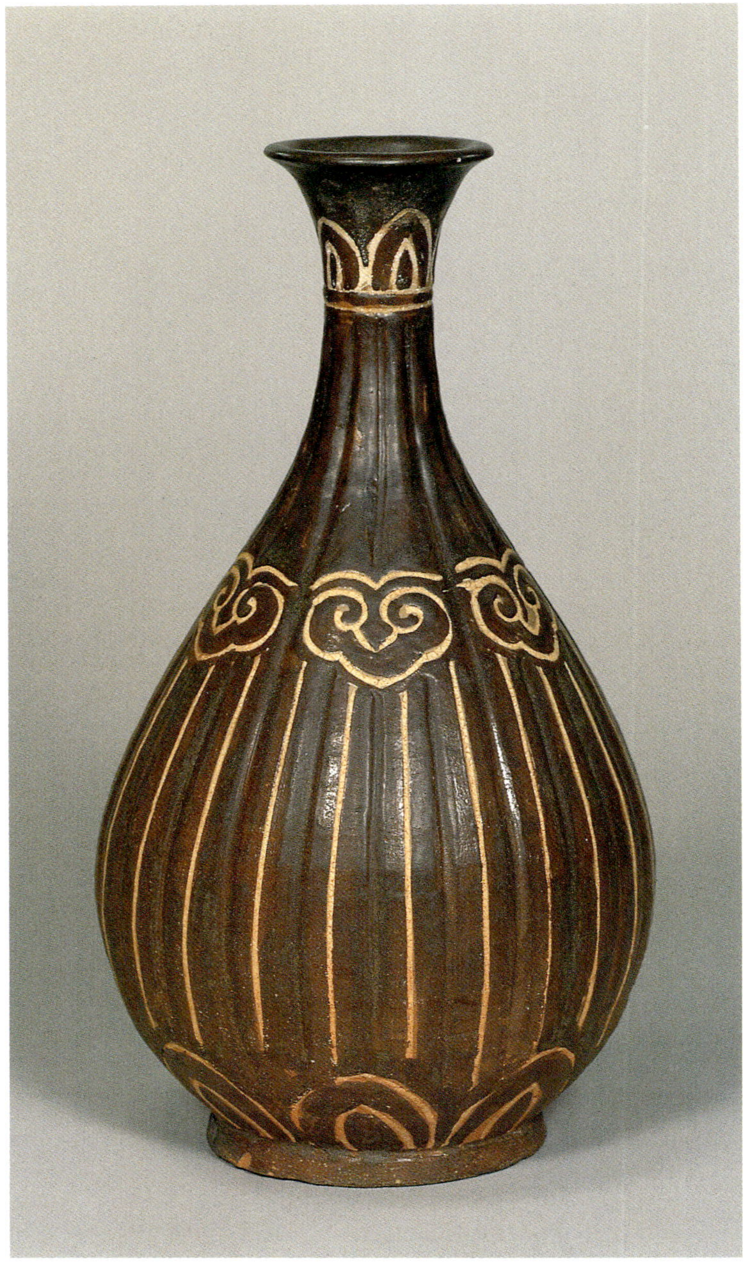

28
Bottle with a carved lappet design

Iron-brown-glazed stoneware

Koryŏ dynasty, second half of the thirteenth century

34 cm

Acc. no. 20785

In addition to celadon ware, white-glazed, black-glazed, and iron-brown-glazed wares were also produced in the Koryŏ period. Iron-brown-glazed ware was fired after a glaze with a high iron content was applied directly to the body; black-glazed ware had a higher content of iron in the glaze. However, extant examples of these two wares are rare. With the exception of the piece exhibited here, iron-brown-glazed bottles, in particular, are found in no other collection. This is the only iron-brown-glazed bottle known.

This bottle has been painted entirely with an iron glaze, except for its base, and distinct areas have been carved away to create surface designs. Its neck and base are decorated with carved lotus petals and its shoulder with a carved lappet design. The elaborate fluting over the remainder of the surface was carved into the body before firing and was carved again after application of the glaze. Iron-glazed ware of this type is thought to have been produced under the influence of Chinese Cizhou ware. Two iron-glazed flasks in the National Museum of Korea, Seoul, and in the Museum of Oriental Ceramics, Osaka, both of which date to the second half of the thirteenth century, support a similar dating for this bottle. The base of the vessel is covered with an iron slip but is unglazed.

This *maebyŏng*, like the one in no. 27, was coated overall with iron pigment, except for its base, and fired after an application of celadon glaze. Since the vessel appears to be black, it was mistakenly thought at one time that a black glaze had been applied to its surface. Many examples of this type, including this vase, have white inlaid decorations, but the inlay method used here apparently differed from the usual one. Typically, a design was carved into the partially dried clay body and was inlaid in white or red clay, after which the vessel was fired. A celadon glaze was then applied, and it was fired again. However, in the case of iron-coated celadon ware, the vessel was fired before the iron pigment was applied and the design was carved and inlaid. Consequently, the edges of the inlaid design are often fuzzy and imprecise.

The inlaid pattern on this *maebyŏng* depicts ginseng-plant leaves. The organization of the shape and the expression of the design are quite elegant, and because of the refined and uniform appearance of the color, the vessel is one of the most representative examples of its type. The shallow, cup-shaped mouth is attached to the shoulder in a flawless manner characteristic of *maebyŏng* at their height. On the basis of this shape, the vase was probably produced in the mid-twelfth century. Glaze was applied to the base, with one area left unglazed.

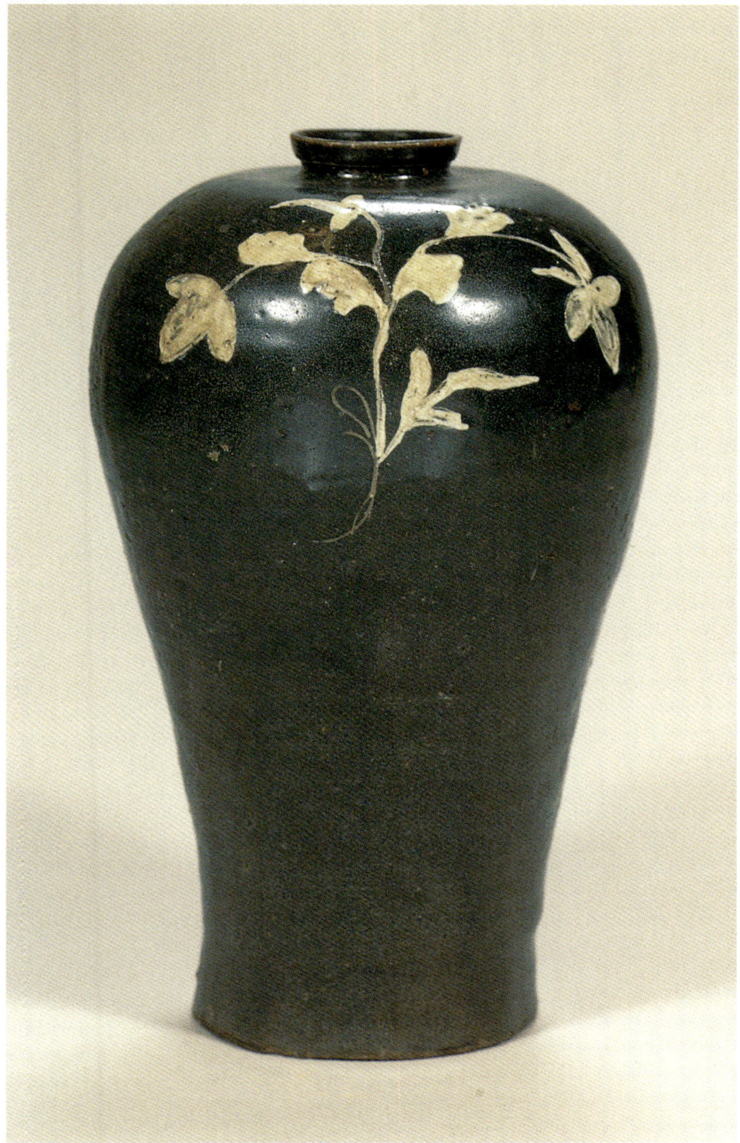

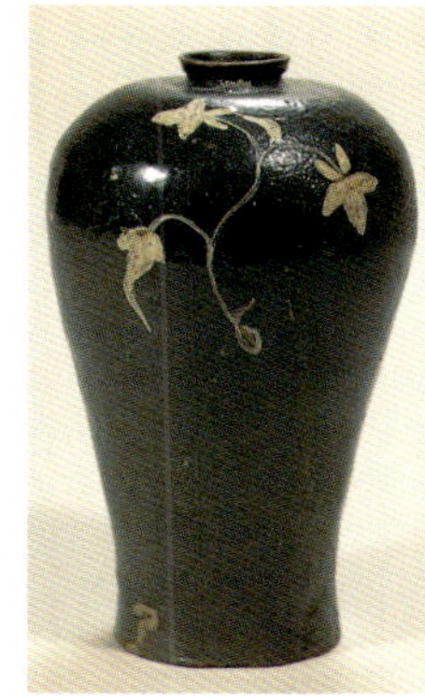

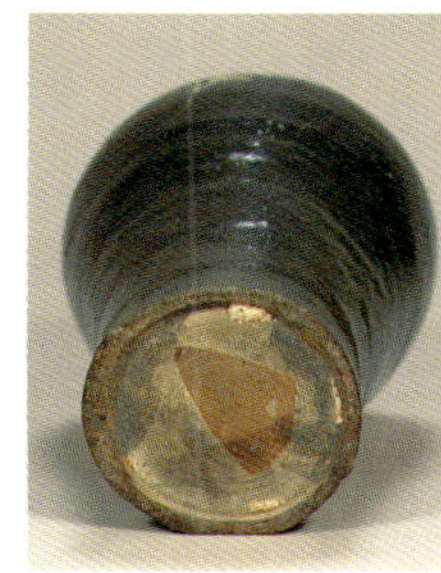

29

Maebyŏng *with an iron-oxidized slip and an inlaid ginseng-plant design*

Celadon

Koryŏ dynasty, mid-twelfth century

26 cm

Acc. no. 20259

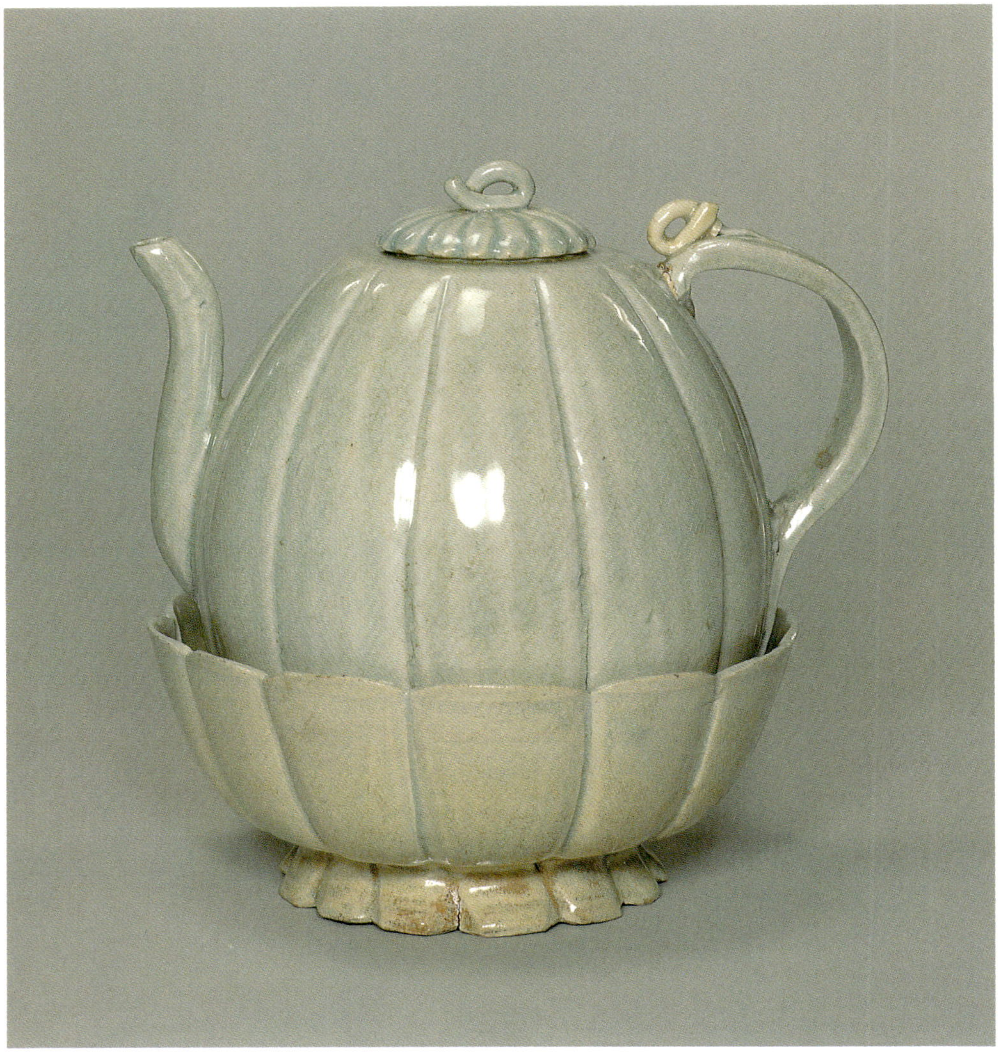

Numerous examples of Koryŏ celadon ware are known, but only a few extant examples of Koryŏ white ware survive. Yet, Koryŏ white-ware shards excavated at Yuch'ŏn-ri, Poan-myŏn, Puan-gun, North Chŏlla province (now in the Ewha Women's University Museum, Seoul) and at Sŏ-ri, Yidon-myŏn, Yongin-gun, Kyŏnggi province, indicate that white ware was produced in large quantities.

This ewer with a lid and warming bowl is rare even for Koryŏ white ware, since many Koryŏ ceramics were stolen from burial mounds and the lids and warming bowls of ewers, for example, were often lost. This ewer assumes the shape of a lobed melon, and the warming bowl, conforming to this shape, is also lobed. Taller and deeper warming bowls can be found in Chinese Ding and Jingdezhen ware of the Northern Song period. The warming bowl was so named in China because it warmed the wine within the ewer. The general Koryŏ practice of drinking wine cold may have prompted the creation of the shallower bowl seen here. The bases of the ewer and warming bowl each have six spur marks.

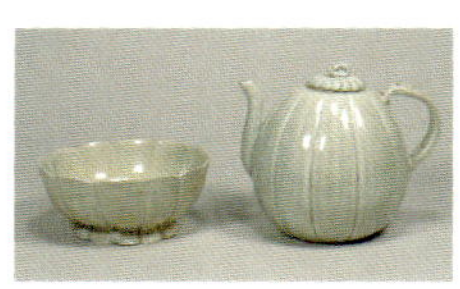

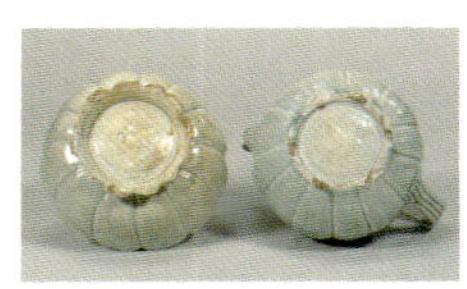

30

Lobed ewer and stand

Porcelain

Koryŏ dynasty, twelfth century

21.9 cm

Acc. no. 20313

Although this *maebyŏng* appears to be decorated with a white inlay, the design was not carved into the unglazed surface but was painted in white slip using a brush. Brush-painted designs are also found on iron-painted celadon, examples of which are far more common than the rare white-slip-painted ware.

The body of this vase is divided into three areas and is decorated with four clouds of fungus shape (Kor. *yongji*; Chin. *lingzhi*). The designs are executed according to a single pattern but differ slightly in size and shape. A certain freedom of expression can be detected in the brush marks on the rim of the mouth, in the varied arrangement of the large and small *S*-shaped designs on the shoulder, and in the carved key-fret pattern on the lower part of the body. This expressiveness contrasts with the tightly organized designs characteristic of inlaid celadon ware at its height. The larger size of this vase and its thicker walls and more pronounced *S*-shaped body all point to additional changes in construction and vessel shape introduced after the beginning of the thirteenth century. The foot is glazed and bears approximately seven spur marks.

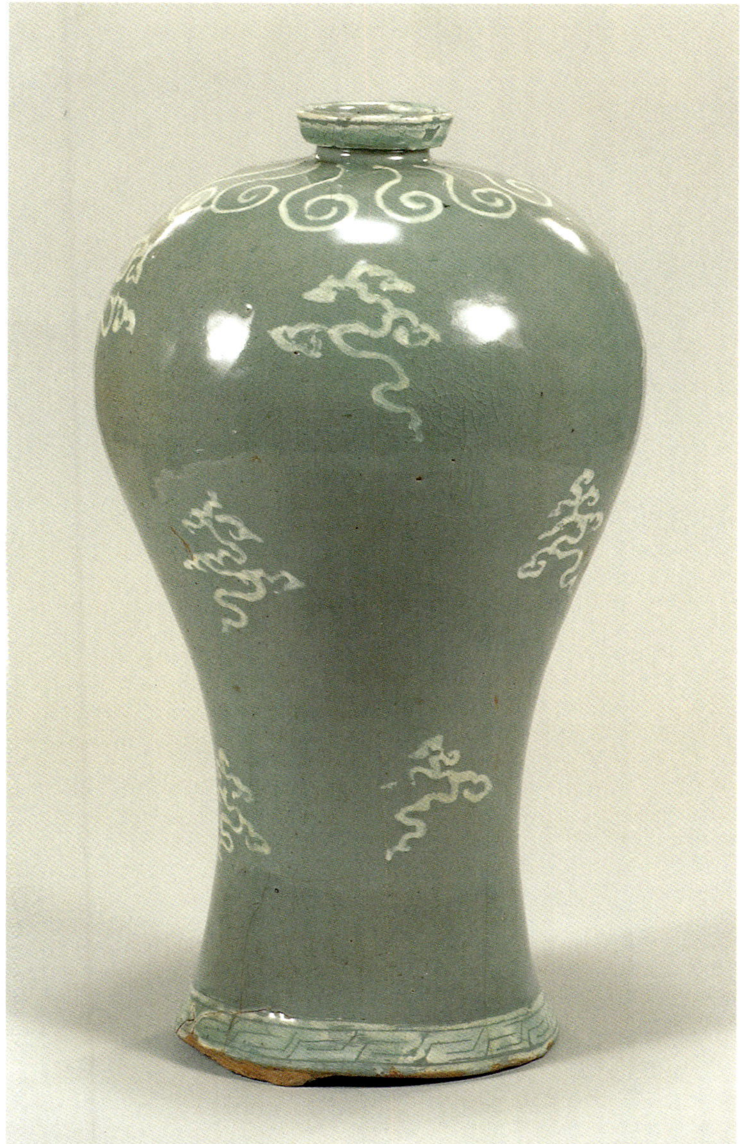

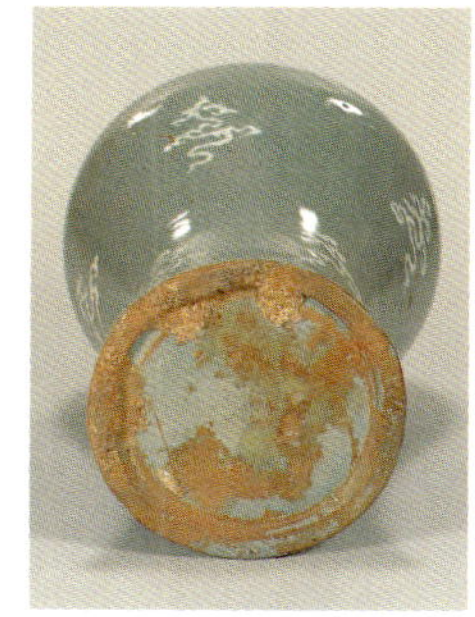

31

Maebyŏng *with a white-slip-painted cloud design*

Celadon

Koryŏ dynasty, first half of the thirteenth century

38.9 cm

Acc. no. 20249

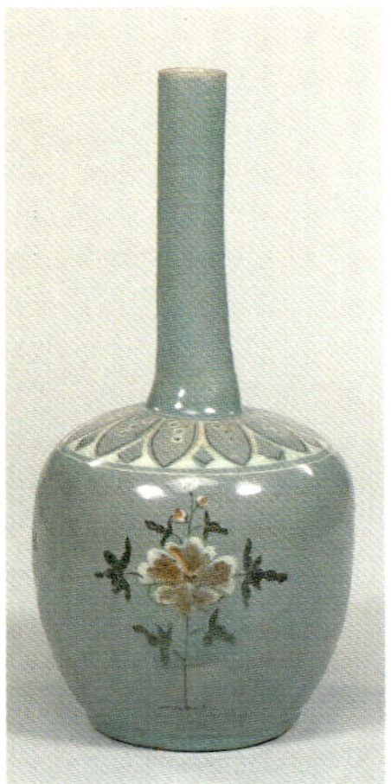

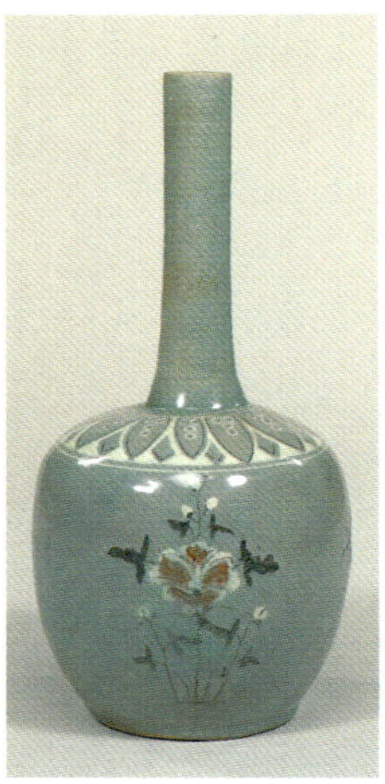

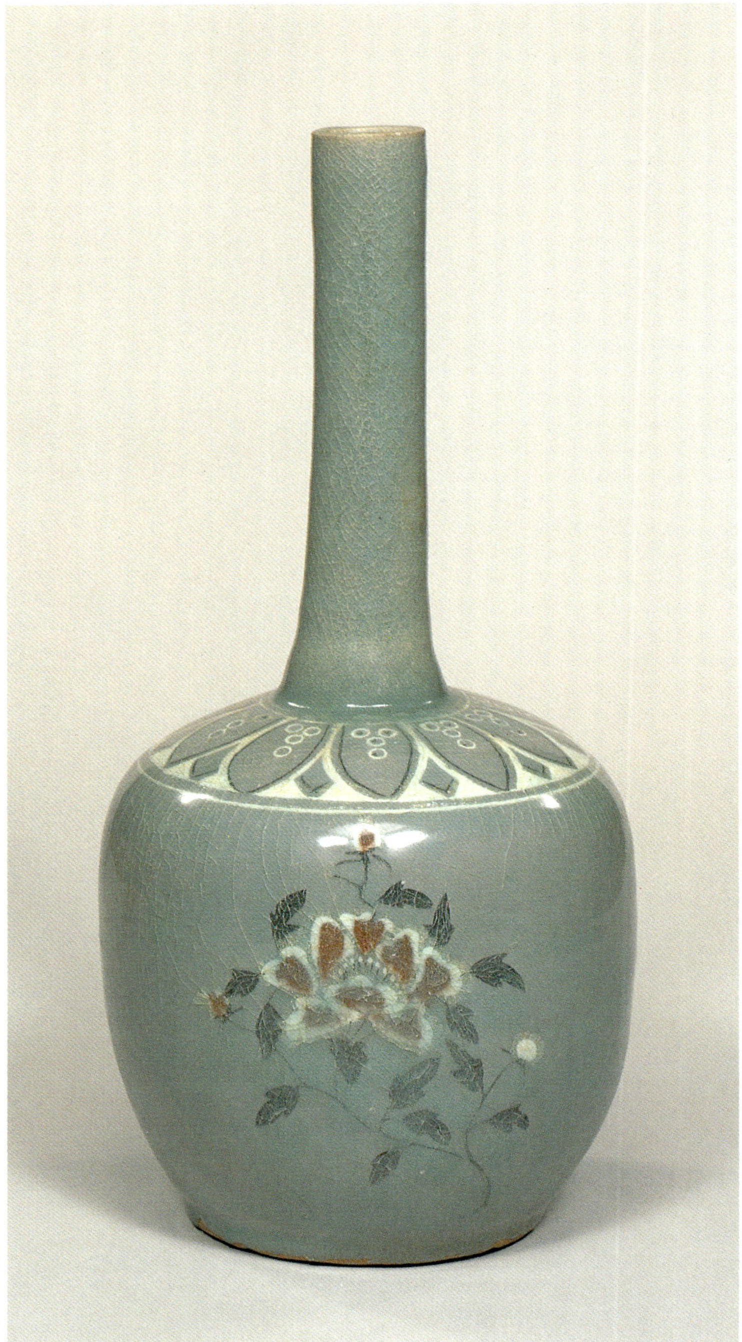

32

Long-necked bottle with underglaze copper-red spotting and an inlaid peony design

Celadon

Koryŏ dynasty, thirteenth century

34.5 cm

Acc. no. 20839

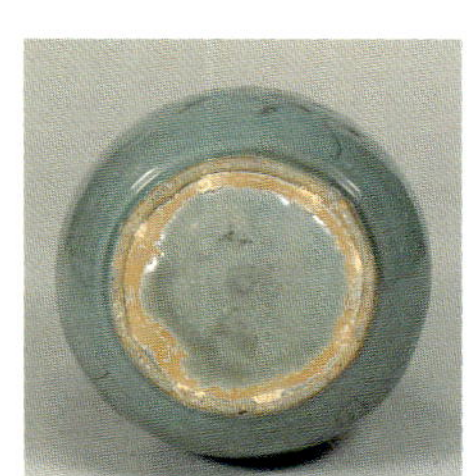

This celadon bottle is decorated with an oxidized underglaze copper-red design. This pigment was applied using one of three decorative methods: the entire surface could be coated with copper pigment; the design could be brush-painted in copper pigment; or the copper pigment could be applied over a white inlaid design. The first two were rarely employed, and most extant examples utilize the third. These decorative methods are never seen in Chinese ceramics but are peculiar to vessels of the Koryŏ period. Nevertheless, an attempt to copy the iron-red glaze of Chinese Ding ware may have been responsible for the Korean discovery of the copper-red glaze. This discovery may have occurred during the first half of the twelfth century, prior to the thirteenth century, when underglaze-red celadon ware flourished. In China, the use of underglaze copper-red pigment on white ware, not celadon, only appeared in the fourteenth century. Thus, in terms of the use of underglaze copper-red pigment, Korea preceded China by two centuries.

This long-necked bottle is decorated with three black-and-white inlaid peonies. Copper-red pigment was applied to parts of the white inlaid buds and flowers to create a refined effect. Among the small number of underglaze-red celadon pieces, there are even fewer that possess the elegant red color of this example, which further enhances its value. The neck of this bottle is not original and has been replaced with one from a similarly shaped vessel. The foot and surrounding area were scraped, and eight spur marks mixed with sand are visible.

This small bowl has a notched rim that turns slightly outward to assume the shape of a five-petaled flower, as is the case with the bowl in no. 34. Its center is ornamented with a chrysanthemum design, and chrysanthemum sprays appear in each of the five adjacent sections. The piece was not thrown on a potter's wheel but was made in a mold; this is apparent from the areas where the design is on the verge of disappearing, which indicates the use of a worn mold. Although the use of molds may be attributed to the influence of Chinese Yaozhou ware, in the Koryŏ period further decoration was added by hand. The five incised chrysanthemum sprays and the carved lines indicating flower petals were all added after the piece was molded. Such workmanship indicates that Koryŏ celadon vessels were not mass-produced but were made as unique objects.

A fine crackle covers the entire body. The glaze was well fired and has a strong greenish tone; it pooled at the center of the bowl, in particular, where it has a jewel-like, translucent, deep green hue. The construction of the bowl indicates a date of the mid-twelfth century. The bottom is unglazed and has three spur marks.

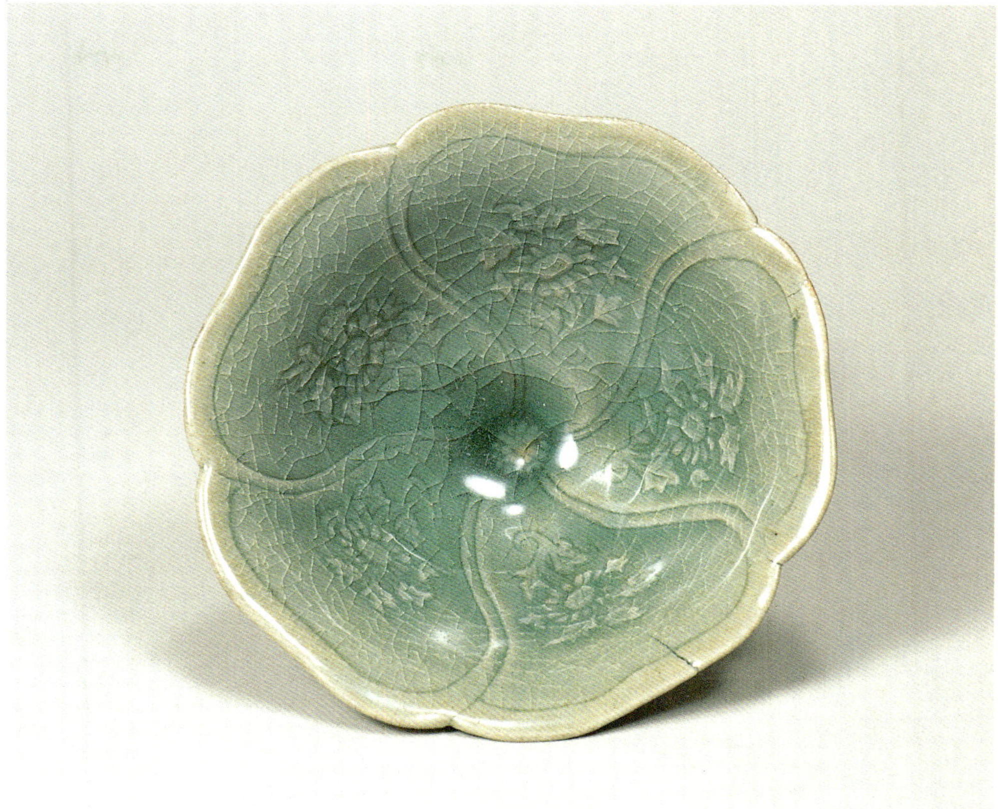

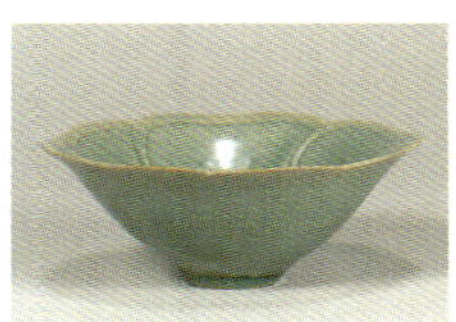

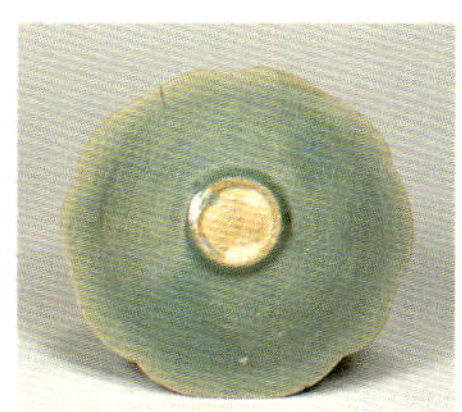

33

Bowl with a stamped chrysanthemum design

Celadon

Koryŏ dynasty, mid-twelfth century

11.7 cm (diam.)

Acc. no. 20350

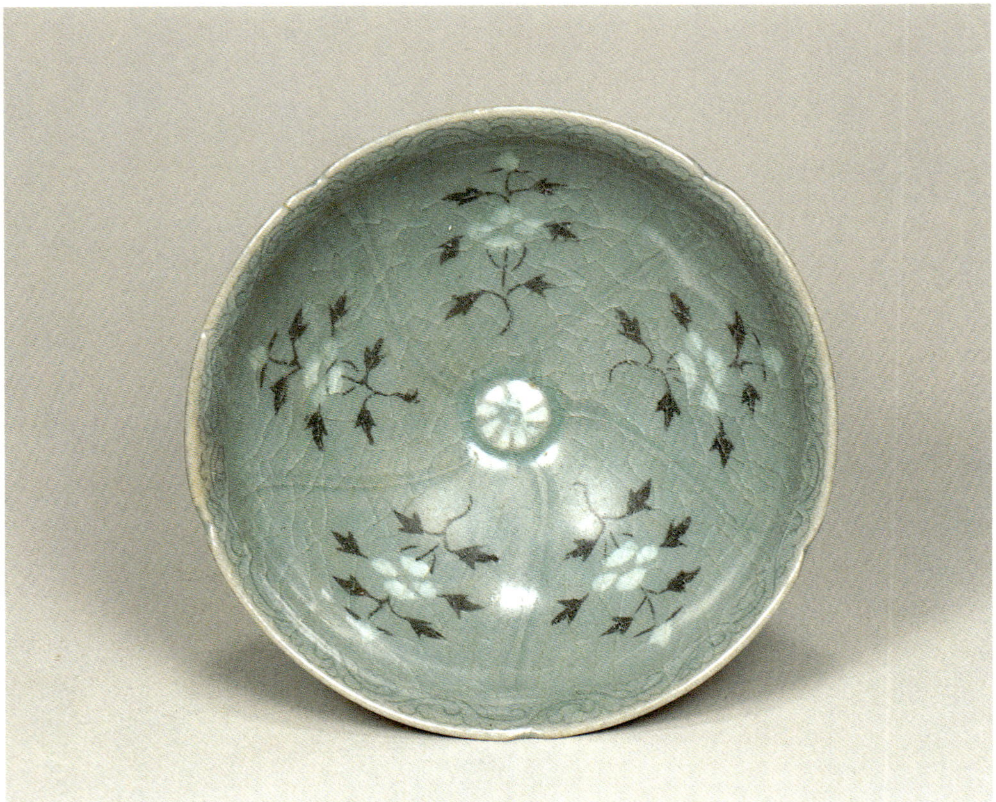

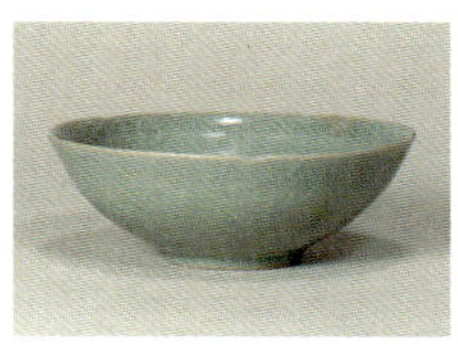

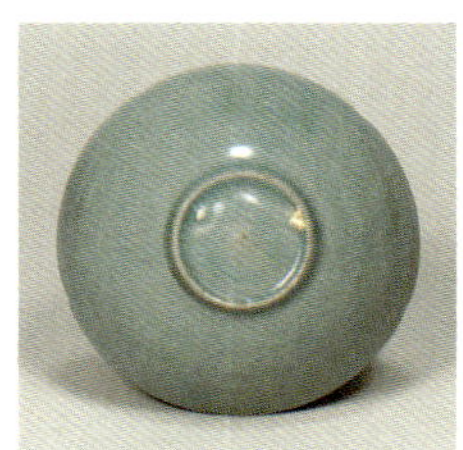

This small bowl may have originally been made as part of a wine cup and stand set. Its notched rim suggests the shape of a five-petaled flower. Its center is decorated with a white-inlaid chrysanthemum design with five black-and-white inlaid peony sprays arranged around it. The peonies are separated by the raised, curved lines delineating the flower petals that define the bowl's overall shape. A finely incised classical scroll pattern encircles the inner surface of the rim. Both inlaid and incised techniques indicate a date of about the mid-twelfth century, just prior to the period when the inlay technique attained its peak. The glaze is a grayish blue, with the blue tone predominating. The foot is glazed and bears three spur marks.

34

Bowl with an inlaid floral design

Celadon

Koryŏ dynasty, mid-twelfth century

10.9 cm (diam.)

Acc. no. 20356

How and when the application of an iron-brown design under a celadon glaze first occurred is still unclear. It has been suggested that this technique originated in response to southern Chinese iron-painted celadon ware; similarities in such elements as the classical scroll pattern and overall design support this argument. However, in the overall design, the influence of Cizhou ware from northern China is also apparent. Among Koryŏ celadons are many examples that indicate that production techniques and surface designs do not necessarily originate from the same source. It should be noted that the iron-painted celadons of the Chinese Yue tradition had a strong effect on the origin of Koryŏ celadon ware. It is quite possible that Yue ware was introduced first in Korea, and decoration originating from both the north and south of China was adopted later.

This bowl, although small, is elegantly designed. Its interior and exterior are each decorated with four peonies surrounded by classical scroll patterns. The veins of the leaves were carved to accent the design. Due to slight oxidation, a brownish hue pervades the glossy and beautiful glaze. Meticulous workmanship suggests a mid-twelfth-century date for the piece. After glaze was applied to the base, it was almost entirely scraped off.

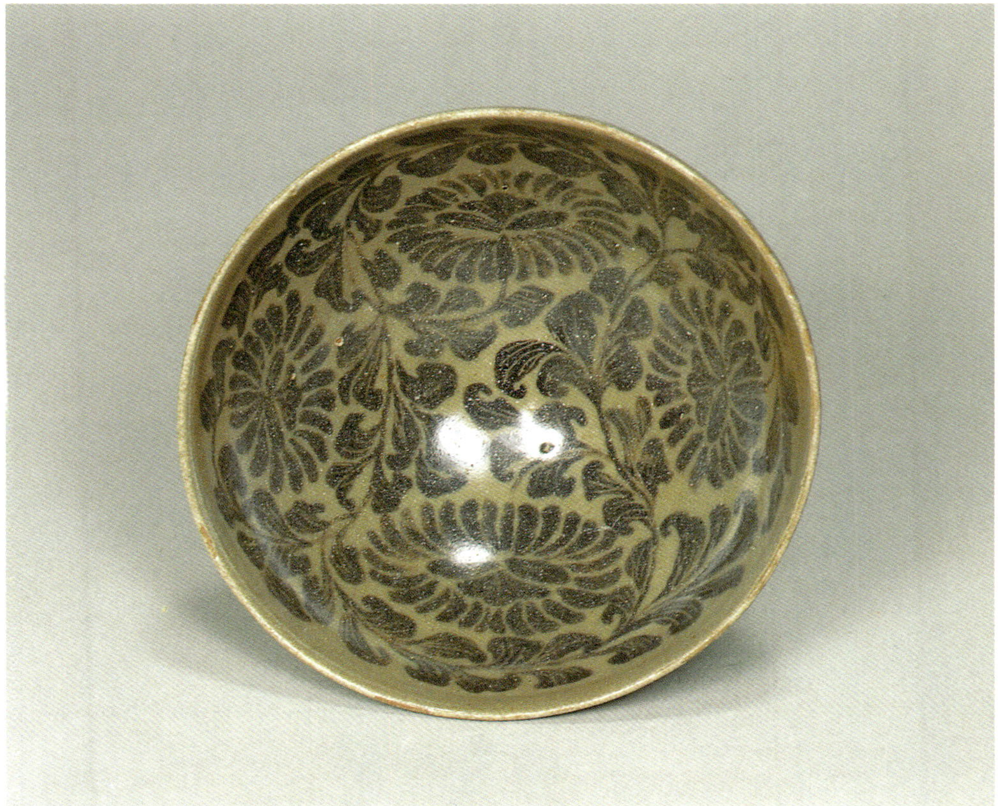

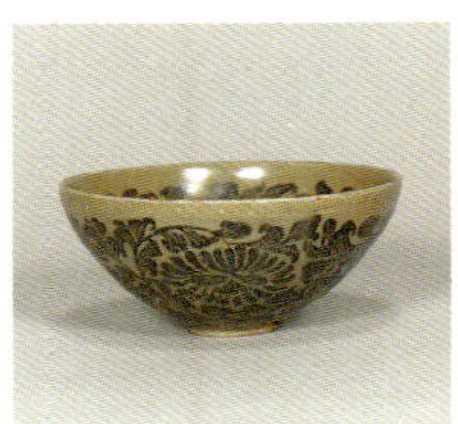

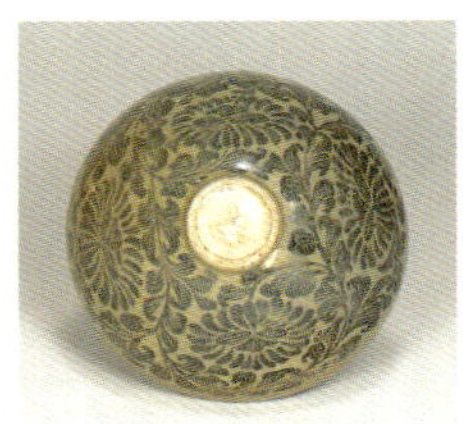

35

Bowl with an underglaze iron-brown peony-scroll design

Celadon

Koryŏ dynasty, mid-twelfth century

10.5 cm (diam.)

Acc. no. 21020

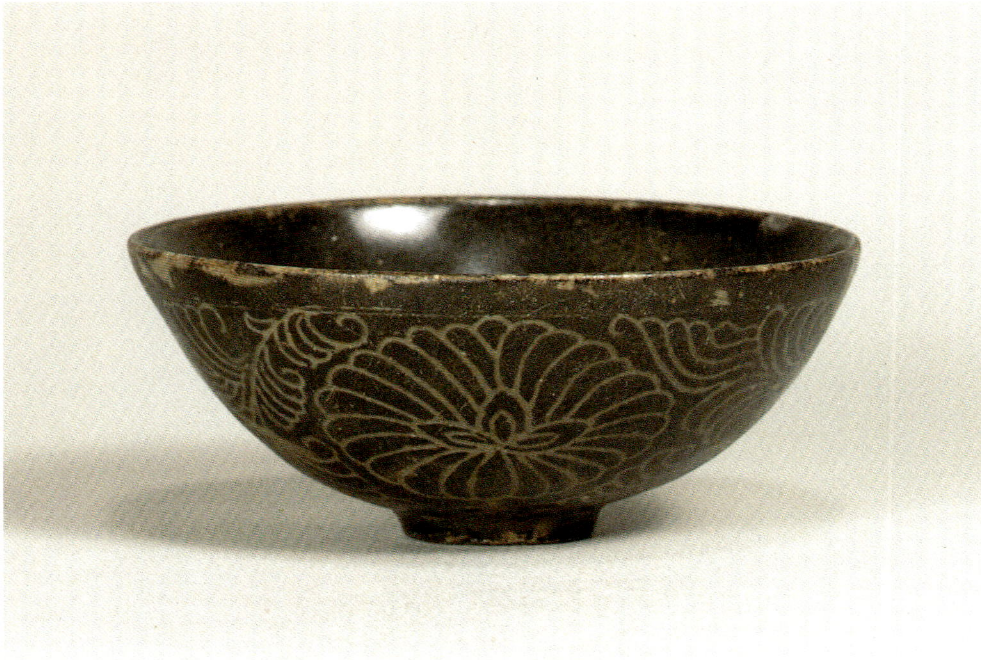

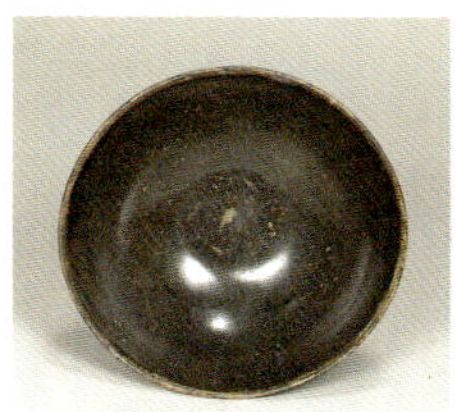

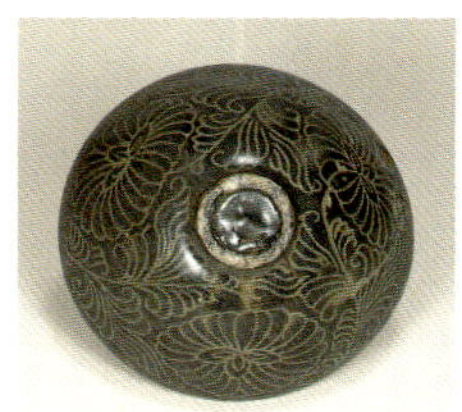

Among Koryŏ ceramics that appear to have a black glaze, it is difficult to distinguish between the rare vessel with a true black glaze and the more numerous examples painted with iron pigment under a celadon glaze. This small bowl is of the second type and was painted with iron pigment under a transparent glaze. After the design was incised, a celadon glaze was applied and the piece was fired. The grayish blue of the celadon glaze can be detected only in the bowl's finely incised designs. Extant examples of this type are rare; among the few known is a famous long-necked bottle in the Fitzwilliam Museum, Cambridge, England.

The design consists of three peony flowers on the exterior of the bowl with peony scrolls filling the surrounding area; the interior is free of decoration. This peony decoration is similar to that of Chinese Yaozhou ware and Xicun ware from Guangzhou province. Shards of a similar type have been found at the kiln site of Sadang-ri, Taegu-myŏn, Kangjin-gun, South Chŏlla province. On the base, the foot rim alone is unglazed. The spur marks are unclear, and it is difficult to cite their exact number.

36

Bowl with an iron-oxidized slip and an incised peony-scroll design

Celadon

Koryŏ dynasty, twelfth century

10.9 cm (diam.)

Acc. no. 20125

Marbled ware refers to vessels made from two or three different kinds of clay that have been stacked on a board and kneaded. The intermixing of the clay produces a naturally irregular design much like that of wood grain or marble. This type of ware was popular in China from the Tang through the Northern Song periods, but there two kinds of clay were mixed and a yellow glaze applied over the surface. This technique differs fundamentally from that of Koryŏ marbled ware, in which three kinds of clay (pure, brown, and white) were used, and the piece was finished with a celadon glaze. A considerable number of Chinese marbled pieces exist, but there are no more than ten extant examples of Koryŏ marbled celadon, including such small pieces as bowls, wine cups, covered boxes, and little bottles.

Marbled ware was discovered at the twelfth-century kiln site of Sadang-ri, Taegu-myŏn, Kangjin-gun, South Chŏlla province; this bowl probably dates to that period. Carefully constructed from its small base to its sides, which extend straight upward, its well-proportioned shape and beautiful glaze indicate that it is an excellent example of its type. The base is unglazed.

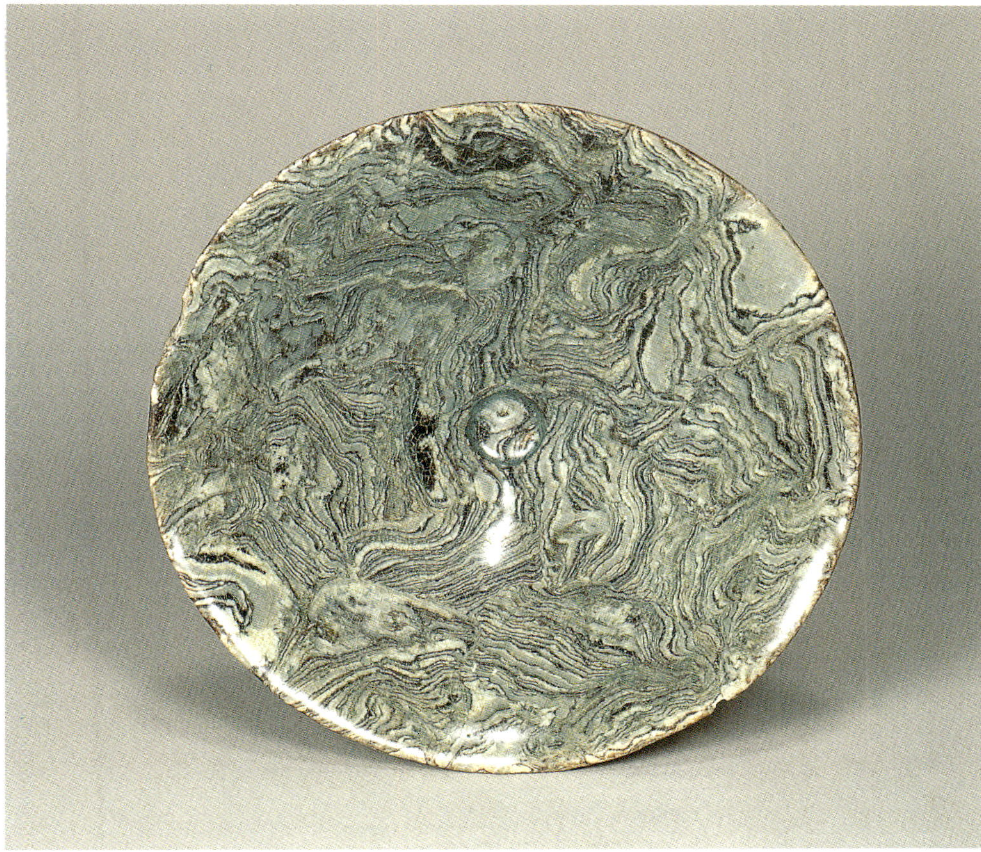

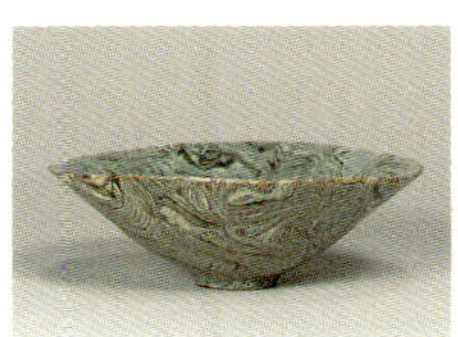

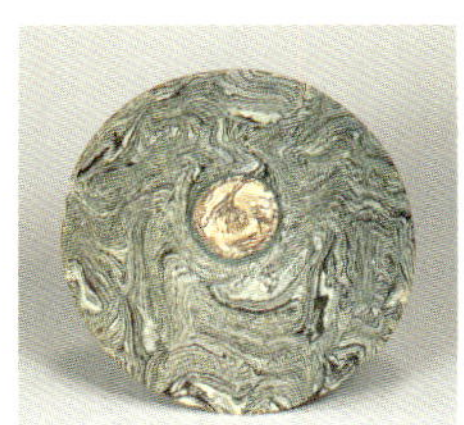

37

Bowl with a celadon glaze

Marbled ware

Koryŏ dynasty, twelfth century

14 cm (diam.)

Acc. no. 20248

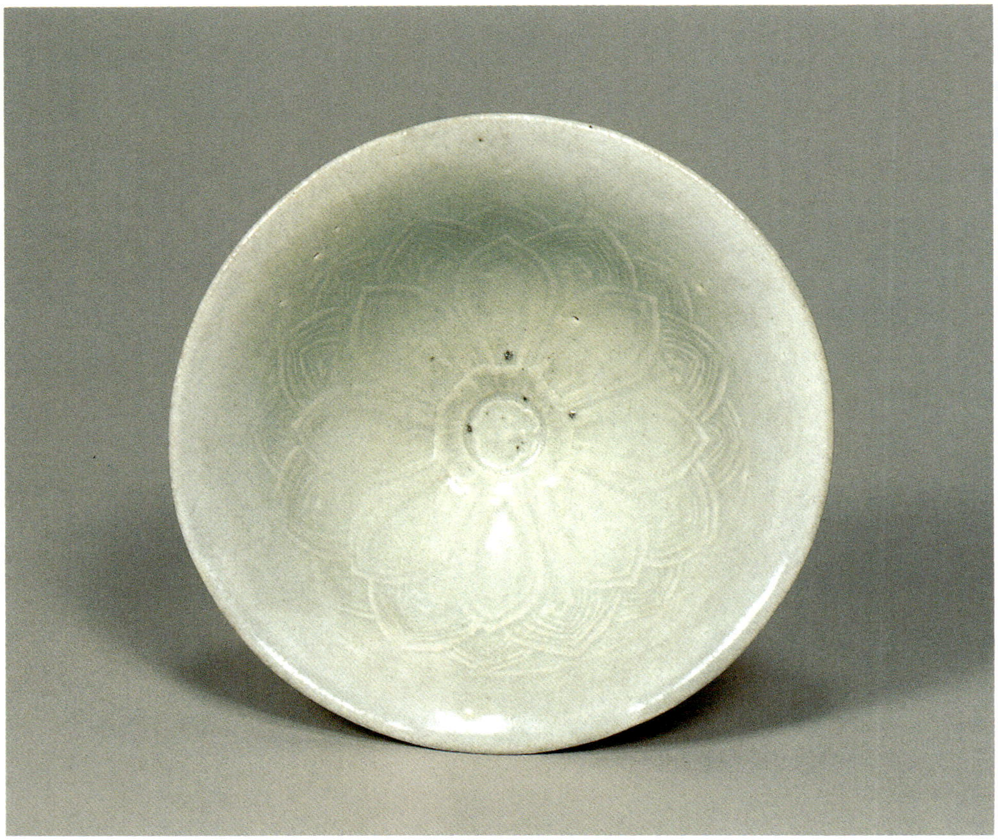

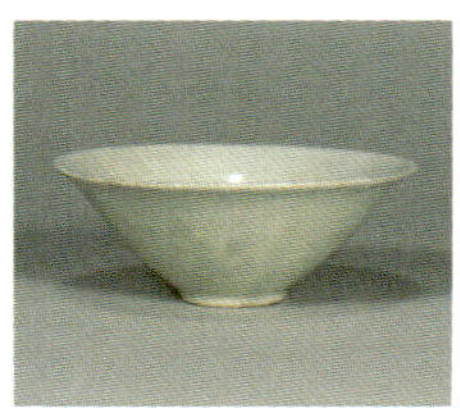

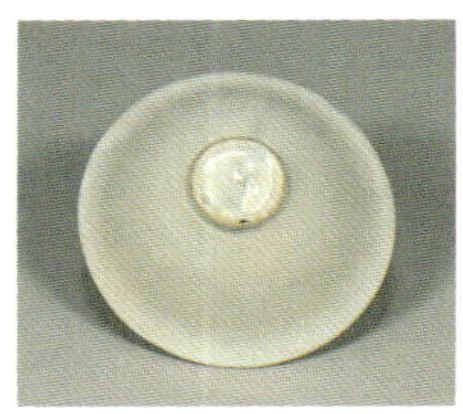

38

Bowl with a stamped lotus-petal design

Porcelain

Koryŏ dynasty, twelfth century

11.7 cm (diam.)

Acc. no. 20553

Of the few surviving examples of Koryŏ white-glazed ware, there is certainly nothing equal to this exquisitely decorated bowl. Koryŏ white-glazed ware may be decorated with incised, carved, impressed, or inlaid designs. This small bowl was manufactured from a mold and is ornamented with a splendidly detailed stamped design of lotus petals arranged in three layers. A Koryŏ celadon bowl of the same size and design is found in the National Museum of Korea, Seoul. Since white-glazed and celadon wares were known to have been made at the same kiln, these two bowls may have been manufactured from the same mold. Many shards of white-glazed ware were discovered at the kiln site of Yuch'ŏn-ri, Poan-myŏn, Puan-gun, North Chŏlla province; and white ware of lesser quality was found at two sites, a provincial kiln at Sŏ-ri, Yidon-myŏn, Yongin-gun, Kyŏnggi province, and another kiln at Sadang-ri, Taegu-myŏn, Kangjin-gun, South Chŏlla province. Although the development of white ware remains unclear, it may have been produced from the first half of the tenth century until the end of the fourteenth century. The base, except for the foot, is glazed.

This small bottle stands only 8.6 centimeters high but is unique in its careful construction and its finished form and design. Its mouth is ornamented with a band of key-fret patterns, its neck with a string of jewels, and on its shoulder and around its base are bands of lotus petals. The main design consists of four flying cranes on a background filled with clouds of fungus (Kor. *yongji*; Chin. *lingzhi*) shape. These are rendered entirely in white inlay with black inlay limited to the edges of the lotus petals and to the cranes' heads and legs. Moreover, the string of jewels and fungus-shaped cloud patterns are stamped and inlaid. The tendency to vary the use of white inlay and stamps is noticeable after the beginning of the thirteenth century. The glaze is perfectly fired, producing a glossy grayish blue with a strong gray tone, and the base is also glazed.

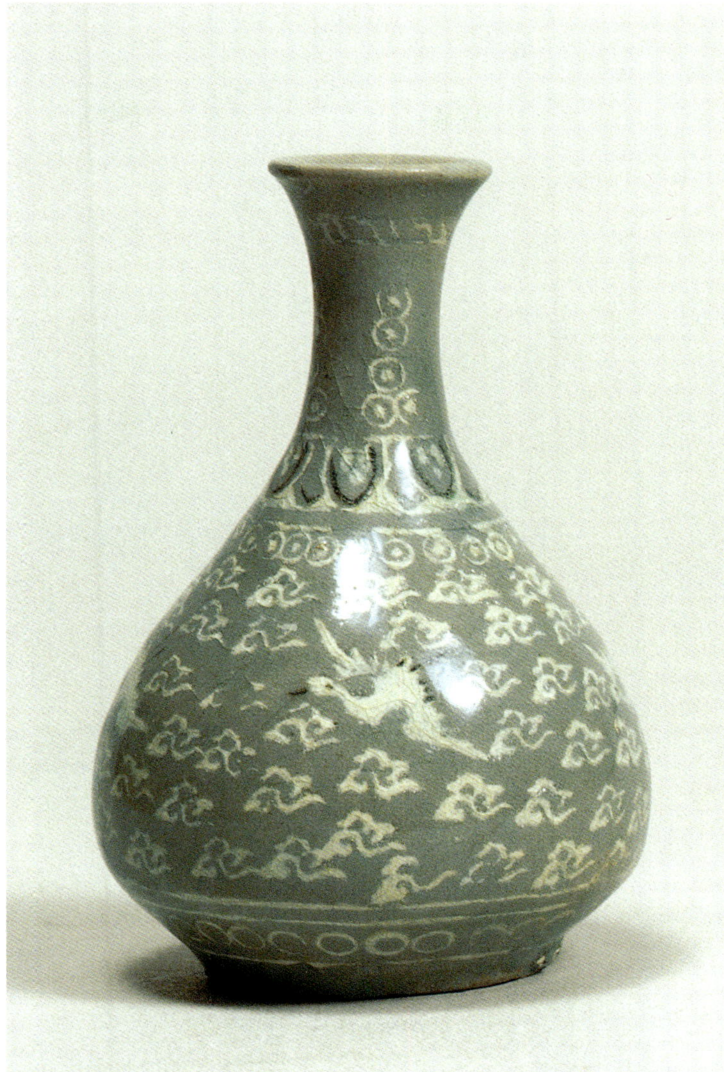

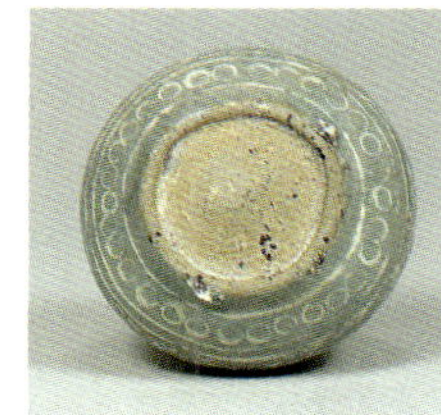

39

Bottle with an inlaid crane and cloud design

Celadon

Koryŏ dynasty, early thirteenth century

8.6 cm

Acc. no. 21030

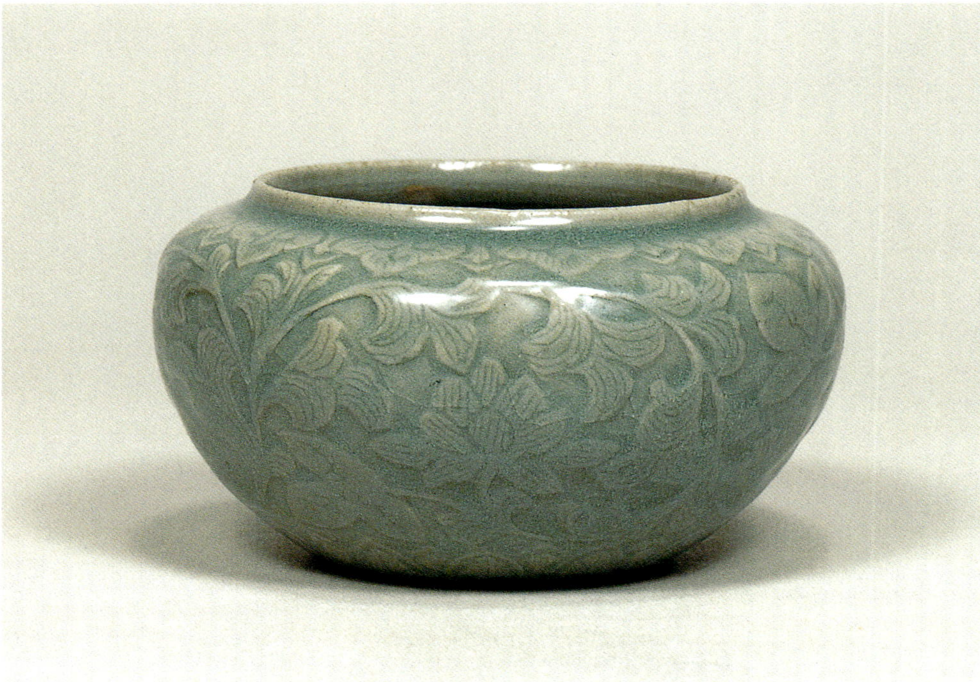

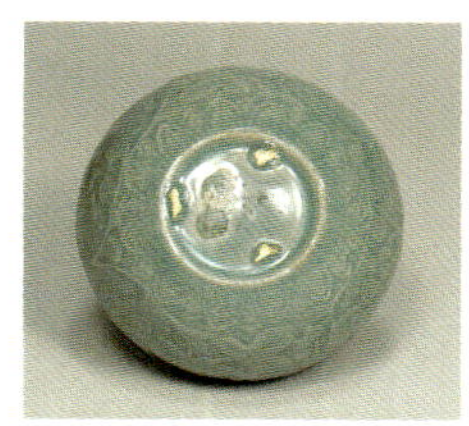

Extant examples of shallow jars with wide mouths, such as this one, are practically nonexistent. A carved band of lappet patterns encircles the rim below the mouth, and carved peonies and peony scrolls adorn the body. The finely incised lines of the flower petals and the veins of the leaves heighten the quiet and elegant effect of the design. Moreover, the beautiful and uniform grayish blue-green glaze indicates a date of the mid-twelfth century, when Koryŏ celadon reached its height and the so-called jade-colored celadon was more valued than gold or silver. In China, such celadon was considered one of the "most precious things on earth," along with the flowers of Luoyang and the inkstones of Duanxi. This small jar serves as one piece of evidence testifying to the truth of that statement. The base is entirely glazed, and three small spur marks are visible.

40

Jar with a carved peony design

Celadon

Koryŏ dynasty, mid-twelfth century

6.2 cm

Acc. no. 20327

Koryŏ celadon-covered cosmetic boxes are of two types: a tall box, used as a medicine or reliquary container; and a flat box, generally used as a container for toilet articles. Examples also exist of a rectangular celadon box into which an oil bottle and a flat covered box fit as a set.

The lid of this box is decorated with a seven-pointed star with seven chrysanthemum flowers placed between the points; these surround an eight-petaled flower. This design, which is unusual and slightly excessive for Koryŏ celadon, may have astronomical significance. Generally, a thick glaze on an inlaid celadon obscures the pattern, while a thinly applied glaze may allow it to penetrate the glaze. This box's clearly delineated design and beautiful glaze indicate that it was fired under ideal conditions. Glaze, design, and construction reveal the careful workmanship with which this piece was made and suggest a date of the second half of the twelfth century.

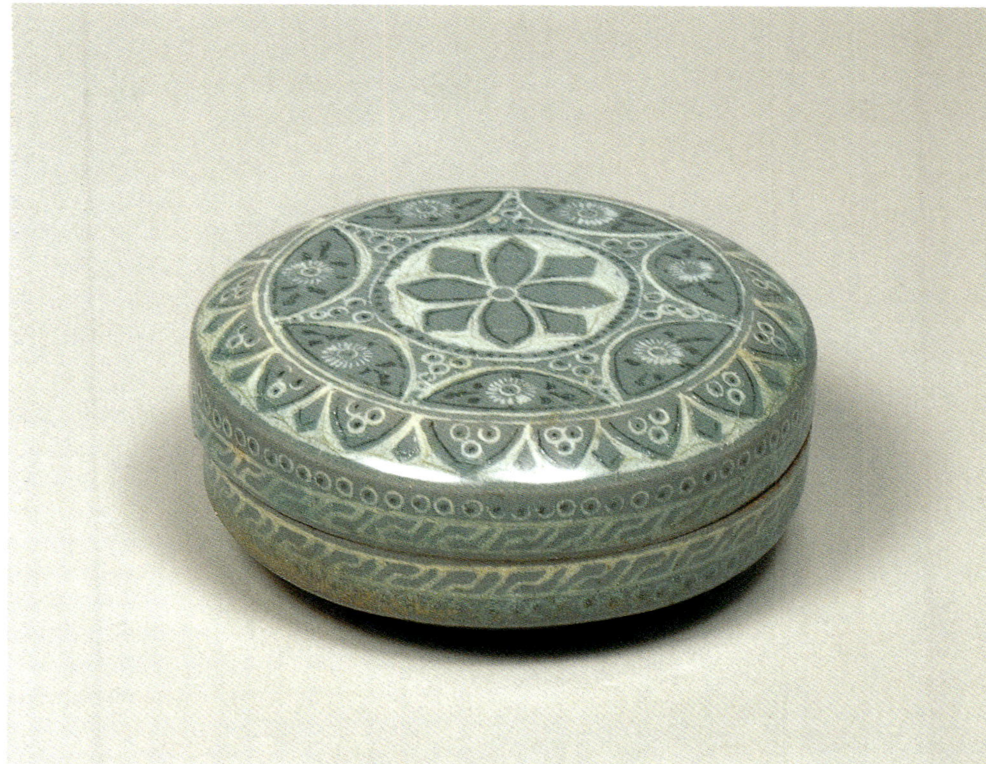

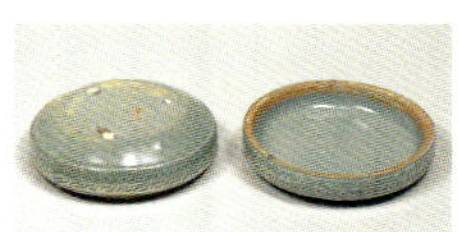

41

Covered cosmetic box with an inlaid floral design

Celadon

Koryŏ dynasty, second half of the twelfth century

4.4 cm (diam.)

Acc. no. 20366

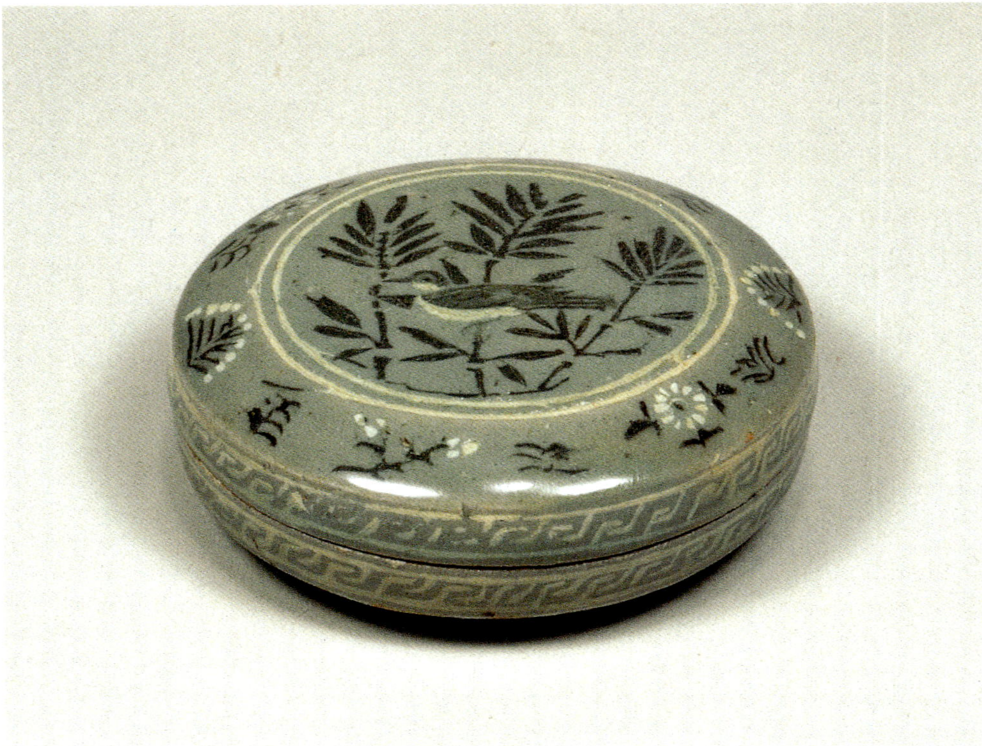

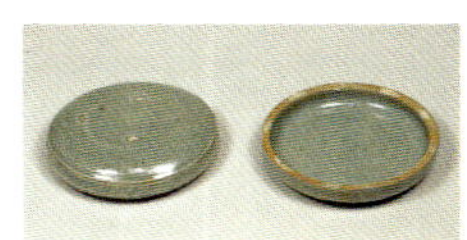

Decoration on the lids of cosmetic boxes generally consists of geometric or floral designs, but a pictorial design such as this one of a bird resting on a branch is seldom encountered. The leaves of the branch are rendered entirely in black inlay and the bird in both black and white inlay, with gray clay inlaid on the head, body, and wings. Celadons in which the inlay is executed in three colors of clay are quite rare. The central design is framed by a double circle, around which are stylized chrysanthemums and flowers in white inlay. The side of the box is decorated with a white-inlaid key-fret pattern, and the base is incised with a combination of stylized key-fret and cloud patterns. The evenly fired, dark grayish-green glaze and a slightly rough, unpretentious design indicate a date of the second half of the thirteenth century. This unique container is characterized by a simplicity whose highly individual appeal is unique among the many known examples of covered boxes.

42

Covered cosmetic box with an inlaid bird and tree design

Celadon

Koryŏ dynasty, second half of the thirteenth century

9.7 cm (diam.)

Acc. no. 20512

The Koryŏ technique of applying a decorative design in gold over a celadon glaze is noted in an entry for the year 1297 in the *Koryŏ-sa* (History of Koryŏ). It states that when a Koryŏ high official presented a gold-decorated vessel to the founder of the Yuan dynasty, an inquiry came from the Chinese leader as to whether the purpose of the applied gold was to strengthen the vessel. After the emperor was informed that it was simply used for decoration, another inquiry was made as to whether the gold could be reused. In response to this query, he was told that a vessel could be easily broken, and if that occurred, the gold would also be unusable. After hearing this, the Yuan emperor ordered that no more gold be used on ceramic ware. On the basis of this entry and extant examples, it is thought that gold-painted celadon was produced beginning in the mid-twelfth century and continued to be made in abundant quantities into the latter half of the thirteenth century.

The lid of this covered box is ornamented with a reverse inlaid classical scroll pattern, with a chrysanthemum design painted in gold at its center. However, only slight traces still remain of the worn gold paint applied over the glaze. On the basis of the design, glaze, and workmanship of the piece, it can be dated to the second half of the thirteenth century.

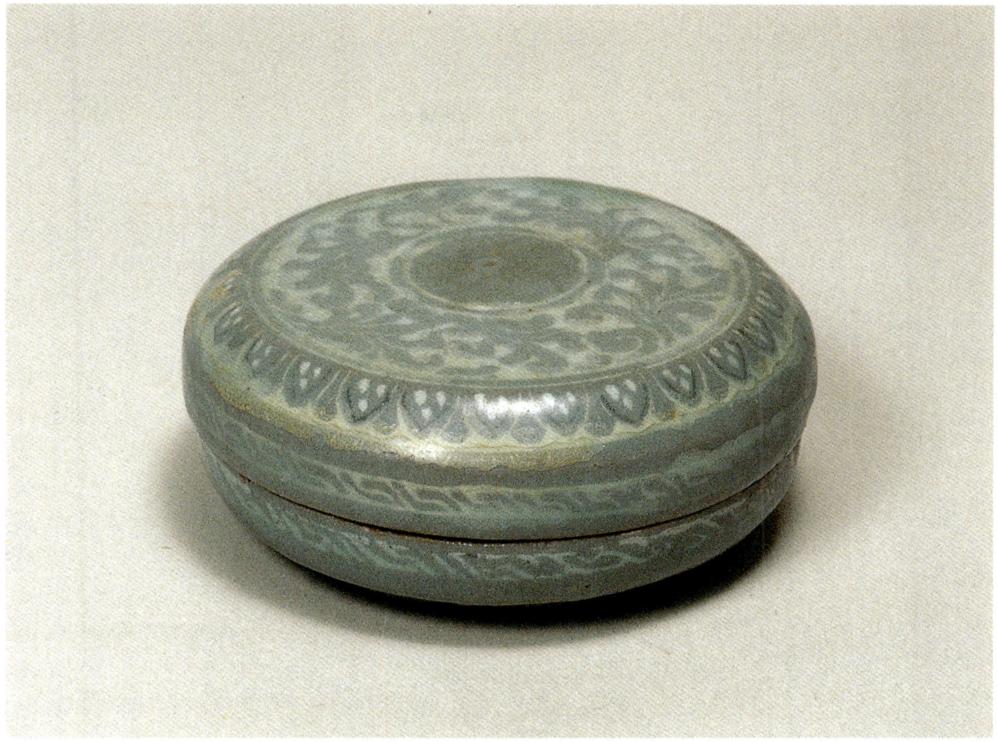

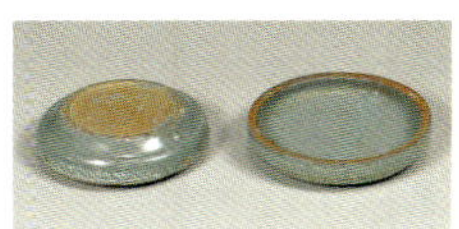

43

Covered cosmetic box with an inlaid scroll and overglaze gilt chrysanthemum design

Celadon

Koryŏ dynasty, second half of the thirteenth century

7.6 cm (diam.)

Acc. no. 20083

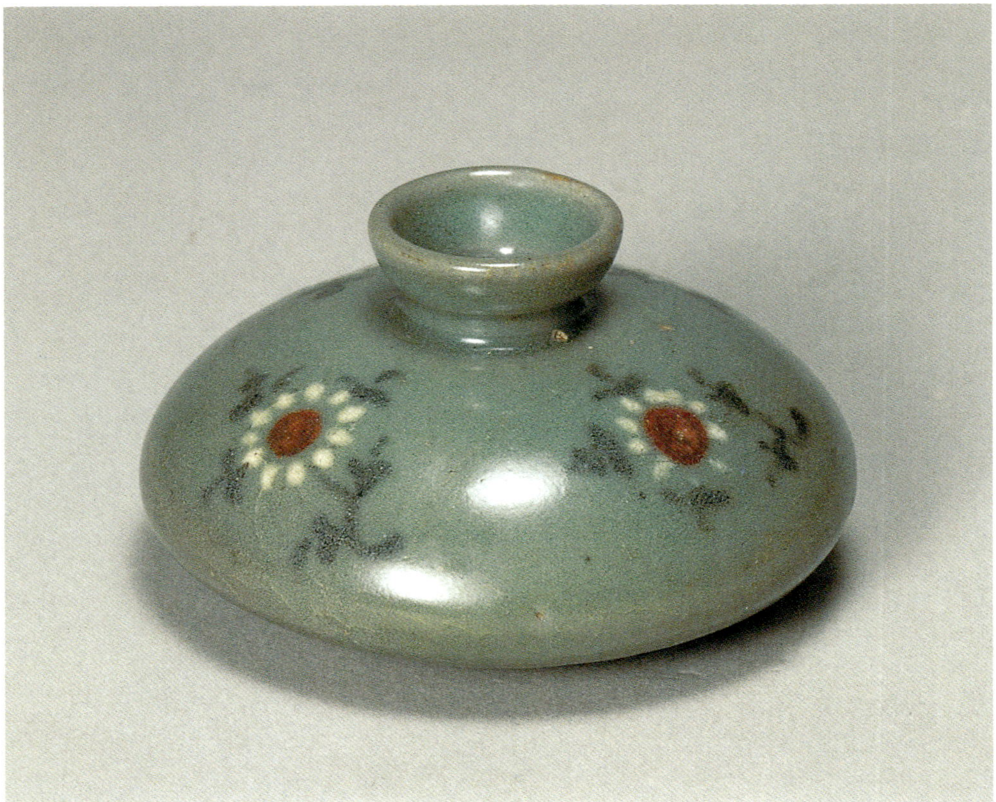

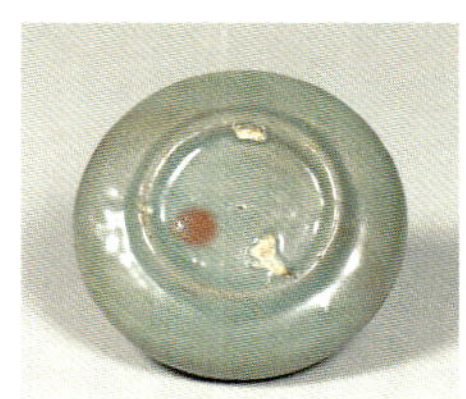

Oil bottles are thought to have held women's hair oil, but this hypothesis has not been proven. This vessel shape, which consists of a round, shallow, disk-shaped body and a small, cup-shaped mouth, first appeared during the United Silla period. The more ancient form is rounder at the top and bottom and thus more globular in character.

This oil bottle is covered with a pattern of four chrysanthemums with black-inlaid leaves and stems, white-slip-painted flower petals, and round flower centers rendered in a rare combination of inlay technique with white slip and copper-red paint. Because it is difficult to stabilize copper pigment, examples of a brilliant red color attained using underglaze copper pigment are rare. The thick glaze obscures the black inlaid design, which is somewhat dull. On the basis of its shape, design, and technique, this bottle was probably produced at the beginning of the thirteenth century. Its base is glazed and has a spill of copper-red glaze.

44

Oil bottle with underglaze copper-red spotting and an inlaid chrysanthemum design

Celadon

Koryŏ dynasty, early thirteenth century

6.5 cm

Acc. no. 20406

The water dropper, one of the accessories found in a scholar's study, was produced in enormous quantities in the Chosŏn period. Most ceramic water droppers were made after the mid-eighteenth century; those dating from the fifteenth and sixteenth centuries are very rare.

This Punch'ŏng water dropper is one of a few fifteenth-century examples that survive. Its jewel-shaped body has a spout in the form of a dragon's head and a handle in the shape of a dragon's body, both popular decorative motifs of the time. Water enters the vessel through a tube or reed inserted vertically into the body through an eight-millimeter hole located on the vessel's bottom; no spilling occurs outside the vessel's body. This system of water intake was handed down from the Koryŏ period. The central area of the body is decorated with a stamped rope-curtain design and the lower section with inlaid lotus petals. Even for such a small Punch'ŏng vessel, these carefully rendered designs represent a refined level of craftsmanship. Although the stamping and inlay techniques of the Koryŏ period continued to be used on Punch'ŏng ware in the Chosŏn period, the content and the expressive methods of that ware, as well as its construction techniques, were completely different. The base of this vessel, with the exception of the foot rim, is glazed.

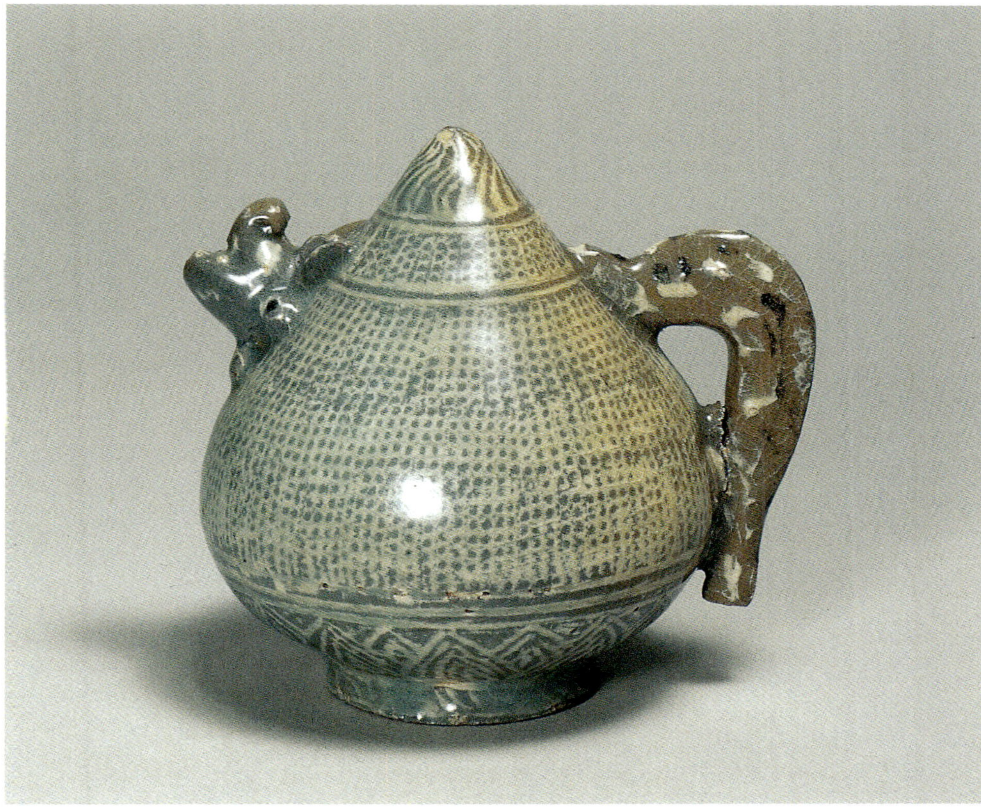

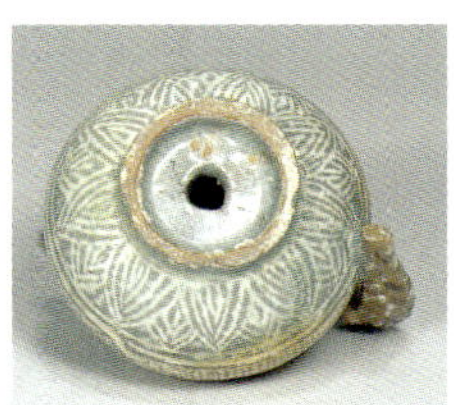

45

Water dropper with a stamped rope-curtain design and a slip-inlaid lotus-petal design

Punch'ŏng ware

Chosŏn dynasty, fifteenth century

9.6 cm

Acc. no. 21204

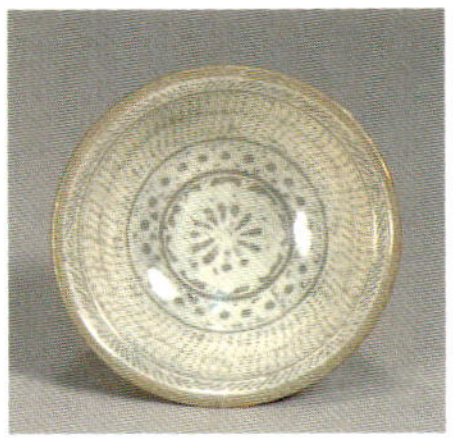

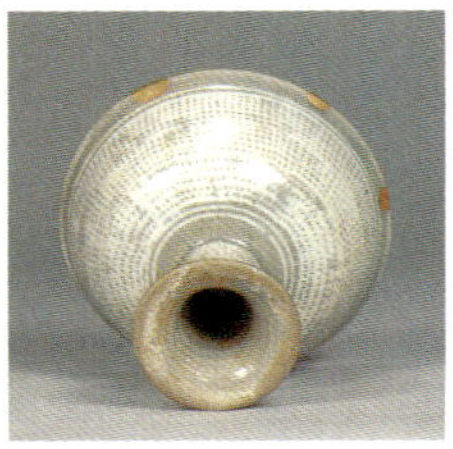

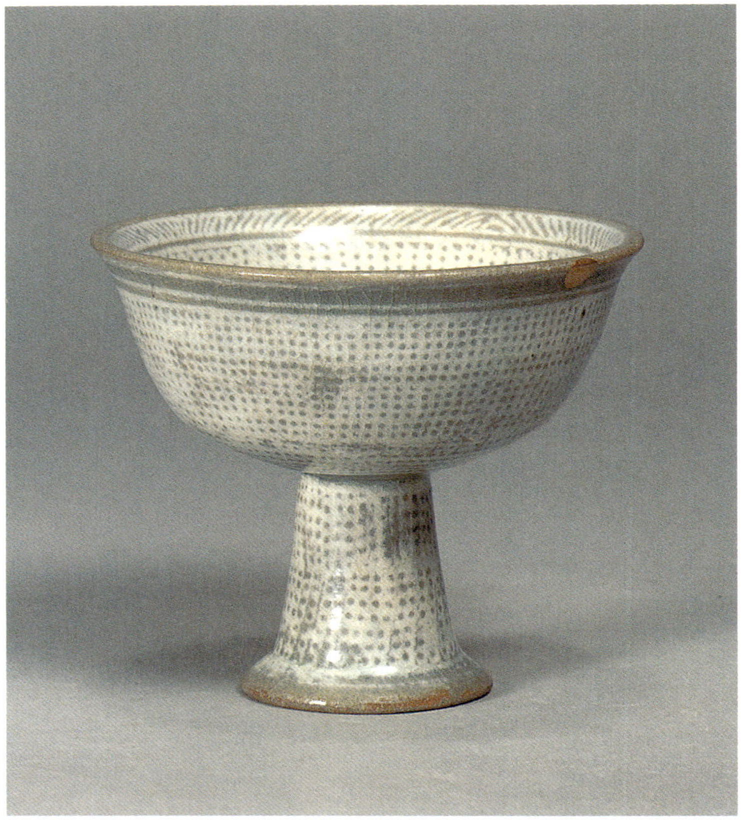

In China, stem cups were often made in white ware and in blue-and-white ware during the Yuan and Ming periods. Perhaps this shape was much appreciated by nomadic peoples such as the Mongols, who founded the Yuan dynasty. On the Korean peninsula, many examples of stem cups are found in Punch'ŏng ware of the fifteenth and sixteenth centuries, the early part of the Chosŏn period. From the seventeenth century on, these vessels were made in white ware, to be used as ritual objects.

Punch'ŏng ware stem cups were generally decorated with stamped or inlaid designs. With the exception of an inlaid grass pattern encircling the rim and a chrysanthemum design located at the center of this small piece, a stamped rope-curtain pattern otherwise covers its entire exterior and interior. Its careful construction and the application of the design and glaze convey perfectly the appeal of Punch'ŏng ware. These elements indicate that this piece should be dated to the mid-fifteenth century, when Punch'ŏng ware flourished.

46

Stem cup with a stamped rope-curtain design and a slip-inlaid grass and floral design

Punch'ŏng ware

Chosŏn dynasty, mid-fifteenth century

7.3 cm

Acc. no. 20631

The Chosŏn dynasty, which was established at the end of the fourteenth century, brought about many institutional changes and a centralization of power during the first half of the fifteenth century. Simultaneously, new vessel forms emerged that differed from those of Koryŏ celadon. One such shape is an elongated jar with four loop handles on the shoulder, a type found in both Punch'ŏng and white ware. This jar, which originally had a lid, was probably produced as a placenta jar, so named because the placenta of a child born to a noble family was placed in it and buried in a specially built structure in the mountains. This custom, which was carried out to insure the happiness of the child, was particularly popular in the fifteenth and sixteenth centuries. Placenta jars consisted of two components, a small inner and a large outer jar; the work exhibited here is thought to be an inner jar.

This vessel is ornamented with a white-inlaid key-fret pattern around its rim and a band of white-inlaid lotus petals around its lower body. The remainder of the body is covered with a stamped chrysanthemum design. Around the central section of the body are four small circles containing the characters *chang*, *hŭng*, *chip*, and *yong*. *Chang hŭng* refers to the Chosŏn ruling house's ceramics and woodworking bureau, to which ceramic ware was sent from all areas of the country. From that storehouse, it was lent out, as necessary, and after being used, was collected again. Inscriptions bearing the names of public offices are found on many extant vessels.

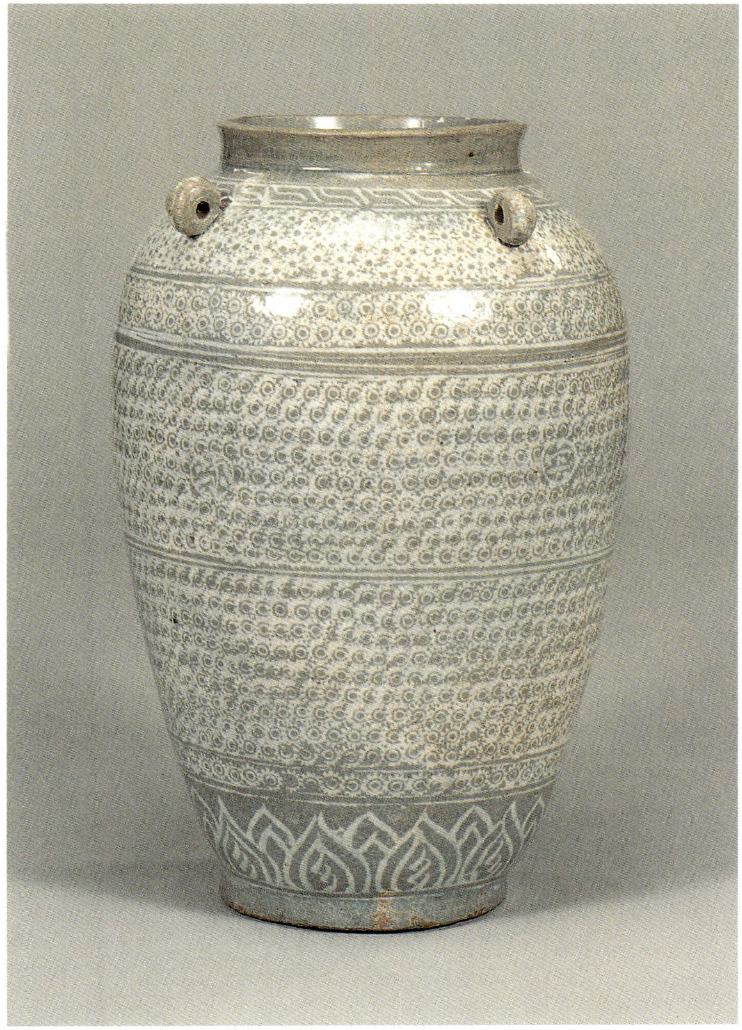

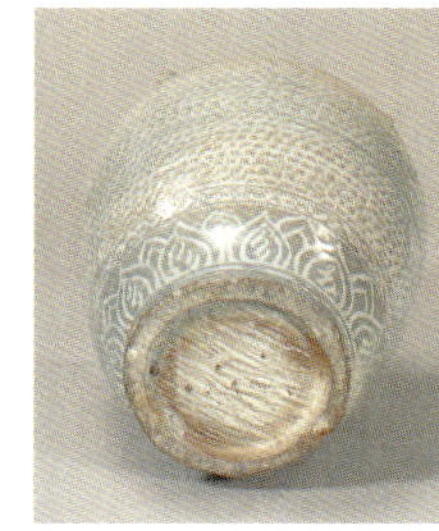

47

Four-handled jar with a stamped floral design, a slip-inlaid key-fret and floral design, and an inscription

Punch'ŏng ware

Chosŏn dynasty, fifteenth century

28.5 cm

Acc. no. 20490

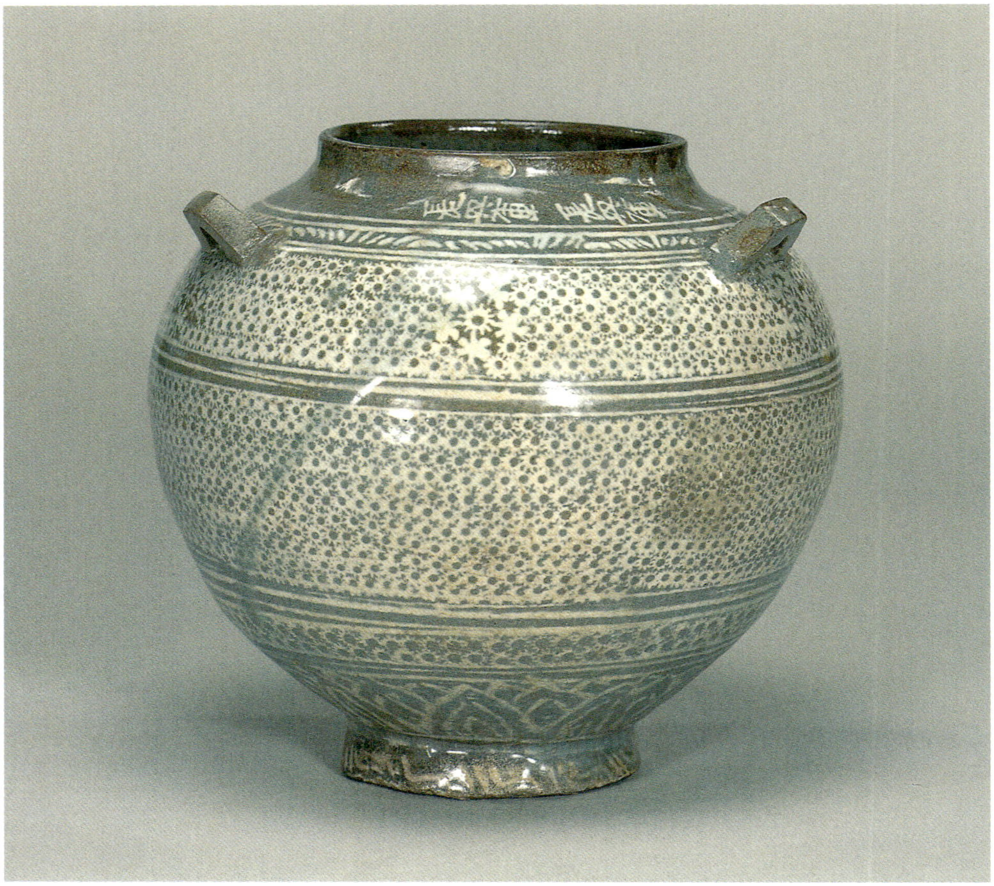

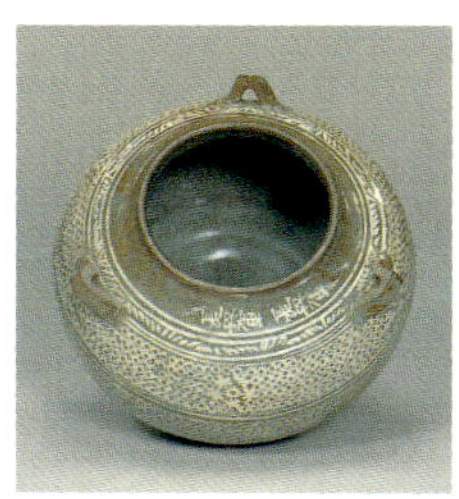

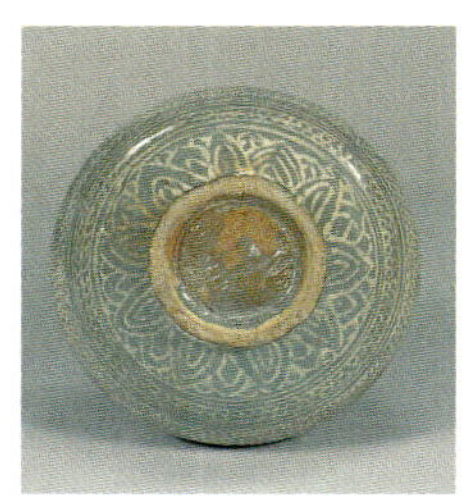

48

Three-handled jar with stamped and slip-inlaid floral designs and an inscription

Punch'ŏng ware

Chosŏn dynasty, fifteenth century

17.1 cm

Acc. no. 20295

This fairly large container was probably also used as a placenta jar, but as such has several special characteristics. First, placenta jars are usually tall and narrow, but this jar, unlike others of the type, is more spherically shaped. Second, instead of four loop handles on the shoulder, there are three in this case. Third, the inscription of the name of the public office is usually written vertically with each character inscribed separately on the central section of the body, but here the inscription *chang hŭng go* is turned on its side and runs parallel to the rim. The careful organization of the form and the application of the design indicate that it may have been produced for a specific nobleman.

The bands of grass patterns on the shoulder, the lotus petals on the lower body, the key-fret patterns on the foot, and the two inscriptions below the rim are all filled with white inlay. As in no. 47, an overall design of small chrysanthemum flowers covers the entire body, with the exception of the upper section, where four large flowers appear. These designs are all stamped and inlaid with a white slip. The base is glazed; only the foot rim is unglazed.

Methods of decorative expression changed considerably at the beginning of the fifteenth century. Orderliness and realism, as well as the use of small units of design, characterize the decoration of Koryŏ-period vessels. In the Chosŏn period, however, as designs were enlarged, greater simplification and abstraction followed. Such freedom in decorative expression also means that the motifs are not always immediately identifiable. This bottle is typical in that it manifests these new tendencies.

A large design, probably of ginseng-plant leaves, covers the central section of the body. Little attempt was made to render the leaves realistically; rather, through the vehicle of this subject matter, an effort was made to convey the activity of life or perhaps to express an irrepressible urge or impulse. Even the shape of the bottle suggests this feeling. The final step in constructing a vessel involves a clean scraping of the surface, but, in this instance, the final scraping was omitted, resulting in a deliberately uneven, rough surface. This avoidance of a perfectly balanced shape is one of the special characteristics of Punch'ŏng ware. The deeply carved lines encircling the lower part of the body impart to it a unique feeling of simplicity and strength. The base is glazed.

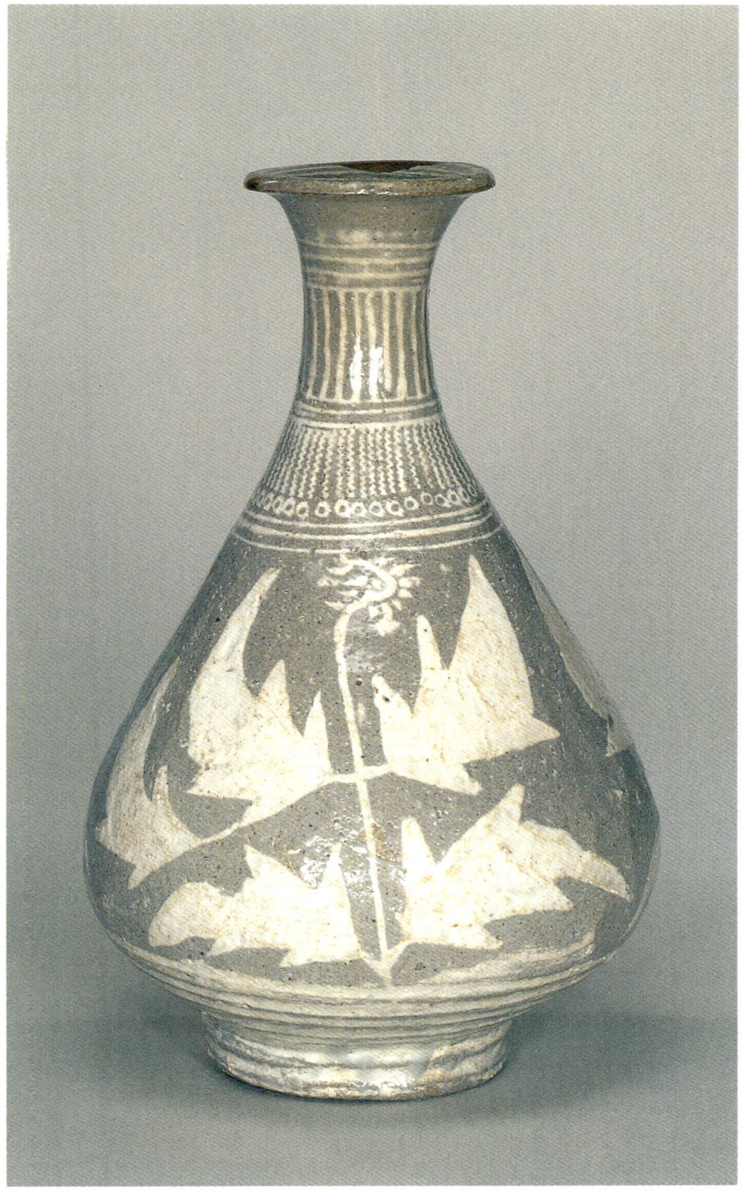

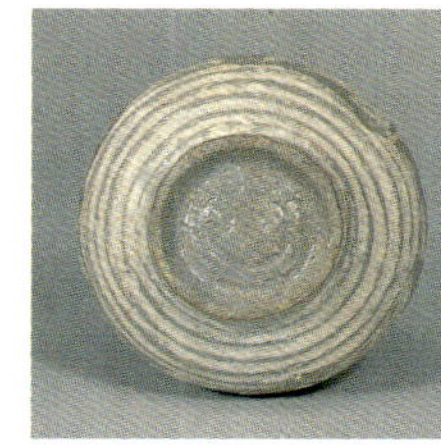

49

Bottle with stamped decoration and a white-inlaid ginseng-plant design

Punch'ŏng ware

Chosŏn dynasty, fifteenth century

29.6 cm

Acc. no. 20397

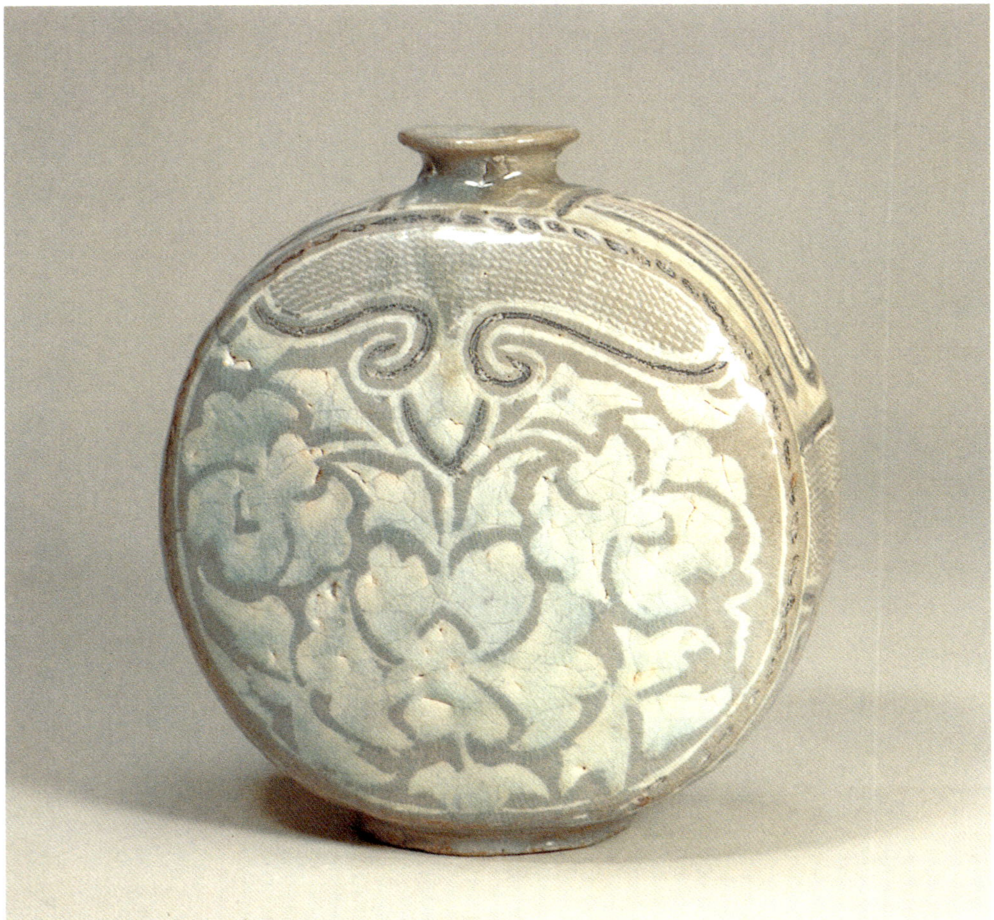

The flask shape, which is a characteristic one in Punch'ŏng ware, first appeared in Koryŏ celadon ware beginning in the latter half of the thirteenth century. Punch'ŏng flasks differ from Koryŏ celadon flasks in several ways. First, the Koryŏ flask has a broad, cup-shaped mouth, whereas the Punch'ŏng flask has a small, trumpet-shaped opening that more clearly reflects its function as a wine bottle. Second, the Koryŏ celadon flask is elongated and swells at the shoulder, extending in an almost straight line to a footless base. The Punch'ŏng flask, on the other hand, is globular and always has a high foot. In the early Chosŏn period, the high foot is not restricted only to flasks but is also characteristic of jars and bottles.

Flasks of a similar type were discovered at the kiln site of Udong-ri, Poan-myŏn, Puan-gun, North Chŏlla province. Among the shards found at the site was a mortuary tablet, dated 1440, with a design similar to the one on this vessel. Thus, it seems appropriate to date this piece to about the same period. The flask, with its well-proportioned shape, superbly expressive design, and strong blue glaze, is an outstanding example of Punch'ŏng ware.

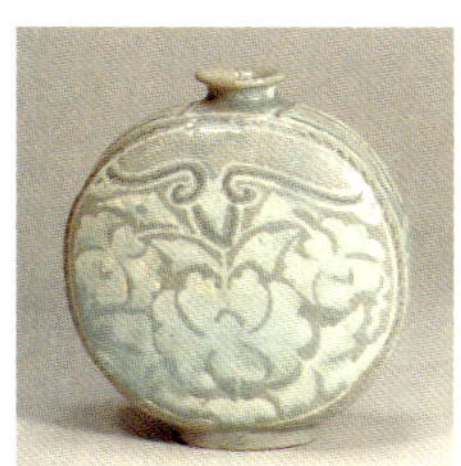

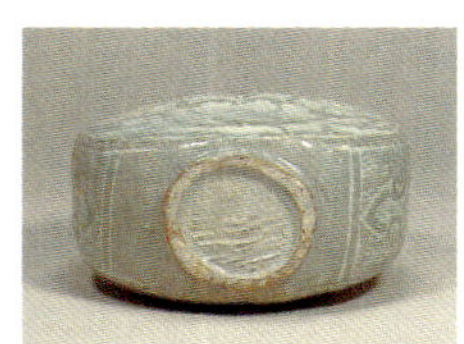

50

Flask with a white-inlaid peony design

Punch'ŏng ware

Chosŏn dynasty, mid-fifteenth century

22.6 cm

Acc. no. 20810

This bottle's distinctive feature is its beautiful, stable shape. Its everted rim and thinly potted walls create a powerful sense of balance. The lower part of the body swells out and turns in toward the base, with its low center of gravity emphasizing this feeling of stability. There are many Punch'ŏng-ware bottles, but examples with this type of shape are relatively rare.

The stamp and inlay techniques used to decorate this bottle are typical of Punch'ŏng ware, thus indicating that it should be dated to the fifteenth century. On the central section of the body is a large-scale inlaid peony design rendered in a fairly abstract manner. Designs on Punch'ŏng ware in general tend to become simplified and stylized to the point where they reflect a greater freedom of expression. In contrast to the expressive quality of the peony pattern on this piece, however, the orderly and refined stamped rope design found above and below it heightens the dignity of the bottle. Moreover, a stamped chrysanthemum pattern has been applied around the outer surface of the foot. Punch'ŏng bottles are usually even decorated on their bases; since this type of bottle was actually used, the decoration could still be appreciated when turning the vessel upside down. The glaze on this piece, which contains a fair amount of iron, has a strong blue hue that is the result of deoxidation during firing. The base is glazed, but part of its surface was burned due to excessive heat in the kiln.

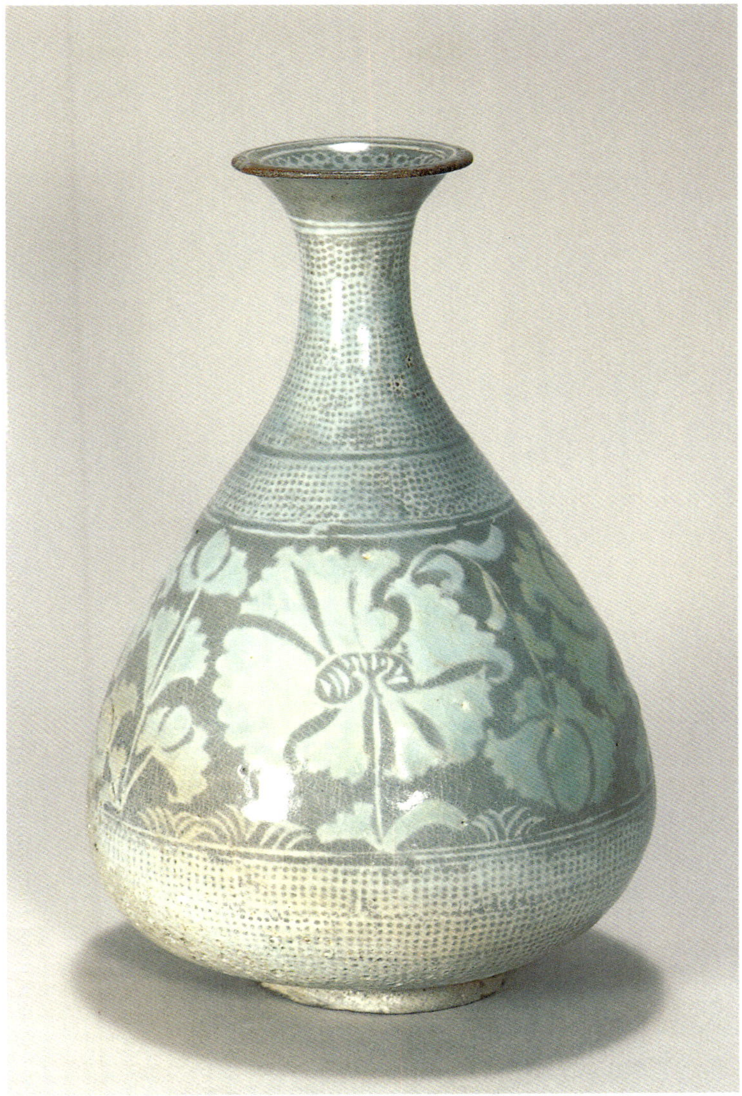

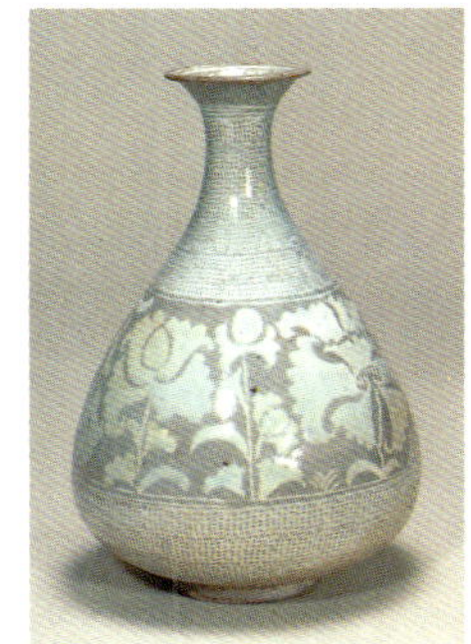

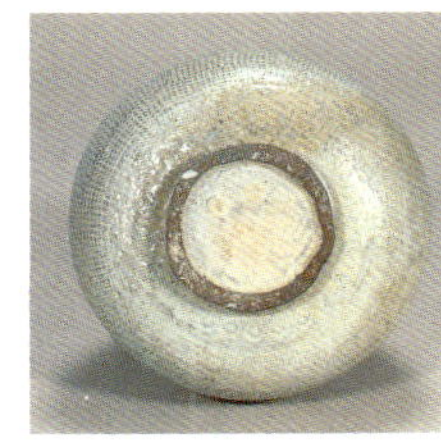

51

Bottle with stamped decoration and a white-inlaid peony design

Punch'ŏng ware

Chosŏn dynasty, fifteenth century

27 cm

Acc. no. 20653

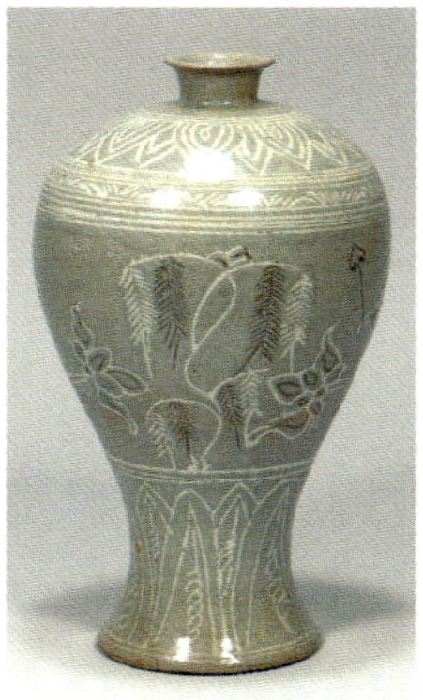

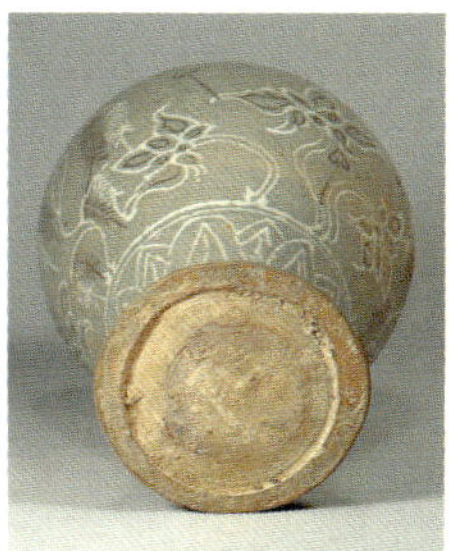

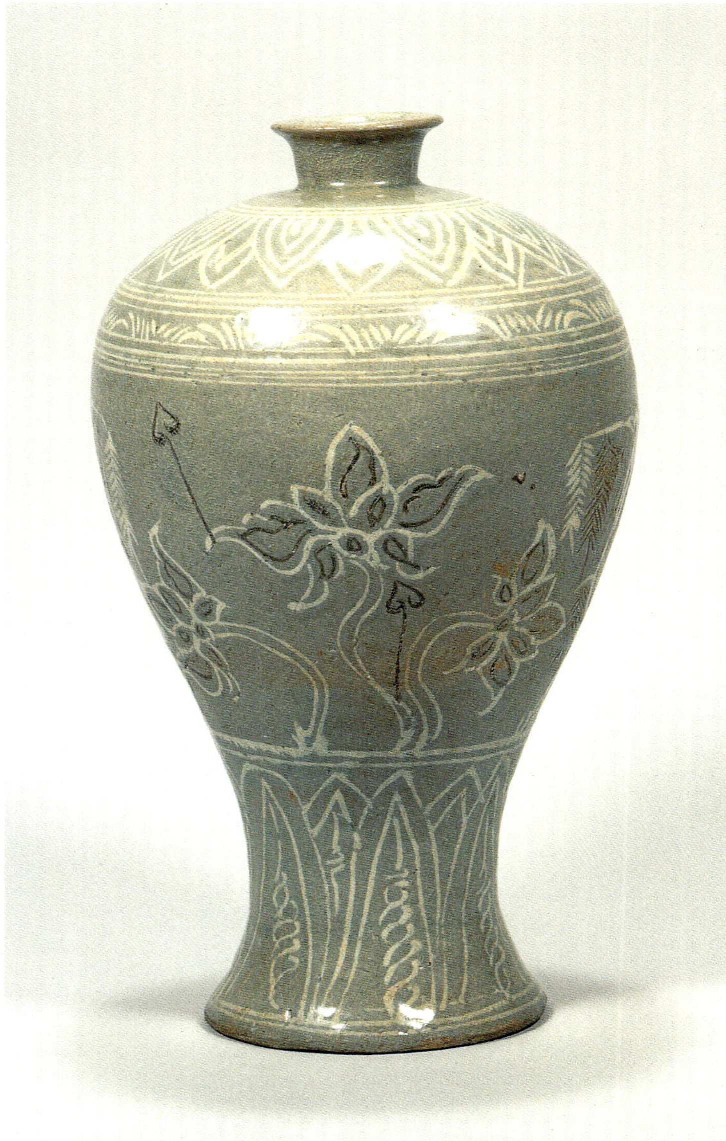

52

Maebyŏng *with a slip-inlaid lotus and willow design*

Punch'ŏng ware

Chosŏn dynasty, early fifteenth century

31.5 cm

Acc. no. 20492

Among several extant *maebyŏng* of the late Koryŏ period, one datable to the mid-fourteenth century was discovered at a kiln site at Sadang-ri, Taegu-myŏn, Kangjin-gun, South Chŏlla province. Fragments of an inlaid celadon dish found with it bore an inscription dated to the period 1341 to 1367. The design motif on the late Koryŏ *maebyŏng* is similar to that of the early fifteenth-century vase exhibited here. This decoration consists of a black-and-white inlaid lotus and willow, with bands of lotus petals encircling the shoulder and lower body. However, this *maebyŏng* differs from the mid-fourteenth-century example in its manner of construction. First, the fourteenth-century example has a gently sloping shoulder which extends down to a somewhat small base, but the fifteenth-century example has a more tense, taut shoulder and a larger base, which reinforces its stability. Second, the lines of the design in the earlier work are weak and lifeless, whereas, in the later piece, they are energetic and lively. Third, the earlier example has a dark gray glaze with an irregular surface, but the later vase has a perfectly fired light blue glaze with a glossy, beautiful surface. These factors seem to indicate that the *maebyŏng* exhibited here was produced under the more stable political conditions of the early Chosŏn period. The foot is shallow and evenly hollowed out and has a thin layer of glaze.

Celadon ware made in the late Koryŏ period is sometimes difficult to distinguish from Punch'ŏng ware made in the early Chosŏn period. Due to the lack of documented materials, the stylistic evolution in this transitional period has not yet been fully clarified. However, generally speaking, at the beginning of the Chosŏn period, designs became more lively, the drawing freer, the glaze clearer, and shapes more stable.

As for this distinctive, horn-shaped cup, it has a perfectly fired, thin Punch'ŏng glaze. Bordered on the top and bottom by bands of slanted lines, the body is decorated on the front and back with inlaid lotuses, which are separated by strings of chrysanthemums. These designs are freely and vigorously drawn. The tip of the cup bears traces of underglaze iron pigment, which perhaps reflect an attempt at decorative effect. Horn-shaped cups were made in the United Silla period and continued to be produced in celadon during the Koryŏ period. They were also made in white ware, as well as in Punch'ŏng ware, at the beginning of the Chosŏn period, but apparently ceased to exist in any form after the fifteenth and sixteenth centuries.

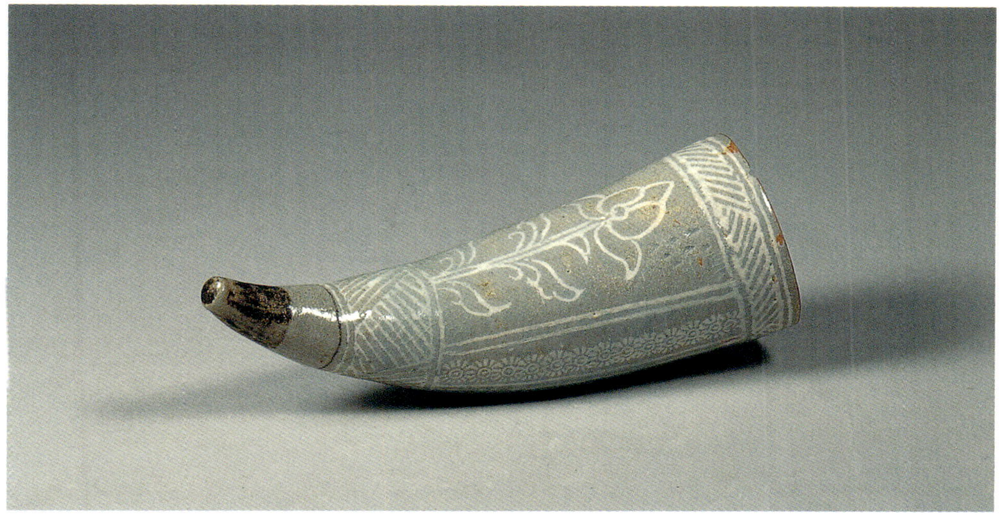

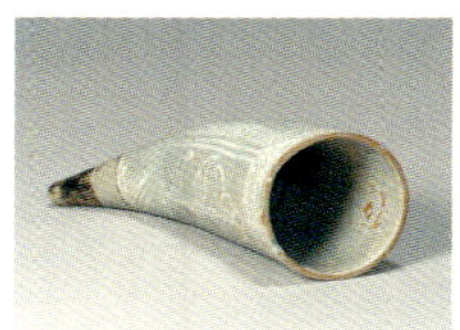

53

Cup in the shape of a horn with a slip-inlaid lotus design

Punch'ŏng ware

Chosŏn dynasty, fifteenth century

14.6 cm

Acc. no. 21000

Important art object

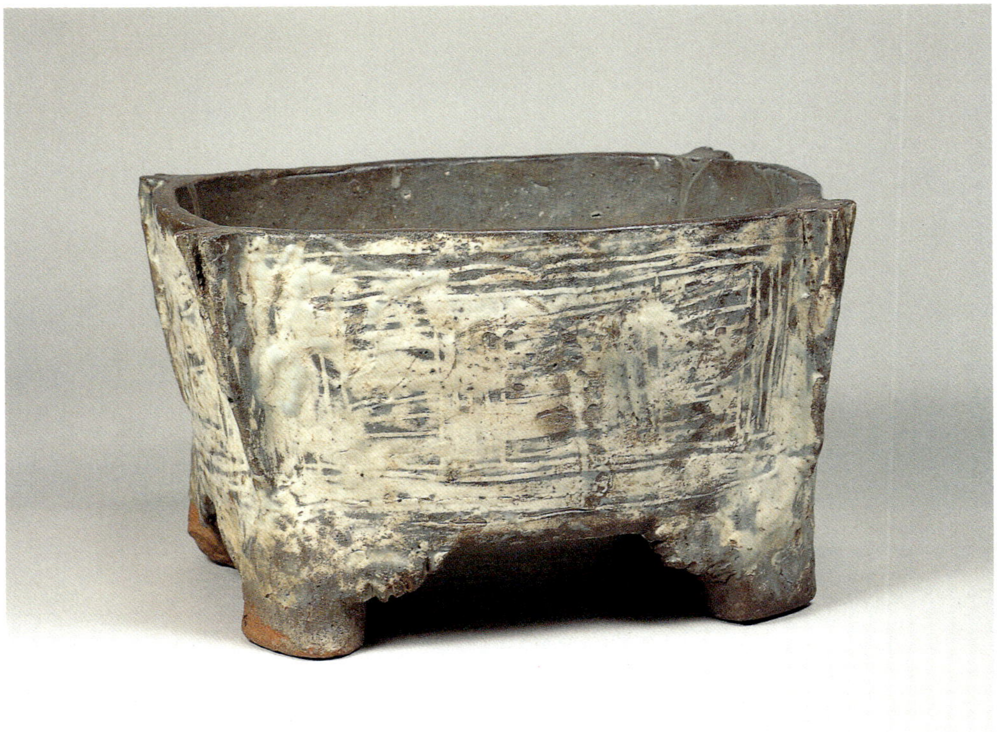

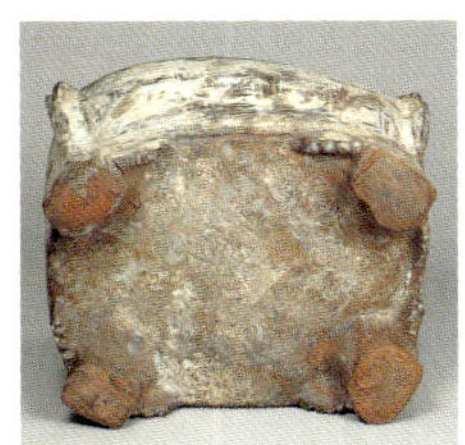

54

Ritual vessel with slip-inlaid and brushed ornament

Punch'ŏng ware

Chosŏn dynasty, fifteenth century

16.2 cm

Acc. no. 20219

At the beginning of the Chosŏn period, Buddhism was rejected, and it became government policy to promote Confucianism in its place. As a result, implements used in Confucian ritual were produced in great numbers. Those made in the fifteenth and sixteenth centuries, during the early Chosŏn period, imitated the forms of Chinese bronzes. The shape of the vessel in this exhibition may have originated with the Chinese bronze *fangyi* (Kor. *bangyi*). Indications of this are seen in the low flanges at the four corners of the body and in the sawtooth decoration of the inner sides of the four legs. Originally, the piece may have had a lid, which is now lost.

This vessel is an example of one of the two trends in ceramics of the early Chosŏn period: on the one hand, classical white ware was produced; on the other, there was a taste for exactly the opposite—simple and rough creations. The shapes of ritual vessels of the latter type are not necessarily well balanced nor are their designs carefully executed. Moreover, their clay contains a large amount of iron and is rough and black. In this case, quite simply, the potter has carefully preserved the intent of a ceremonial vessel and, in so doing, has achieved something spiritual. Consequently, what is expressed is the imposing dignity and solemnity that the vessel possesses. Irregular lines are inlaid on the body, both horizontally and vertically, and are covered with a white slip. The potter's genuine motivation to create and express is akin to that of many avant-garde artists of today.

This jar, which is similar to the ritual vessel in no. 54, is made with a clay that was not carefully selected and contained many irregularities. Consequently, the vessel is dark in color and has a rough, coarse surface. The lower body is covered with a white slip, but its uneven application leaves an impression of crudeness. While this kind of roughness is emphasized, the overall shape is nevertheless balanced and the vessel carefully finished.

The large floral design in the central area is rendered in a simplified and stylized manner, which makes it difficult to identity the flower precisely; it is probably a peony. The design is deeply incised on the white-slip-painted body and the ground around it is scraped off in a sgraffito technique. The shoulder is ornamented with a band of incised lotus petals. The shape of the vessel recalls that of the placenta jar, but this piece lacks the three or four loop handles that are of special use on the placenta jar; therefore, its function is unclear. Because the sgraffito technique was developed mainly in Chŏlla, this jar is probably a product of that area. A jar excavated from a reliquary monument erected in 1430 for Kobong (d. ca. 1428), a monk at the Songkwang-sa in South Chŏlla province, provides a date for this vessel. The perfection of the excavated jar makes it clear that the technique of scraping reached the height of its development well before 1430. Thus, this vessel is probably a work of the early fifteenth century. The base of the vessel is flat and unglazed.

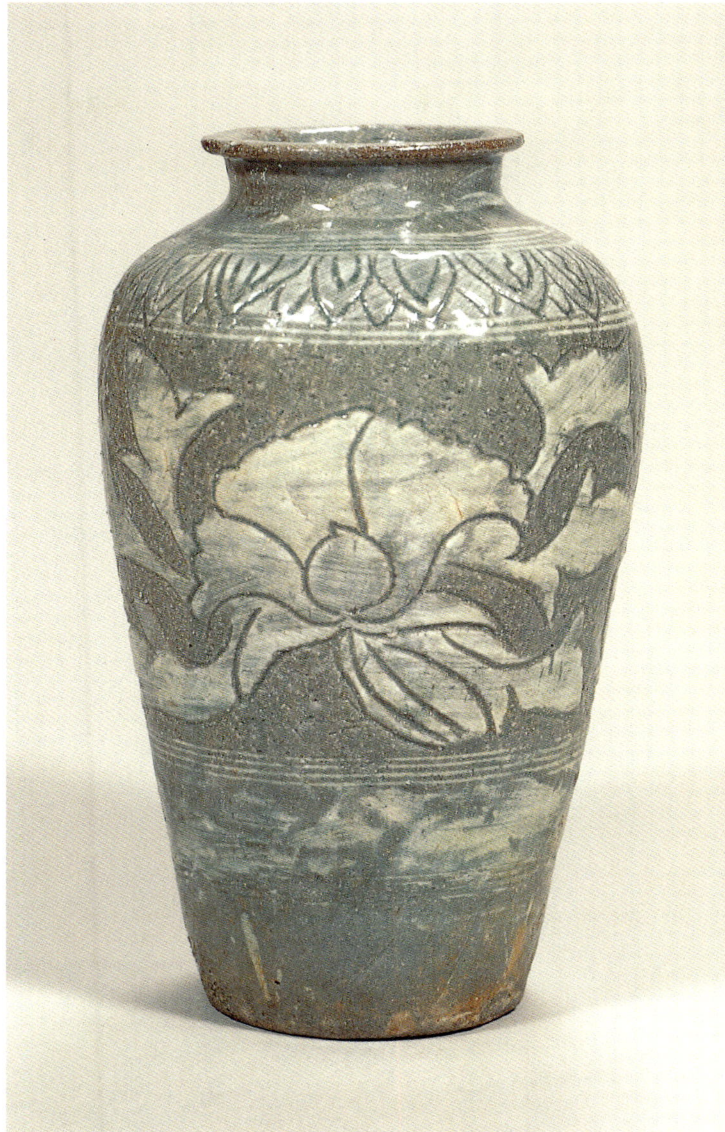

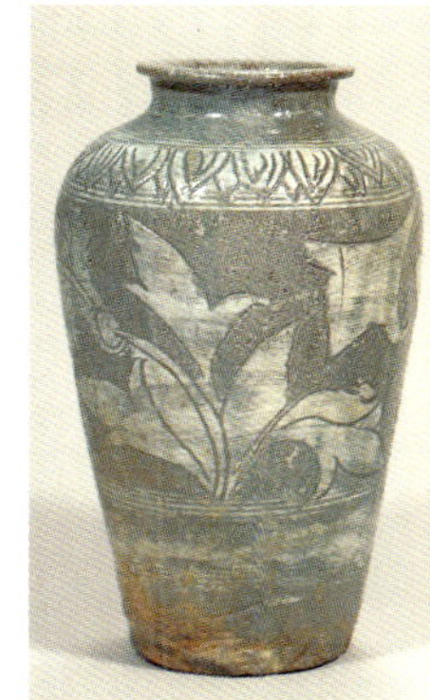

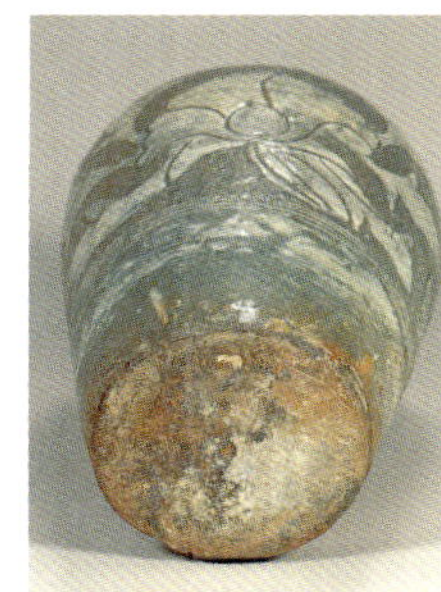

55

Jar with a sgraffito peony design

Punch'ŏng ware

Chosŏn dynasty, early fifteenth century

33.5 cm

Acc. no. 20846

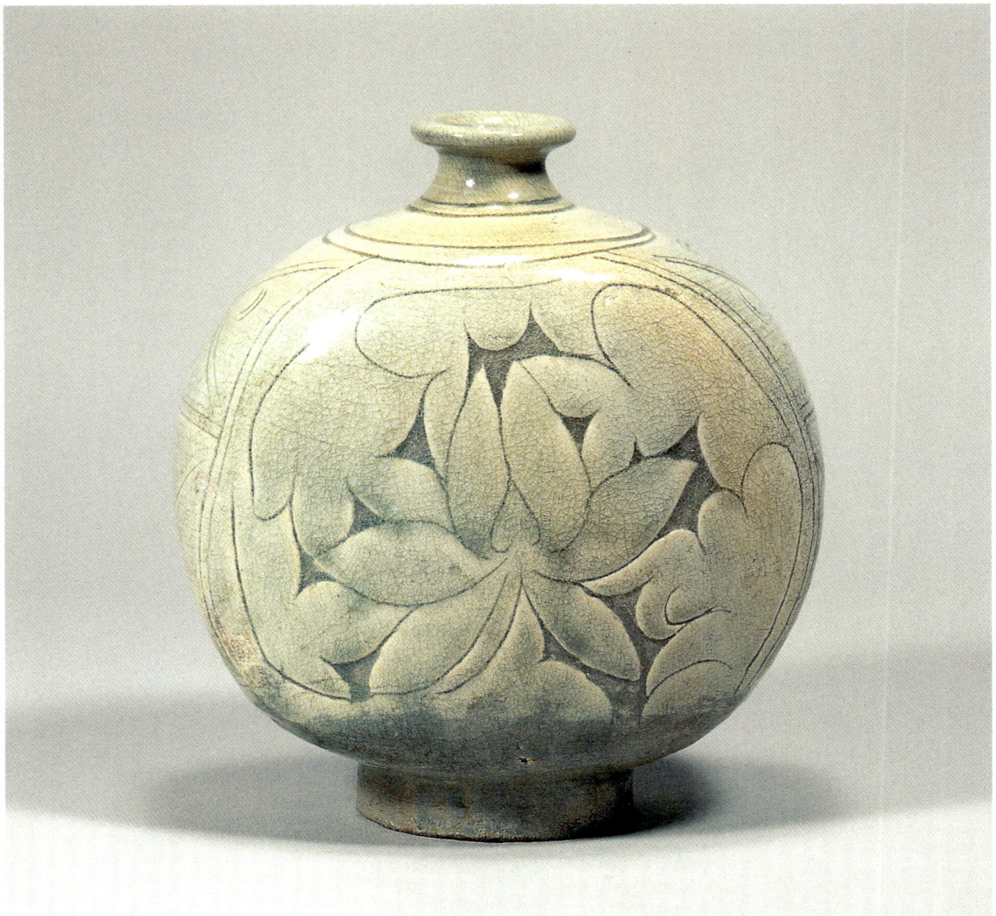

In the Chosŏn period, numerous vessels were made of dark gray clay with a high iron content; their bodies were decorated with a white slip through which lines were carved to form a design. Around the contours of the design, the white slip was scraped off to reveal a raised pattern. This technique was developed particularly in South Chŏlla province and flourished in the fifteenth century.

This flask seems to be decorated with a lotus flower. Patterns on Punch'ŏng ware of the early Chosŏn period are simplified and stylized; in their deviation from the realistic rendition of nature, their abstract sense of design becomes stronger. This floral design, which is neither realistic nor abstract, has a splendidly impressive decorative effect. The other side of the vessel has almost the same design. In firing, the iron content in the glaze produced a grayish-green hue. Especially in those areas where the glaze has pooled, the translucent, dark-green color lends an indescribable appeal to the piece. The base has been carefully scraped flat and is glazed.

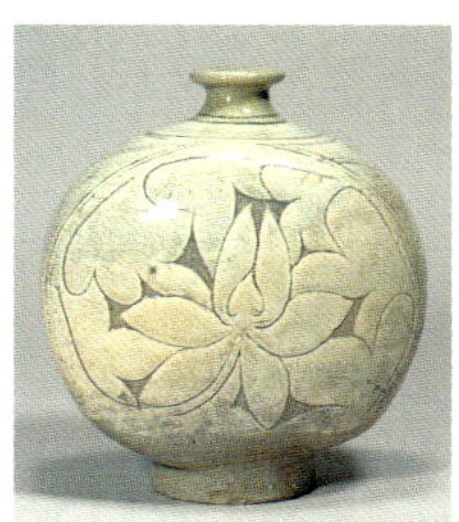

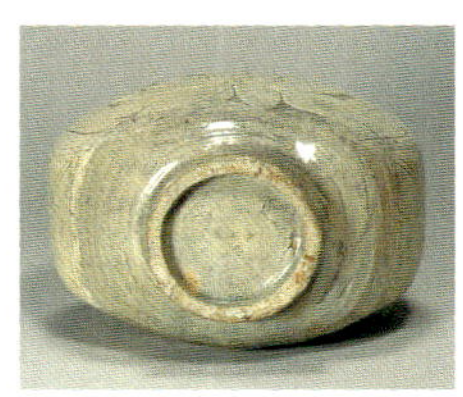

56

Flask with a sgraffito lotus design

Punch'ŏng ware

Chosŏn dynasty, fifteenth century

22.3 cm

Acc. no. 20555

The *maebyŏng*, one of the popular shapes in the Koryŏ period, underwent several changes at the beginning of the Chosŏn dynasty. First, the cup-shaped mouth characteristic of the Koryŏ period completely disappeared and was replaced by a trumpet-shaped mouth. Second, the curve that extended from the shoulder to the lower part of the body became a more pronounced *S* shape and flared out at the foot of the vessel. Third, the manner of decorative expression changed from the realistic to the more abstract.

This *maebyŏng* is ornamented with patterns of rope curtains and strings of jewels on the shoulder and an inlaid design of young plant sprouts on the lower part of its body. The white-slip-painted body is decorated with a raised dragon design revealed by carving away the white slip around it. This type of design, which is not found at all in the Koryŏ period or in China from the Yuan to the early Ming dynasty, is a special creation of the early Chosŏn period. The dragon's head, in particular, is fragmented into various shapes that spread across the vessel. While the dragon motif itself is quite advanced, not much attention was paid to how the final result was to be expressed. In such a case, bold and free forms of the type seen here are created. The areas painted in iron slip serve to emphasize the dragon design and avoid monotony. Glaze was applied to the base, but most of it has worn off.

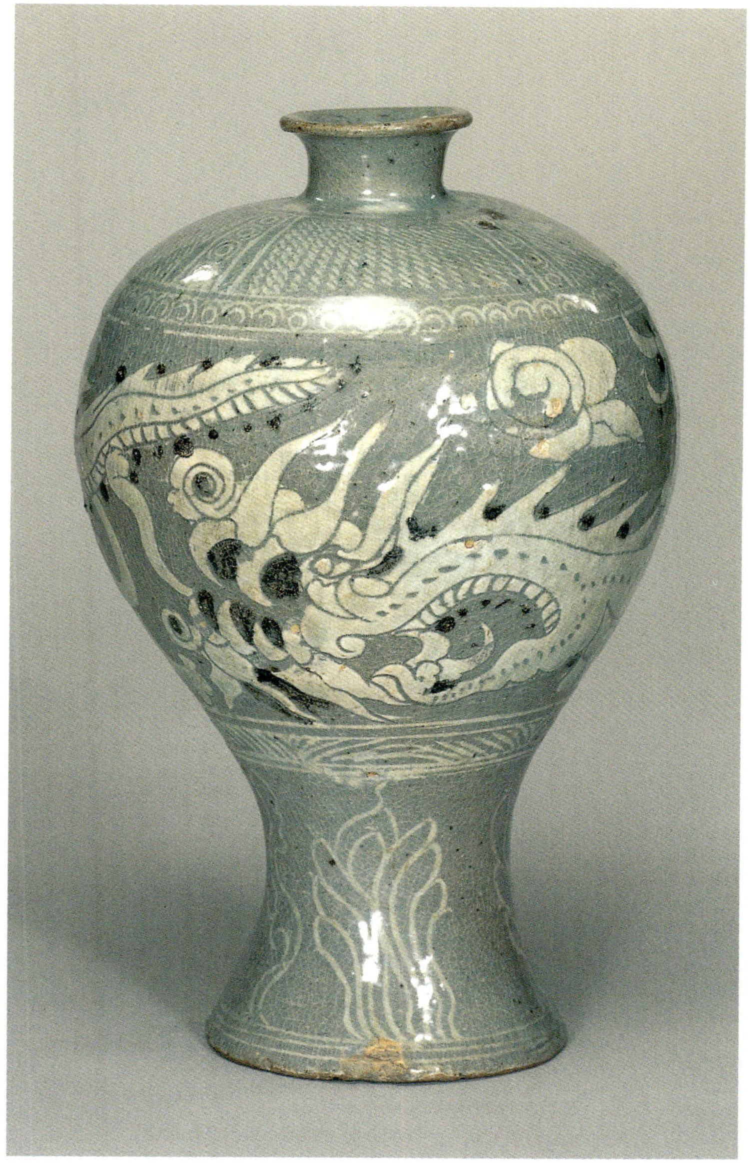

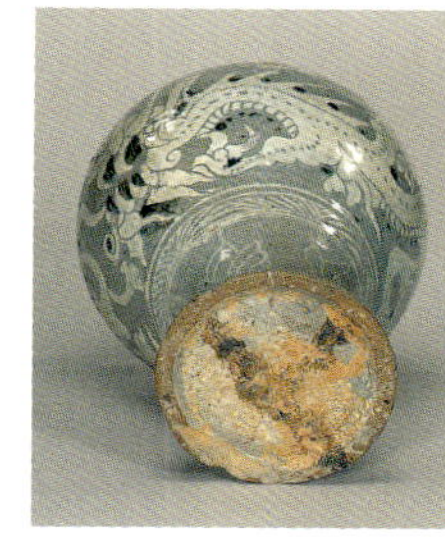

57

Maebyŏng *with a sgraffito dragon design*

Punch'ŏng ware

Chosŏn dynasty, fifteenth century

26.4 cm

Acc. no. 20061

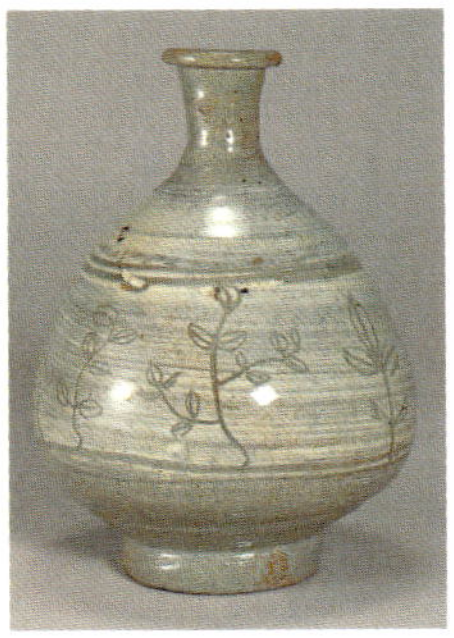

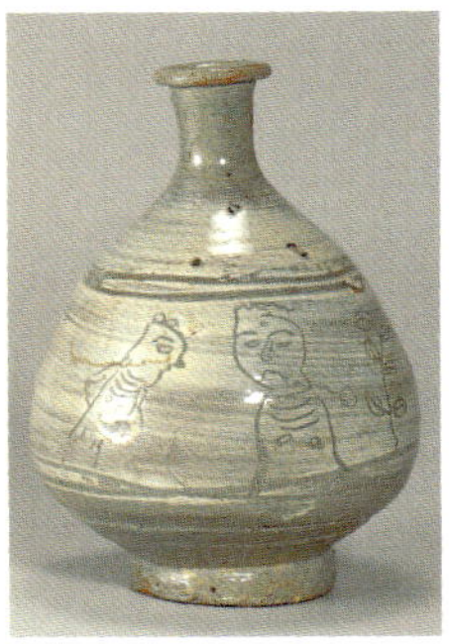

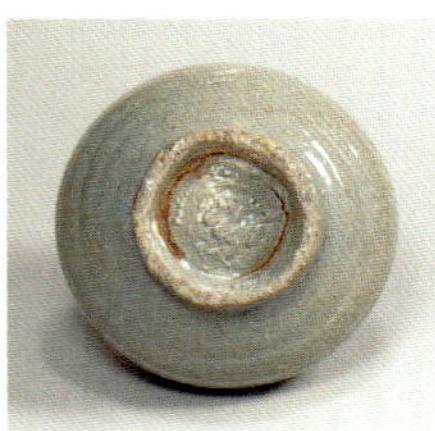

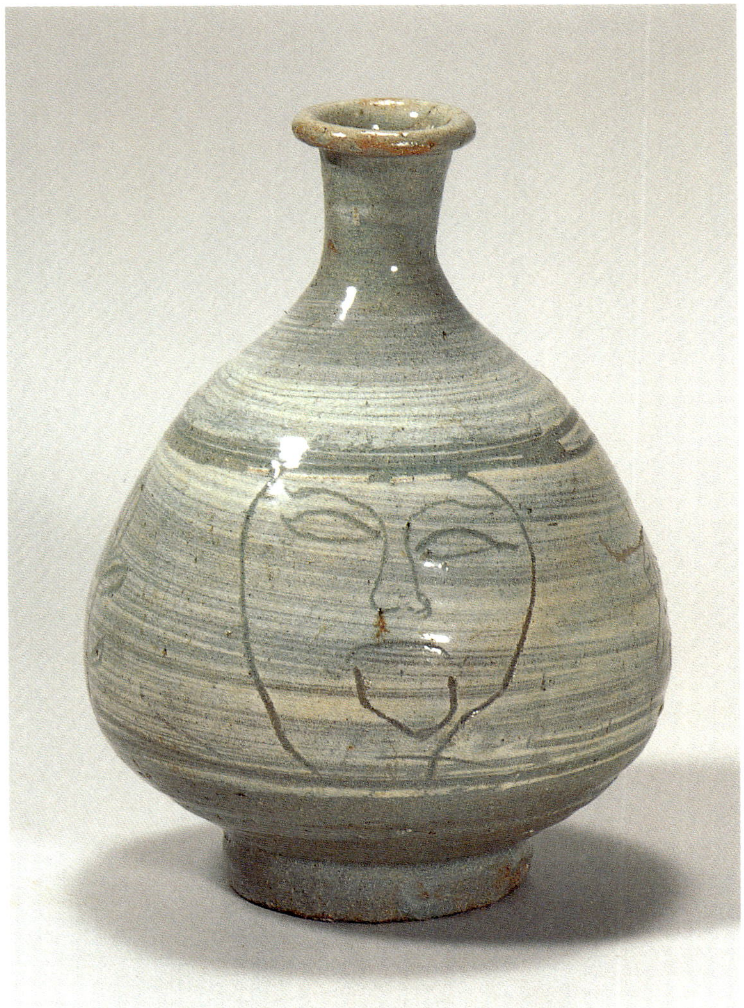

Punch'ŏng ware, which at first glance appears awkward, will occasionally exhibit surprisingly bold and fresh designs such as the one on this bottle or the fish design on the flask in no. 59. The decorative technique for this bottle first called for a white slip to be painted on the body with a brush. This brushed white slip of itself produced a kind of decorative effect. A band made by scraping off the white slip around the shoulder contrasts with the painted areas remaining on the lower body, whereas the neck is hardly painted at all. This arrangement of white slip strengthens the vertical feeling of the bottle. Represented on the central section of the body is a large, incised human face, with a wide-open mouth, which perhaps is meant to be shouting; the expression of the eyes and eyebrows clearly do not suggest a smiling face. Incised human faces and animal bodies among grasses and flowers are found on the other side of the vessel. What these designs symbolize is unclear, but they may possess magical significance. Punch'ŏng ware was not only made for daily use but for ritual and mortuary purposes as well. This bottle may have been made for the latter function. The base was roughly scraped and glazed.

58

Bottle with a sgraffito figure design

Punch'ŏng ware

Chosŏn dynasty, fifteenth century

22.8 cm

Acc. no. 20268

Rice-bale-shaped bottles first appeared in the Three Kingdoms period, mysteriously disappeared during the Koryŏ period, and reappeared in the early Chosŏn period. Extant examples of this form in Punch'ŏng ware are numerous. Such bottles may or may not have a foot; those that do have another, original foot. That is, after the bottle was thrown on the wheel and the original foot was formed, it was placed on its side, and a mouth and second high foot were made separately and attached to what had originally been the sides. When the flask was actually used, it was held horizontally with its mouth facing upward and the second, high foot supporting the bottle. These kinds of bale-shaped flasks appear to have been made for ritual use as wine vessels.

A white slip was painted over the body of this bottle, with the exception of its bottom and mouth, and a fish design was incised through the slip. The fish, drawn upside down in a seemingly naïve fashion, constitutes a simple and powerful statement. The unpretentiousness and deliberate primitiveness of the design are characteristic of ceramics of the period. The original base (i.e., the vertical bottle's base) is unglazed.

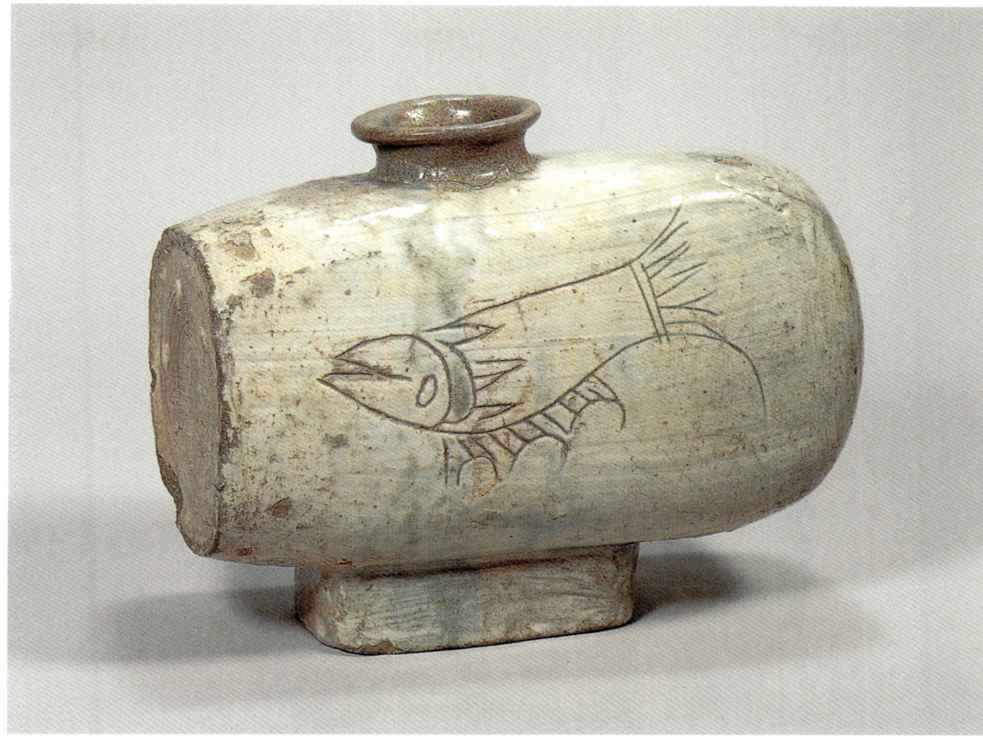

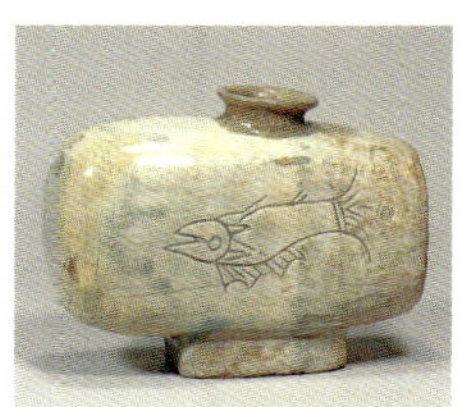

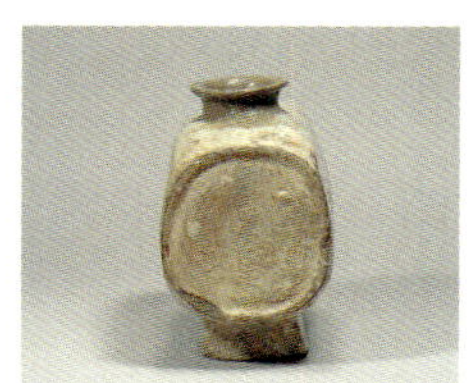

59

Bottle in the shape of a rice bale with a sgraffito fish design

Punch'ŏng ware

Chosŏn dynasty, fifteenth century

19.5 cm

Acc. no. 20378

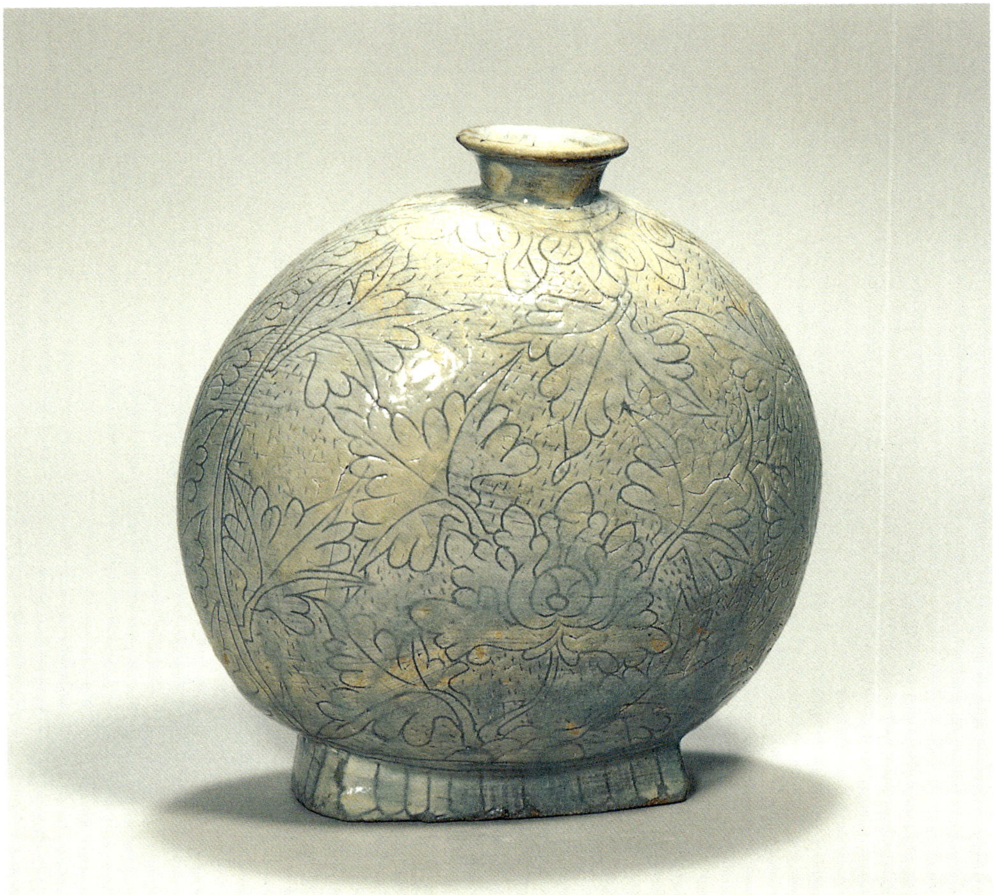

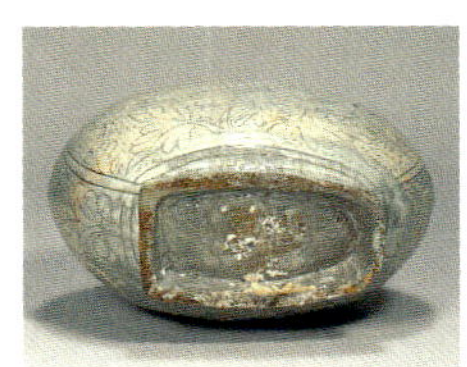

60

Flask with a sgraffito peony design

Punch'ŏng ware

Chosŏn dynasty, fifteenth century

21.9 cm

Acc. no. 20859

Punch'ŏng-ware flasks were constructed in two ways. One method called for throwing a bottle-shaped vessel on the potter's wheel and then hitting the sides, as was the case with the flask in no. 56. A study of extant pieces indicates that this method was more the common one. The second technique consisted of hand-building the vessel by placing two dishes with vertical sides together, as was the case with no. 50. This method was employed to construct this somewhat unusual flask. However, the two hand-built dishes that were put together to make this example were without vertical sides. Thus, the flask, which lacks flat sides, has the shape of a globular, disklike ovoid. Vessels made in this manner are rare, but an inlaid white-ware flask of the same type does exist. The white-ware flask was found with a mortuary tablet which serves to date it before 1466; however, stylistically and technically, this vessel is surely datable to first half of the fifteenth century. The flask exhibited here, while not directly related to the white-ware flask, can also be dated to the fifteenth century.

A peony design is carved into the white-slip-painted body; dotting fills the surrounding space. Both the shape and the incised lotus-petal design on the foot are unusual. A similar piece is found in the National Museum of Korea, Seoul. The bottom is glazed.

The technique used on this jar is characteristic of a very small ceramic-producing area at the foot of the Kyeryong Mountains, located in the middle of the southern part of the Korean peninsula; it is this region that gives its name to the ware, popularly known as Kyeryongsan. A white slip was brush-painted over the entire body of this vessel, except the foot, and a design was applied over it with iron pigment. Kyeryongsan clay has a high iron content that appears dark gray after firing. When a white slip is painted over it, the contrast between the light slip and the dark clay is striking. The iron pigment then painted over the slip, which becomes lacquer black, further heightens the decorative effect.

The iron-pigment design covering the upper body of this jar is one of the most outstanding examples of Kyeryongsan ornamentation. The highly stylized plant design, which is difficult to identify, is of a type often seen on Kyeryongsan ware. Examples of round jars such as the one exhibited here are relatively scarce, but this vessel's small, constricted foot is typical of Kyeryongsan ware. The base is unglazed.

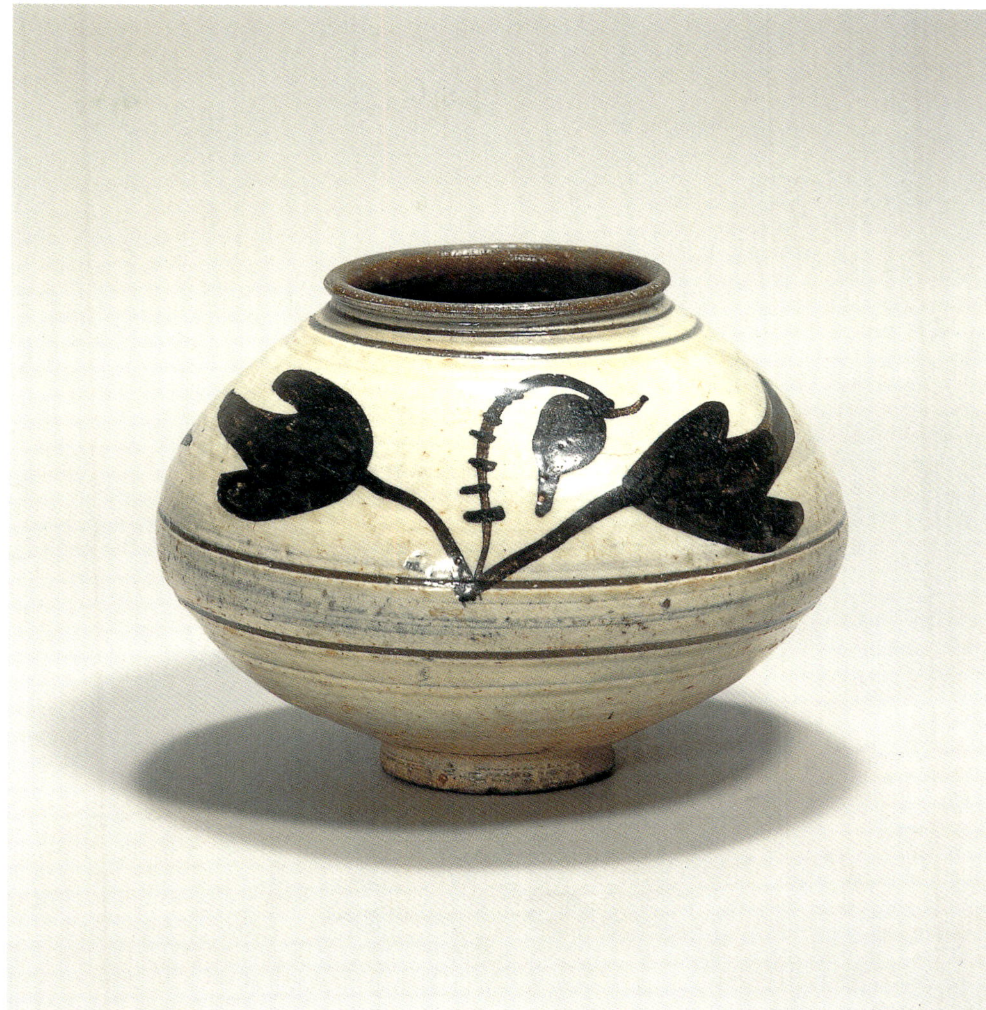

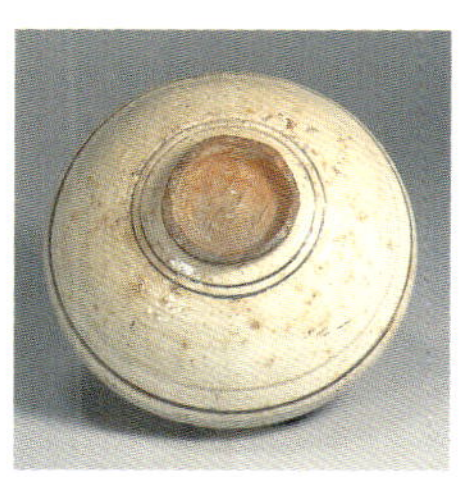

61

Jar with an iron-brown plant design

Punch'ŏng ware, Kyeryongsan ware

Chosŏn dynasty, fifteenth/sixteenth century

14.3 cm

Acc. no. 20943

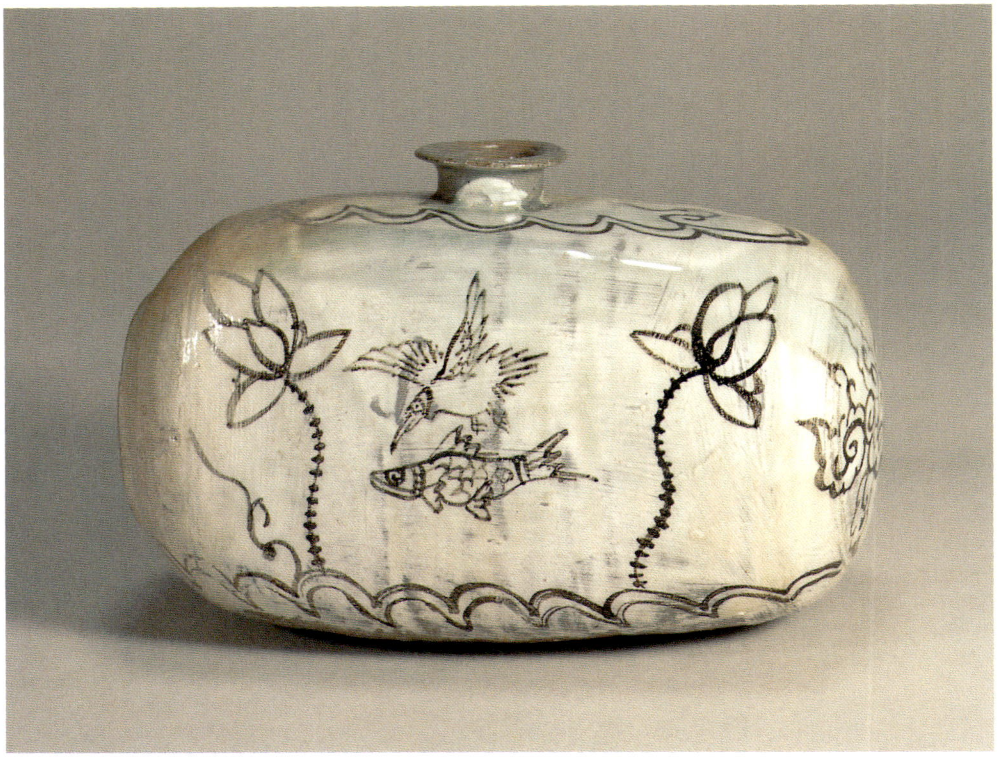

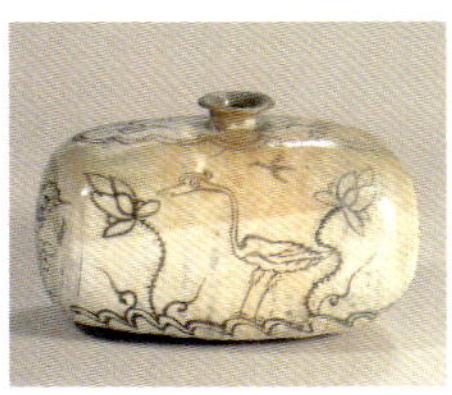

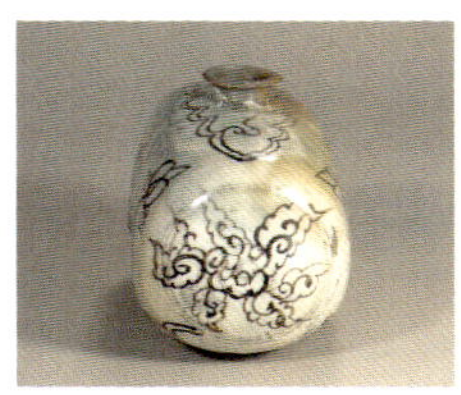

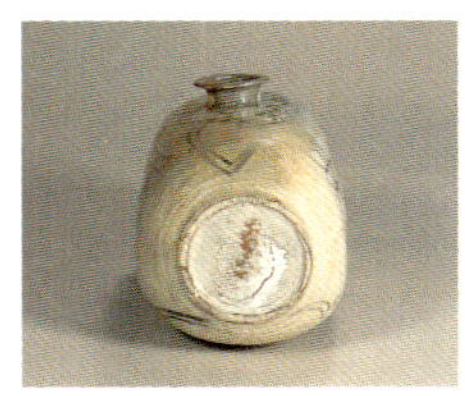

62

Bottle in the shape of a rice bale with an iron-brown bird and fish design

Punch'ŏng ware

Chosŏn dynasty, fifteenth/sixteenth century

14.4 cm

Acc. no. 20773

This rice-bale-shaped bottle, like no. 59, was originally formed using the base on the side of its body, and evidence indicates that it was fired vertically on that base. The finished flask, unstable without a foot, was probably supported by a circular stand made of wood or straw.

The body was brush-painted with white slip, with the exception of the mouth and foot, and over that a design was applied using iron pigment. The iron-painted design depicts a kingfisher swooping down to snatch a fish swimming in a pond where lotuses are flowering. On the other side, a heron rests among flowering lotuses, while a large bird flies serenely in the distance. Each scene captures a single moment executed with crisp and concise brushwork and, at the same time, conveys a serenity that suggests painting of Sŏn (Chin. Chan; Jap. Zen) inspiration. Support for this notion may be derived from the fact that, at the beginning of the Chosŏn period, Buddhism was strongly suppressed and Buddhist monks fled to the mountains to lead a sequestered life. Perhaps this drawing can be attributed to such an exiled Sŏn monk. This piece's renown as one of the most superb examples of Punch'ŏng ware is based upon its finely executed painting and well-structured form.

Bottles with a trumpet-shaped mouth, short neck, flared lower body, and low foot clearly show the influence of vessels popular in China during the Yuan and early Ming periods. This form was very popular in Punch'ŏng ware.

The brush-painted white-slip body of this vessel is decorated with a design painted in iron pigment indicating that it is Kyeryongsan ware. One of the particular characteristics of this ware is that, regardless of the vessel's shape, be it a bowl, bottle, or jar, its foot is always small and constricted, creating an impression of lightness. Four lines encircling the vessel divide it into sections; the main one contains a broad scroll design which, in its semiabstract manner, resembles modern design. In the flowing movement of the fine brush strokes, one senses passionate feeling. The only known documentation for this style at present consists of mortuary tablets dating from the second half of the fifteenth through the sixteenth centuries. One theory suggests that this ware belongs to an even later stage in the development of Punch'ŏng ware, but until further research has been conducted, lack of evidence currently leaves much about its origin unclear.

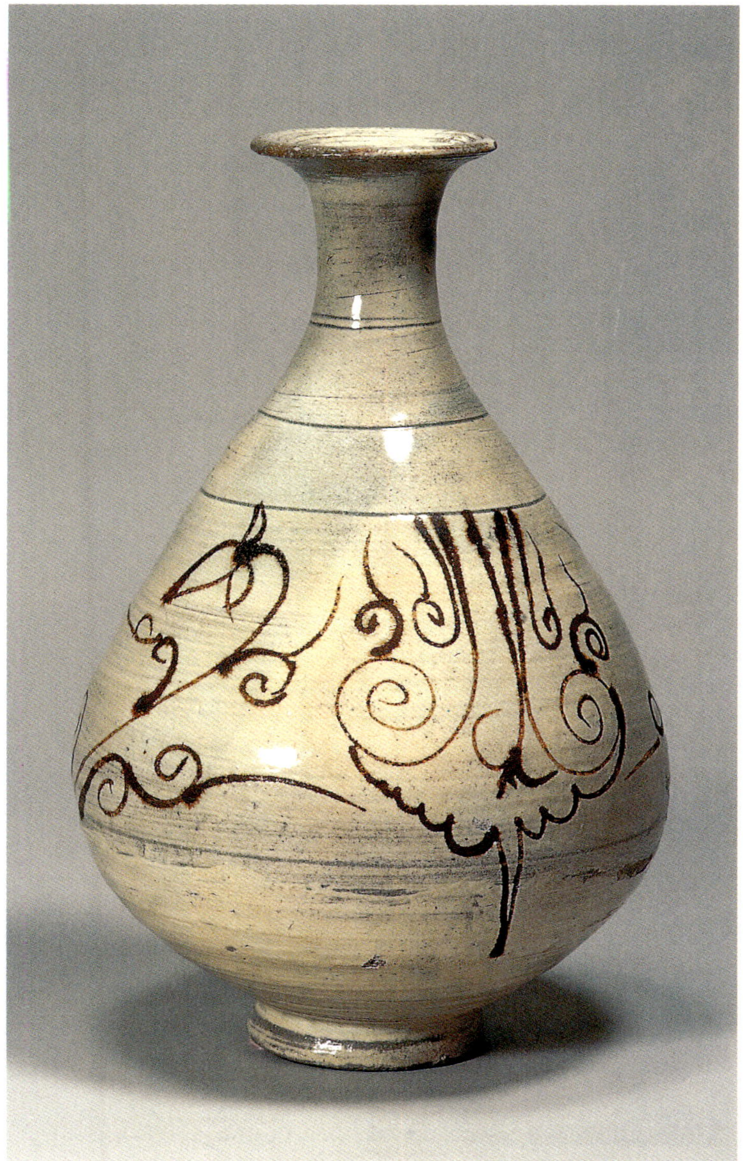

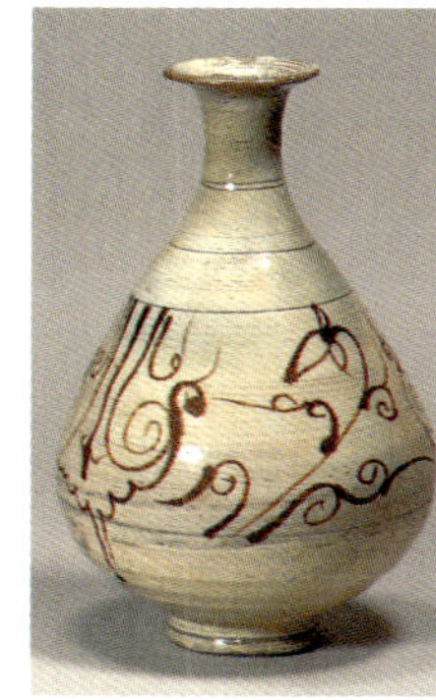

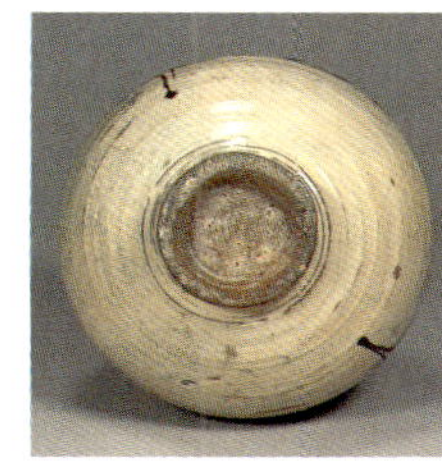

63

Bottle with an iron-brown scroll design

Punch'ŏng ware, Kyeryongsan ware

Chosŏn dynasty, fifteenth/sixteenth century

29 cm

Acc. no. 20292

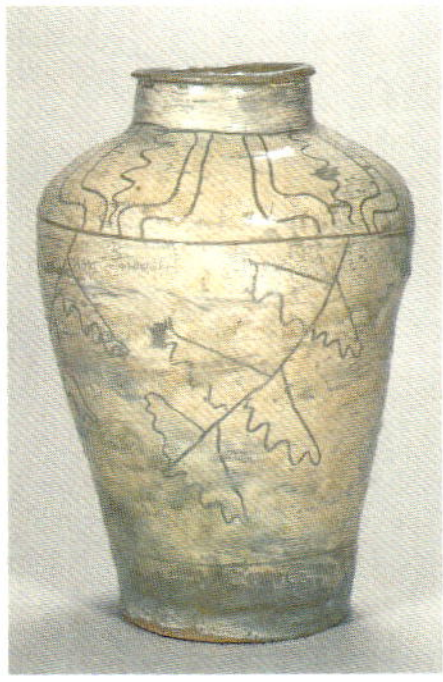

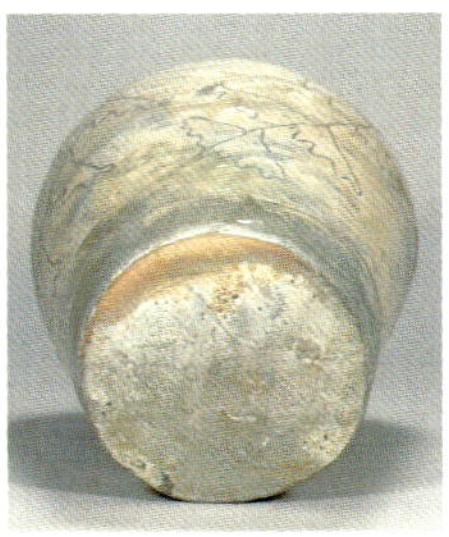

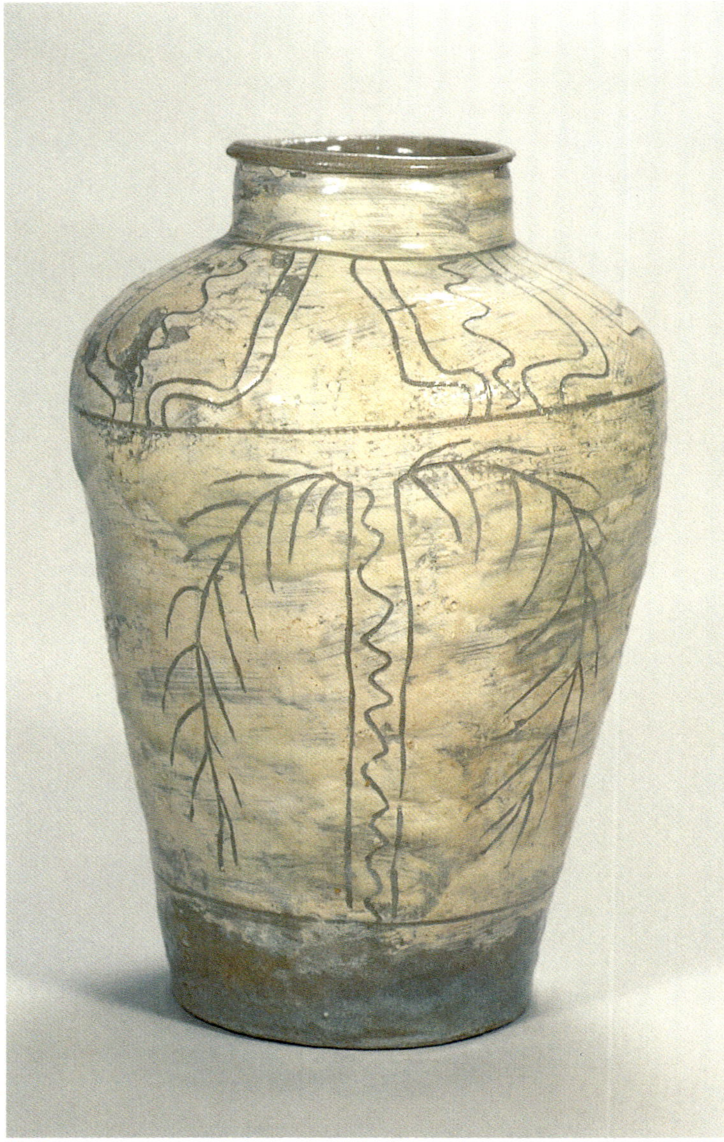

64

Jar with a sgraffito plant design

Punch'ŏng ware

Chosŏn dynasty, early fifteenth century

43.8 cm

Acc. no. 20294

The shape of this jar, with its vertical mouth, straight contour line extending from its shoulder to the lower part of its body, and flat, footless base, is often seen in Punch'ŏng ware. But it is the particular type of scraping technique used on it that is its datable feature. The technique is similar to that of a jar discovered in the reliquary monument of the monk Kobong of the Songkwang-sa in South Chŏlla province. The monk's death in about the year 1428 provides that jar with an early fifteenth-century terminus ad quem; therefore, the vessel exhibited here can be similarly dated.

This thin-walled and exceptionally light jar was constructed using a coiling technique; its body is subtly curved without any modification on the potter's wheel. The entire vessel is covered with an uneven and rough application of brushed white slip. A line drawn as a border encircles the lower body, but paint was applied below it, indicating that the border was ignored. An incised lotus-petal design is found on the shoulder; and incised line drawings, apparently of ginko and willow leaves, decorate the body. These are not executed carefully or neatly, but rather in a deliberately careless, unpretentious manner. Nonetheless, this jar, a typical example of the early Chosŏn aesthetic, forcefully conveys its essentially bare form.

This object assumes the shape of a helmet and is one of five or six such pieces unearthed to date, all of somewhat different size and design. Although their function is not clear, one theory holds that they served as covers for vessels placed in tombs. We cannot be absolutely certain of this, however, since a complete set, including a container together with a helmet-shaped cover, has yet to be discovered.

The piece is decorated with concentric bands of various designs. A zigzag, key-fret pattern encircles the rim. Above this, in order, are a band of lotus petals, a section of patterned waves, and another band of lotus petals. Encircling the top of the dome is yet another wave pattern in the middle of which, at the very top of the dome, is a small, gem-shaped knob. The bands containing the lotus petals were executed by carving away the surface so as to leave the petals in relief, whereas in the other sections lines were incised to create the wave designs. The inside of the helmet cover is glazed, and twenty-three clay spur marks are spaced at regular intervals around the perimeter. The piece was probably fired on a cylinder-shaped pedestal. Its unusual shape and graceful decoration give it particular importance.

The techniques used to produce this kind of Punch'ŏng ware are derived from earlier Koryŏ celadons; in a sense, these wares represent the ultimate extension of that tradition. One special feature of this piece is that the wave patterns were done in such a way that they recall stamen. Interestingly, shards resembling this piece have been discovered at kiln sites at Sanch'ung-gun, South Kyŏngsang province, and the shards are decorated with designs similar to those of Koryŏ celadons.

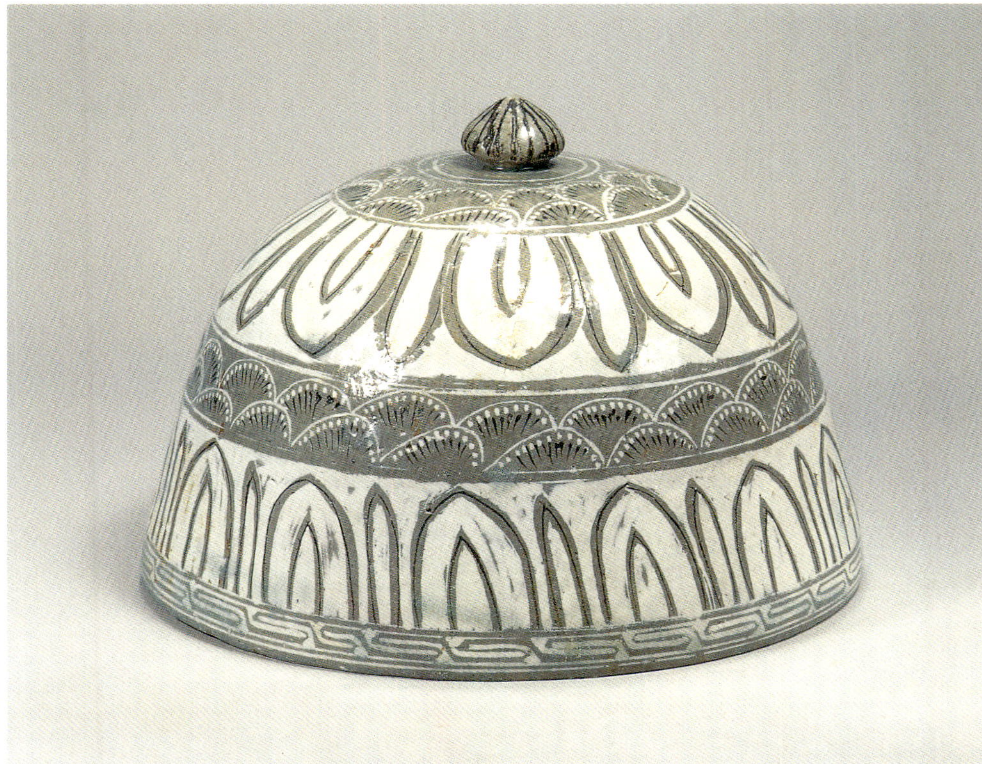

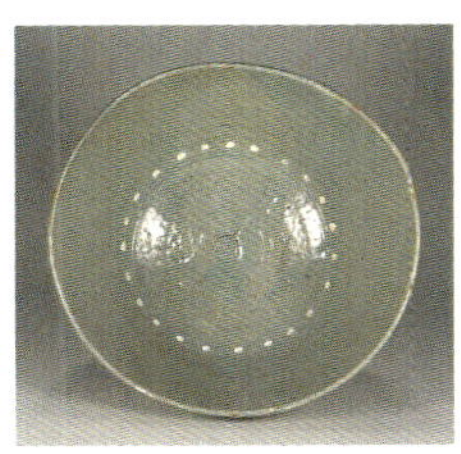

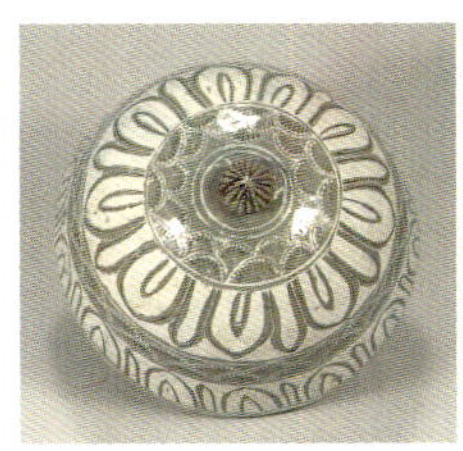

65

Vessel cover with a white-inlaid lotus-petal design on white slip

Punch'ŏng ware

Chosŏn dynasty, fifteenth century

18.3 cm

Acc. no. 20913

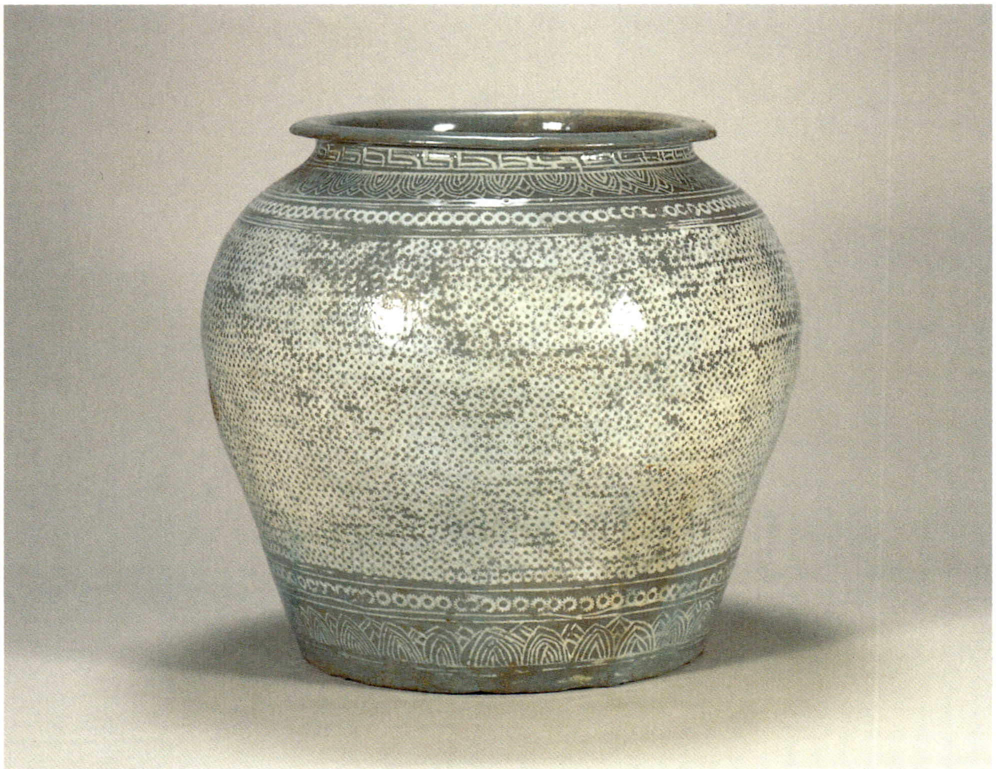

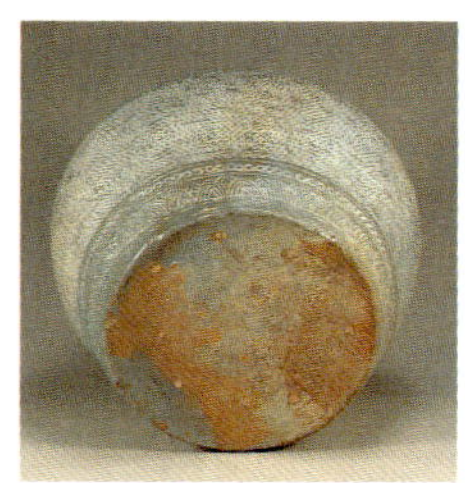

66

Jar with a stamped and slip-inlaid flower-blossom design

Punch'ŏng ware

Chosŏn dynasty, fifteenth century

37.6 cm

Acc. no. 20357

This jar has the wide mouth and ample, rounded body characteristic of Punch'ŏng ware. The shape is related to the function of such jars: they served as the outer containers for smaller vessels used to preserve the placenta. Originally, these jars had lids, but few such covers exist today.

Three bands of decoration encircle the top of this piece, which is somewhat larger and more imposing than most jars of this type. The first band, around the rim of the jar, is done in a key-fret pattern, and the next two represent lotus petals and chrysanthemum flowers, respectively. The chrysanthemum and lotus designs are repeated at the base of the jar. The base is flat, and the glaze ran unevenly in one place. Nevertheless, the designs were executed with great care, demonstrating the importance attached to these jars.

The expanse of the jar's body between the bands is filled almost entirely with impressed designs of chrysanthemum blossoms. To make these, the potter stamped the vessel with a wooden pattern cutter. The indentations were then filled with a white clay, or slip, and a thin, transparent glaze was applied over the entire piece before firing. Although the stamped-flower technique was widely practiced from the thirteenth century on, this kind of design, in which the floral pattern covers almost the entire surface of the jar, did not become common until the Chosŏn period. By the mid-fifteenth century, such pieces were being produced in many provinces. Archeological examples reveal that the stamping and inlay techniques coexisted in the same period.

The term *punch'ŏng* is used for a ware that is covered with white slip. The slip could be applied in any of several different ways. One technique was to incise or, more often, to stamp a pattern into a piece and apply a coating of slip to fill in the indentations. Then, a thin blue-green glaze would be applied and the piece would be fired. Punch'ŏng ware made in this fashion was sometimes decorated with unusual curvilinear and geometric designs. Later, a second technique became more common; the slip would simply be brushed onto a piece and, in place of the stamped design, some decoration might be painted on the surface before firing. Yet another technique involved merely dipping a ceramic piece into a large crock of liquid white slip called "powder." Then, a transparent glaze would be applied and the piece would be fired. A pot executed in this manner might either have been dipped completely into the powder or coated just to the edge of its base. Pieces given an overall coating are called *punin* (Jap. *kohiki*; literally, to apply powder), and they are especially prized by the Japanese. This dipping technique produced a thicker layer of slip than did brushing, which often resulted in chipping of the slip after firing.

This bottle is an example of *punin* ware that was probably brought into Japan at the end of the sixteenth century. It has a very natural shape; its contours, together with the indentations near the mouth, indicate that it was used as a decanter for rice wine. The glaze covers the entire piece, except in places on the shoulder where the slip was chipped off and repaired with lacquer. Long years of use have resulted in stains. These add character to the piece and are greatly admired by the Japanese.

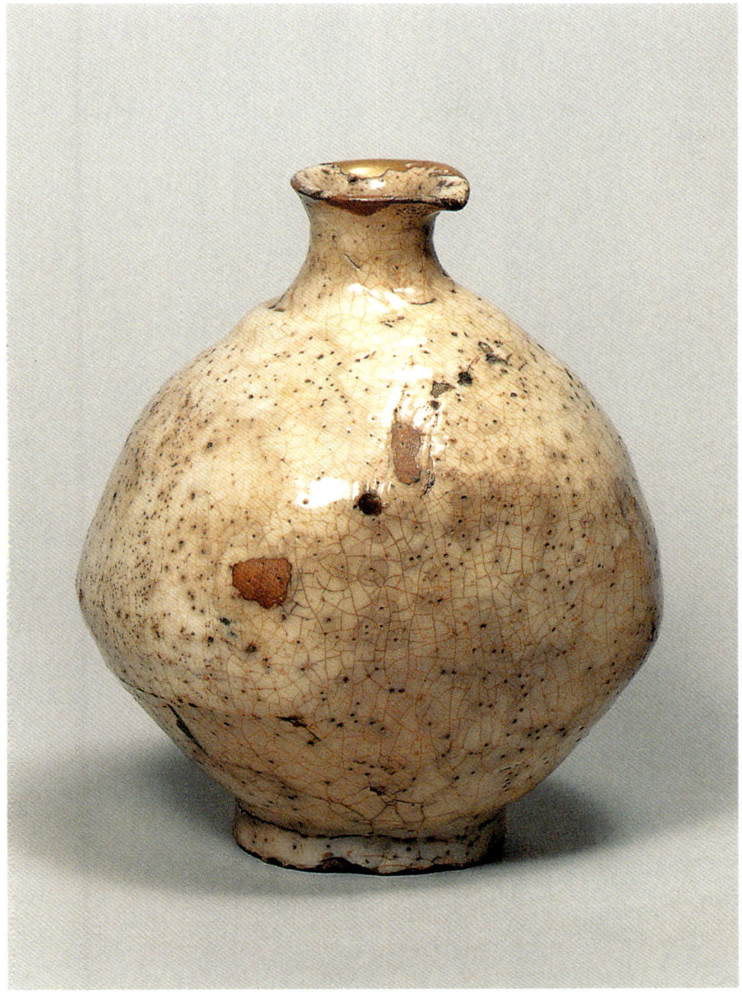

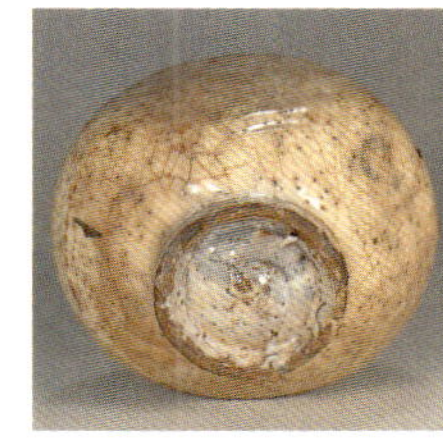

67

Bottle with an overall white-slip coating

Punch'ŏng ware

Chosŏn dynasty, fifteenth/sixteenth century

18.1 cm

Acc. no. 20204

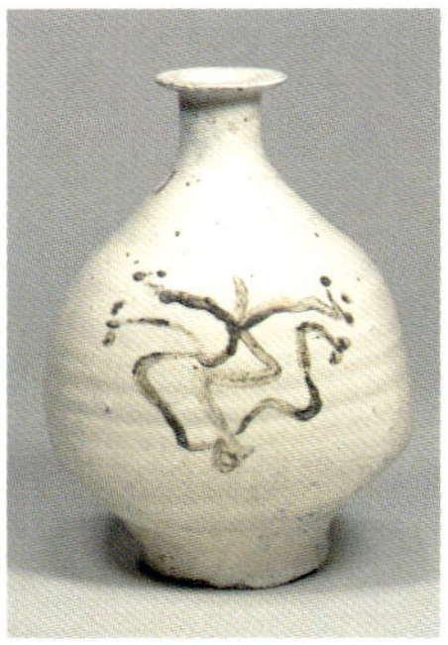

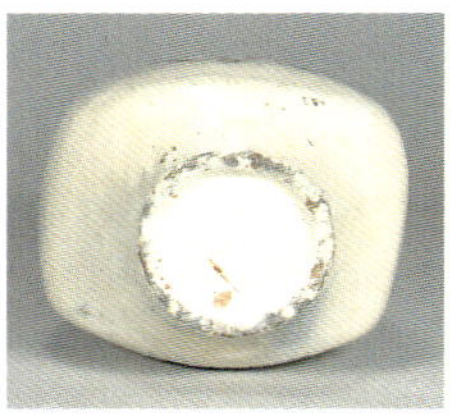

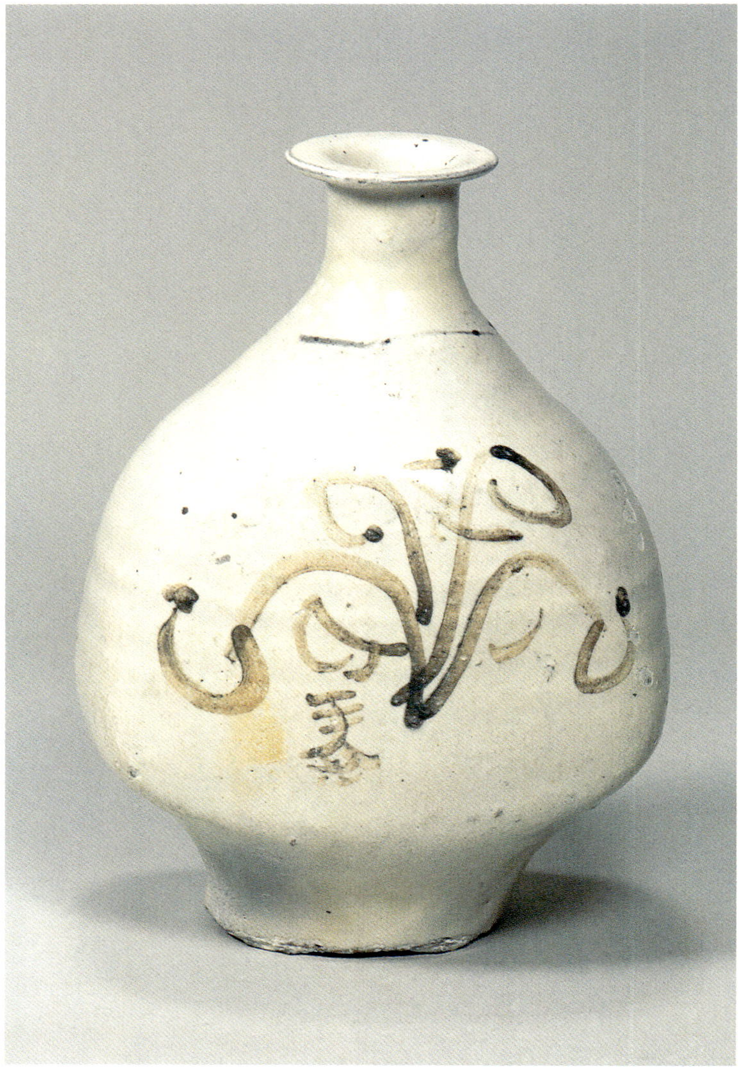

68

Bottle with an iron-brown flowering-plant design over a white-slip coating

Punch'ŏng ware

Chosŏn dynasty, fifteenth/sixteenth century

21.1 cm

Acc. no. 20702

This flask, like no. 67, is an example of *punin* (Jap. *kohiki*) ware made by dipping the entire piece in a crock of white slip so that the slip coated even the underside of the vessel. In this case, a design was drawn on top of the white slip with iron pigment. A transparent glaze was then applied and the piece fired. In Japan, this kind of flask is called a "painted *kohiki*" and is valued for its rarity.

The decoration painted around the middle of the bottle is probably a floral design of plants and flowers, although this is difficult to confirm since the pattern is so simplified. Similarly, it is impossible to determine whether the design under the floral motif is an extension of that pattern or a stylized written character. In any case, the motif fits the bottle effectively and enhances its exquisite shape. The flask's powerful mouth opens in the form of a trumpet, and the curved line that runs from the shoulder flows smoothly into the deeply concave base, which concludes naturally in the vessel's foot. This shape is very rare, and the flask was probably made in South Chŏlla province. The sort of floral motif used here, which is inspired by nature, can also be found on iron-painted porcelain ware produced in the seventeenth century. Consequently, one theory regards the *punin* technique itself as a later development in Punch'ŏng ware.

Black-glazed ware was not produced in large quantities until the Chosŏn period, although some examples dating to the late Silla period have been unearthed from tombs on Ullŭng Island and other pieces have been discovered in Koryŏ-period kiln sites. Almost all extant black-glazed wares are flasks or bottles, and their shapes are practically indistinguishable from such pieces done in the Punch'ŏng tradition. Indeed, black-glazed ware has been discovered at kiln sites known for Punch'ŏng ware in Chŏlla, Kyŏngsang, and Ch'ungch'ŏng provinces. Moreover, some examples have even been found alongside porcelain. These latter examples, however, should be distinguished from the black-glazed wares produced from the seventeenth century on, in which the glaze was applied over a white porcelain body.

This flask is somewhat smaller than most, and its shape is slightly unusual as well. That is, the mouth is exceptionally tiny, and the body immediately swells out, imparting a strong sense of solid structure. The glaze has a deep black luster and covers the base. This piece is one of the finest known examples of a black-glazed flask.

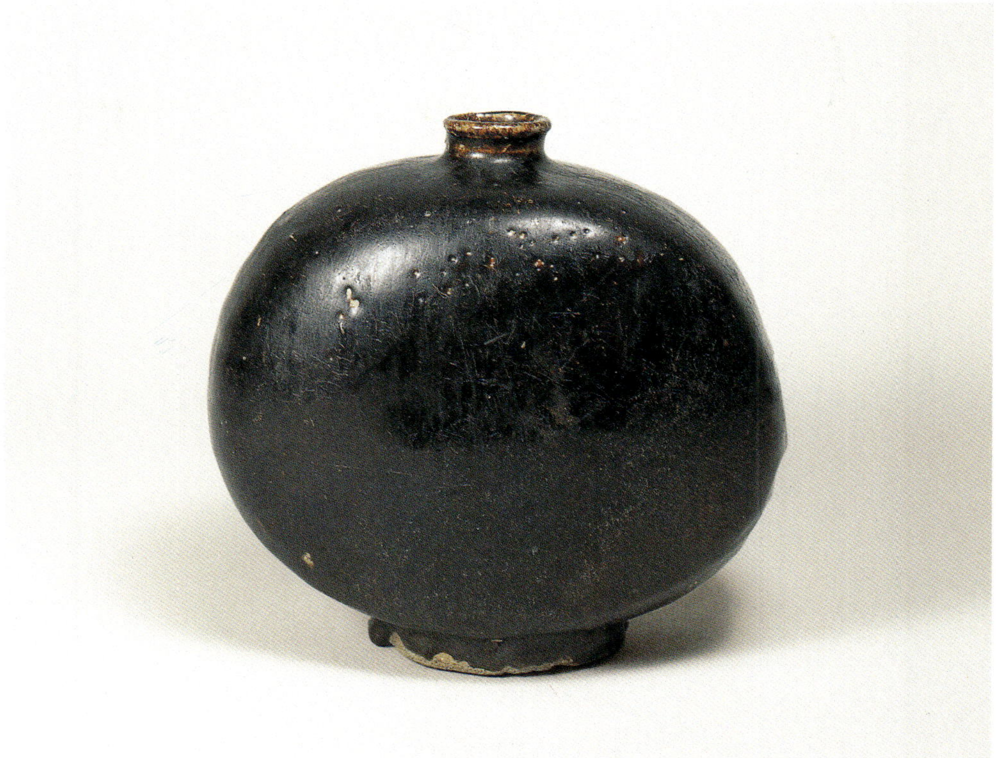

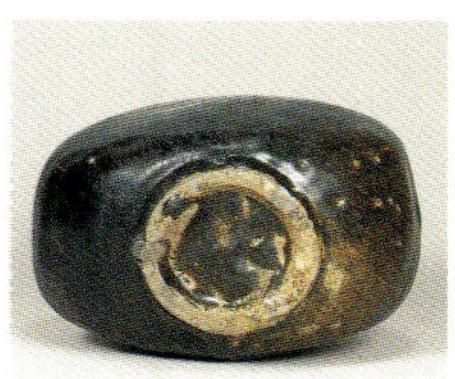

69

Flask covered with a black glaze

Stoneware

Chosŏn dynasty, fifteenth century

17 cm

Acc. no. 20774

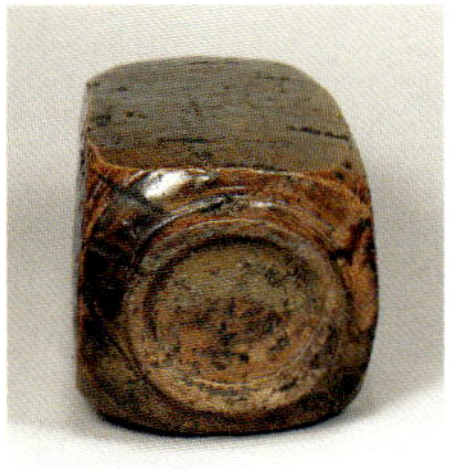

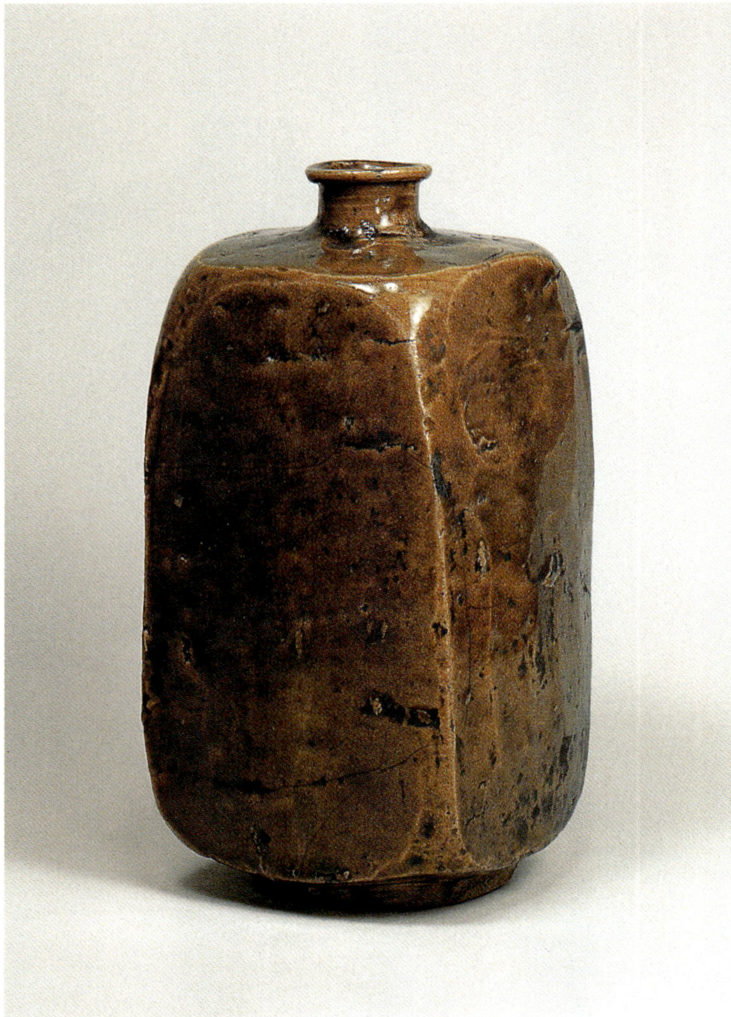

Both caramel and black colorings are produced by iron contained in the glaze. In general, higher iron concentrations yield blacker results. Caramel glazes were especially common during the second half of the nineteenth century, and many examples of such jars and bottles survive today. The jars, it is thought, were used as honey containers.

The bottle included in this exhibition dates to the seventeenth century, making it a very early example of caramel-glazed ware. The piece has an unusual shape as well. In the Chosŏn period, most square bottles were made in molds, but this piece was first thrown in the form of a round flask and then squared off with a spatula. Traces of this process are evident on the shoulders, which have obviously been shaved in order to moderate the squareness of the design. The potter used the spatula with skill and confidence, producing a piece that is at once simple yet exhilarating. The base is not glazed.

70

Square bottle covered with a caramel glaze

Stoneware

Chosŏn dynasty, seventeenth century

26.6 cm

Acc. no. 21021

Porcelain jars were produced in great numbers during the first half of the Chosŏn period. The shallow, everted lip found on the jar included in this exhibition is characteristic of such ware, and thus one might suppose that it dates to the fifteenth century. However, the piece has several unusual features that cast some doubt on this dating. First, the inside of the jar was entirely glazed to prevent liquid from leaking out of it. In addition, the glaze has a slightly reddish tint, due to the iron content of the clay. Moreover, in contrast to most examples dating to the early Chosŏn period, the base of this vessel is not glazed, and it has an exposed foot and an unexpectedly thick and robust wall. Finally, a network of fine crackles extends over the entire surface. Taken together, these characteristics indicate that it might be more appropriate to date this piece to the sixteenth century. Its gentle and rounded shape, which resembles Tang-period porcelain, and its serene, yet lustrous, glaze make this an extremely fine example of a Chosŏn-period jar.

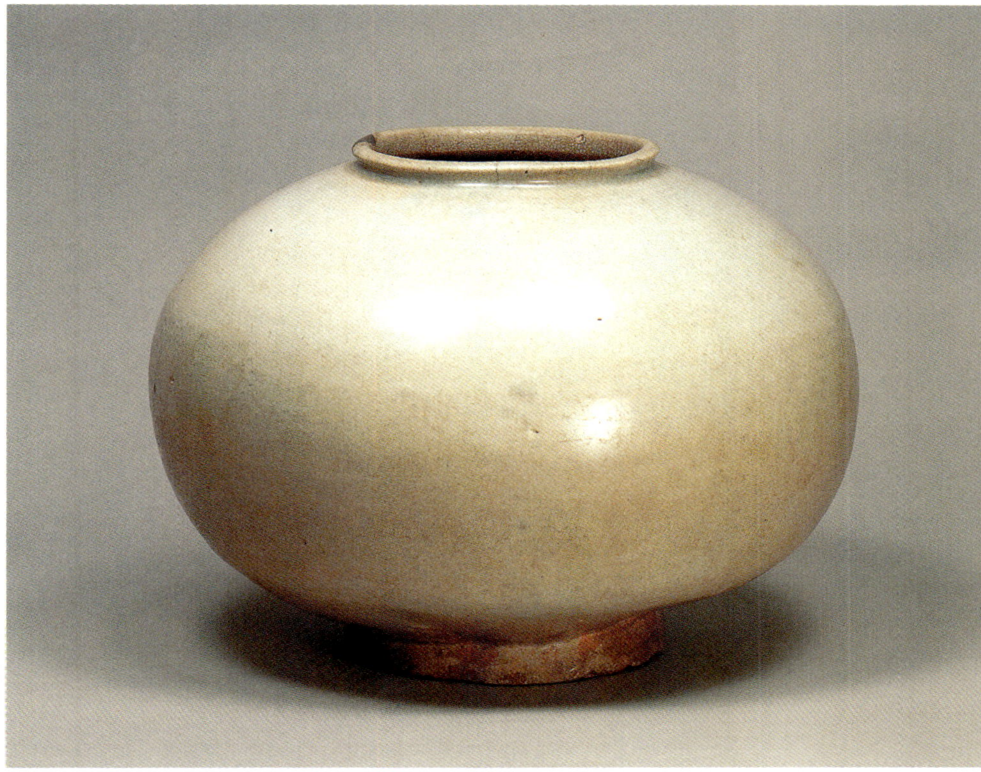

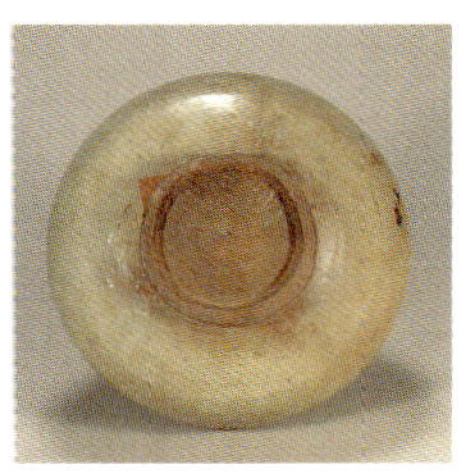

71

Jar

Porcelain

Chosŏn dynasty,
sixteenth century

23 cm

Acc. no. 20197

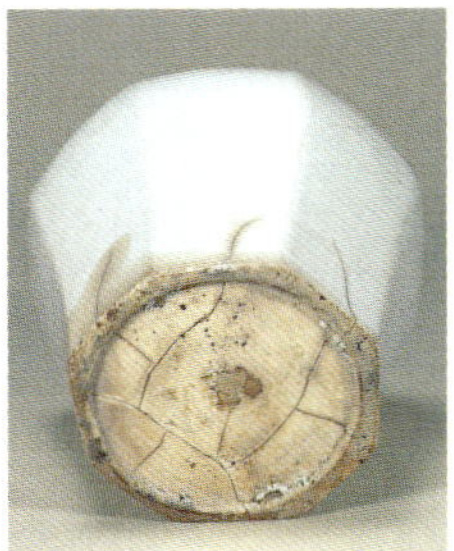

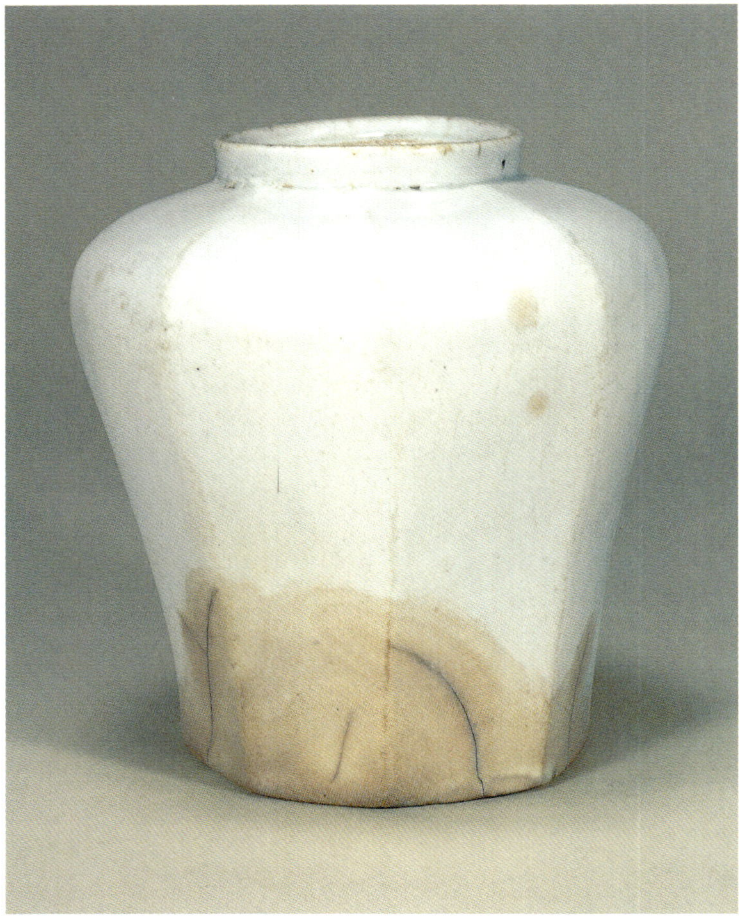

The technique of beveling was already known in the Koryŏ period. Most Koryŏ pieces, however, were copies of Chinese prototypes and were executed only in a limited variety of designs, with long-necked bottles constituting the most common shape. During the Chosŏn period, people came to favor the kind of clean, straightforward design that could be achieved with beveling, and vessels incorporating this technique became very common from the first half of the eighteenth through the latter half of the nineteenth century. Many examples have been unearthed at official kiln sites at Kwangju, Kyŏnggi province.

This jar was beveled into eight sides. The line extending from the shallow mouth to the high, broad shoulder, which then slopes downward on an inward curve to the base, is firm and confident. The shape and glaze do not suggest the feeling of cold perfection found in so many Chinese porcelains, but rather shroud any hint of severity with a mood of gentleness. This jar has many features characteristic of pieces made in the eighteenth century. The base is trimmed slightly and covered entirely with glaze.

72
Beveled jar
Porcelain
Chosŏn dynasty, eighteenth century
21.9 cm
Acc. no. 20101

In general, two methods were used to make square porcelain bottles. In the first of these, the potter employed molds to create each of the four sides individually, joined them together, and then added the top and the base. In the second method, the potter used a different type of mold to shape a lump of clay into a solid, square-shaped pillar and then hollowed out the inside, leaving only the outer shell. Scholars are divided as to which of the methods was used to make square porcelain bottles in Korea in the Chosŏn period.

Square bottles may have either a round or square spout; it is not clear whether one of these styles preceded the other chronologically or whether one predominated at a particular time. Given all of these uncertainties, it is difficult to date any particular piece. Taken all together, the shape, glaze, and type of clay used suggest that the piece included in this exhibition was executed in the first half of the eighteenth century and thus may be counted among the oldest extant square porcelain bottles. It should be noted that the shoulders were all trimmed off diagonally. The base is glazed.

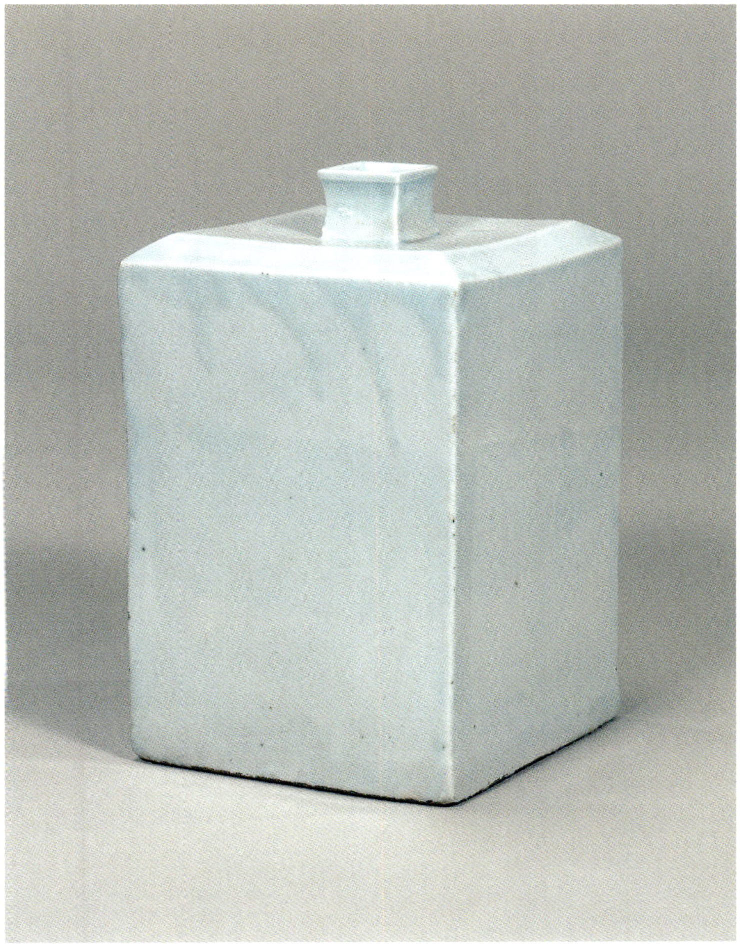

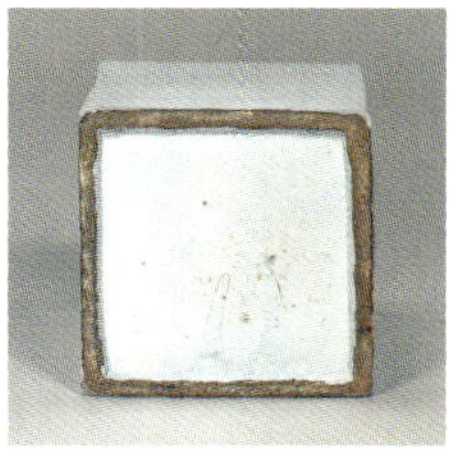

73

Square bottle

Porcelain

Chosŏn dynasty, first half of the eighteenth century

21.3 cm

Acc. no. 20652

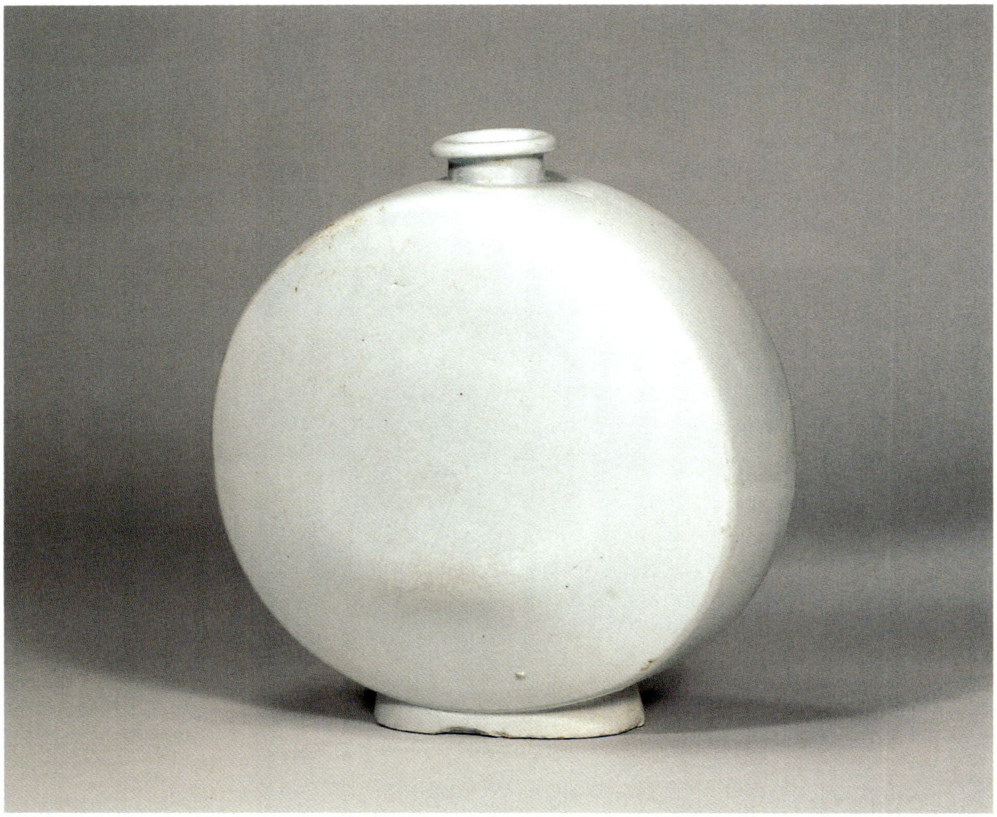

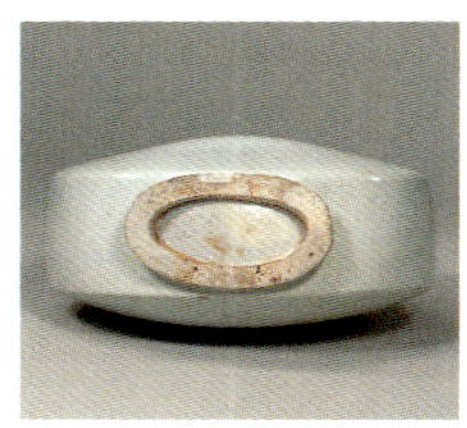

74
Flask
Porcelain
Chosŏn dynasty, seventeenth century
25.7 cm
Acc. no. 20198

This kind of flat-sided shape was very popular for flasks of the Chosŏn period, and people used them as liquor containers. Similarly contoured flasks were done in Punch'ŏng style in the fifteenth and sixteenth centuries, in white as well as in underglaze iron-painted porcelain beginning in the seventeenth century, and finally in blue-and-white ware from the eighteenth century on. Generally, a Punch'ŏng flat-sided flask was thrown on the wheel in the form of a globular jar, and the potter then tapped the sides in order to flatten them. Porcelain flasks, however, were executed differently: two dish-shaped plates were joined together to form the body, to which the mouth and foot were then attached. The shape of the foot changed over time. The feet of pieces made in the seventeenth century assume an oval shape, but, from the mid-eighteenth century on, the four corners were cut diagonally to make the sort of octagonal foot shape seen clearly in no. 79.

This piece probably dates to the seventeenth century, when many underglazed, iron-painted flasks of similar design were produced. The glaze on this particular flask also displays the hard, white finish characteristic of seventeenth-century ware. The foot was made separately and attached to the body later. The base is glazed.

In the early fifteenth century, Korean potters employed techniques developed in early Ming China to create pieces imbued with classic shapes and luminous, pure-white finishes. The collection of essays entitled *Yongjae ch'onghwa*, written by Song Hyŏn in the mid-fifteenth century, informs us that court utensils in the reign of King Sejong (1419–50) consisted almost entirely of porcelain, a telling indication of how prized such wares were at that time. Yet, another historical document claims that the Korean court commissioned some especially elegant pieces for presentation to the emperor of China in 1424.

Horn-shaped cups were made in Korea during the era of the Three Kingdoms and the Koryŏ period. They can also be found among Punch'ŏng wares. Despite this history of nearly continuous production, horn-shaped vessels in porcelain are extremely rare, and extant pieces date to the first half of the Chosŏn period. Apparently, none was produced after the seventeenth century, perhaps because the type lost its place in court rites when the ceremonial wares used in such rituals changed dramatically. The horn-shaped cup included in this exhibition has the graceful, classical form favored at King Sejong's court.

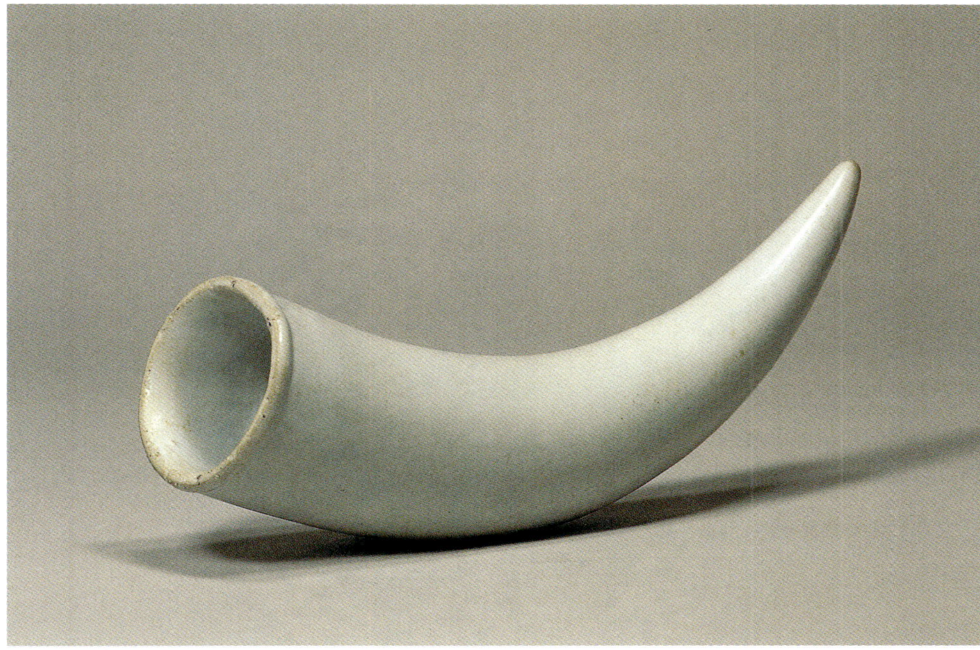

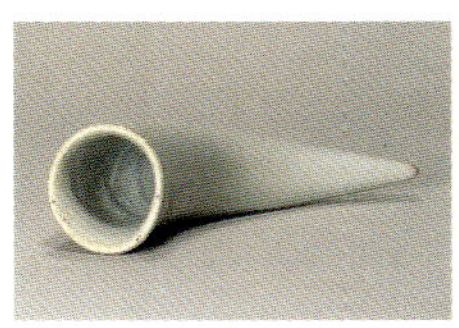

75

Cup in the shape of a horn

Porcelain

Chosŏn dynasty, first half of the fifteenth century

18.8 cm

Acc. no. 20730

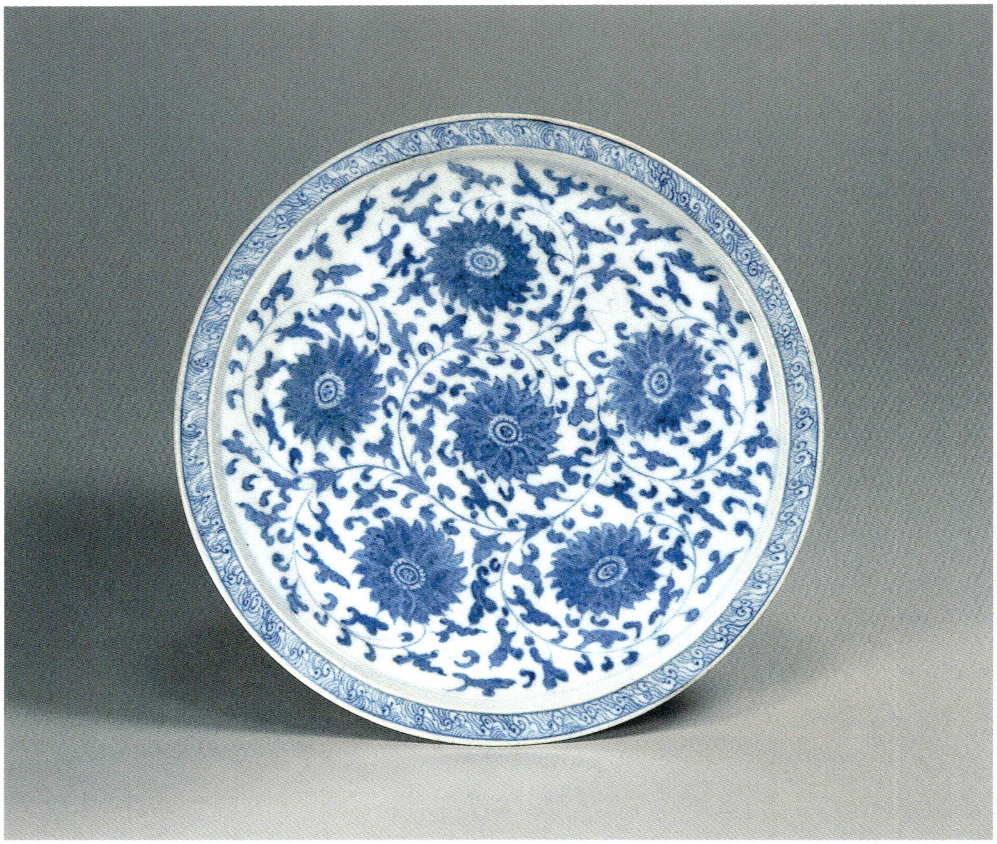

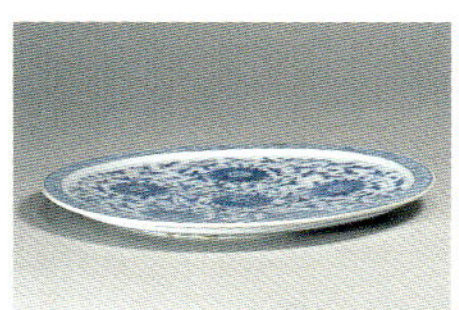

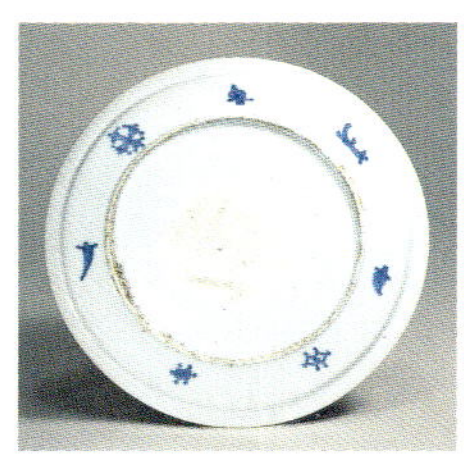

76
Dish with an underglaze floral-scroll design
Blue-and-white ware
Chosŏn dynasty, second half of the fifteenth century
22.7 cm (diam.)
Acc. no. 20427

Although there is a consensus today that the appearance of blue-and-white ware in Korea was influenced by the development of such a tradition in early Ming China, scholars are still not certain exactly when blue-and-white pieces were first made in Korea. One indication that blue-and-white ware was being produced by the late fifteenth century can be found in a document written in 1464, which states that Korean domestic blue cobalt was being utilized for underglaze painting because there was a shortage of the Chinese pigments customarily used to make blue-and-white ware.

This dish is a good example of Chinese influence on Korean blue-and-white pottery. The shape is probably modeled after dishes made in China during the Yongle (1403–24) and Xuande (1426–35) periods; and the shallow, low foot may also imitate Chinese examples. The floral (Kor. *bosanghwa*; Chin. *baoxianghua*) scroll design, however, is more reminiscent of pieces produced during the Chenghua period (1465–87), and so it is probably safer to date this piece to the latter half of the fifteenth century. One other, similar dish has been unearthed at a kiln site at Doma-ri, Kwangju. The two dishes have an "eight-jewel" motif in common, and part of the glaze inside the foot has been chipped off in each case as well. Large diameters and small, shallow feet characterize blue-and-white dishes made during the fifteenth century.

This jar is executed in a manner strikingly different from its Chinese counterparts. For instance, the common Chinese lotus-petal motif is not present here. Rather, the jar is decorated on one side with a painted pine and small birds, and on the other with plums and small birds. Moreover, this sort of arrangement, in which the motif begins at the base and moves up along the shoulder of the body, is not found on blue-and-white ware produced in Ming China. The designs here are exquisite, and we know from the written record that they were painted by court-appointed artists. The underside has been trimmed flat and is glazed.

Perhaps the jar's most unusual feature is that superior-quality cobalt of great transparency was used on the pine needles and small birds, whereas the outline of the tree trunk and its bark were drawn utilizing a muddy cobalt with a high iron content. Cobalt with a high degree of purity was imported into Korea from China. However, supplies became increasingly scarce, and documents indicate that Koreans went to great lengths to discover domestic sources. Cobalt was first mined in Korea in 1463, and we can assume that this jar was made around that time. The shape, glaze, and design of this jar are similar to those of no. 85; the two pieces may well have been made at the same kiln at approximately the same time.

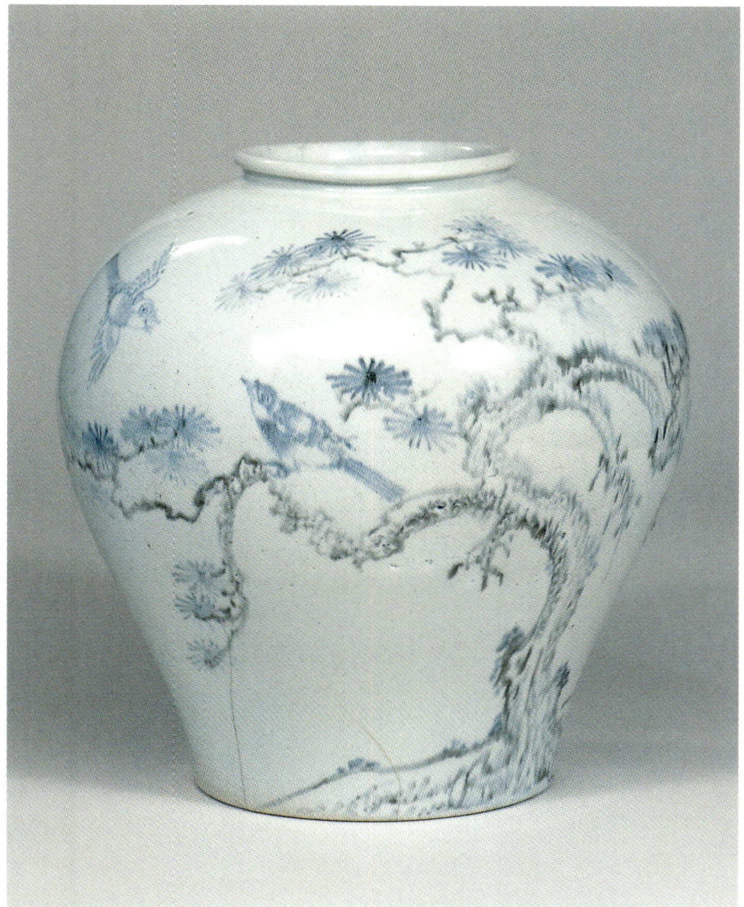

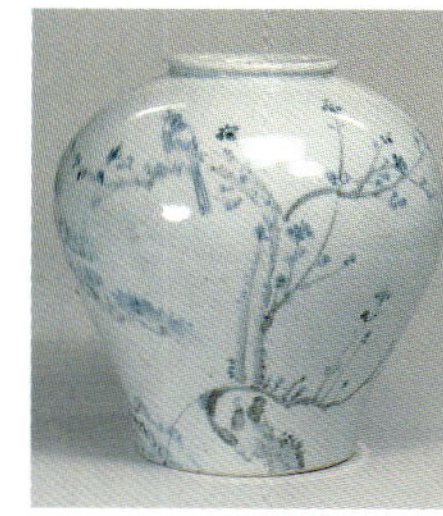

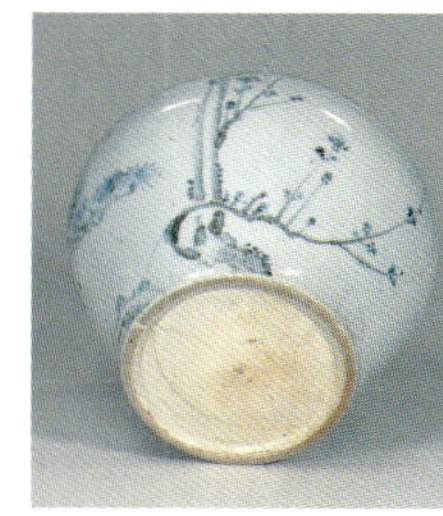

77

Jar with an underglaze bird and pine-tree design

Blue-and-white ware

Chosŏn dynasty, mid-fifteenth century

27.3 cm

Acc. no. 20578

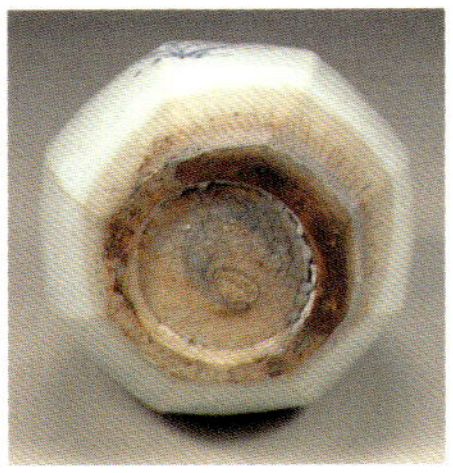

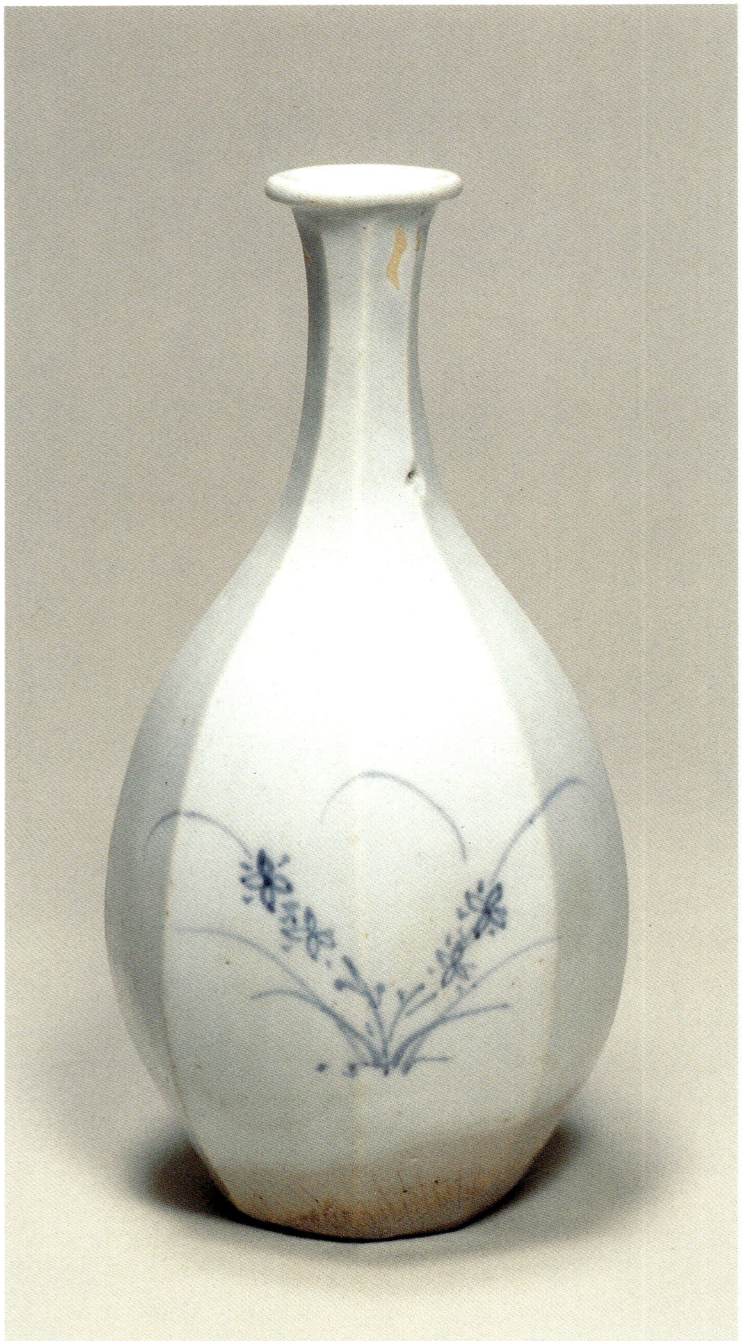

78

Beveled bottle with an underglaze floral-plant design

Blue-and-white ware

Chosŏn dynasty, first half of the eighteenth century

24 cm

Acc. no. 20415

Size and weight combine with a delicate silhouette to make this an unusual example of an eight-sided beveled bottle. This piece also differs from other large beveled bottles in that the footless base has a concave underside. The mouth is tiny, and the lip folds out and around to form an astragal-like shape (Kor. *okyŏn*; Jap. *tamabuchi*). This piece was first thrown in the shape of a round jar on a potter's wheel, and a spatula was then used to cut from just below the mouth in order to form the eight beveled sides. The beveling was done in a skillful, natural manner enhanced by a spatula mark left near the mouth and by the cracks in the glaze at the base of the bottle. The base is glazed.

A simple motif of meadow orchards was painted on each side of the bottle. The design flows with smooth and easy gracefulness, and its delicate size is set off exquisitely by the surrounding white space. Such a restrained pattern is characteristic of blue-and-white ware of the early eighteenth century. This "autumn-grasses" (Kor. *chucho*) motif, as it is called, is regarded as extremely charming in Japan. This bottle has gained considerable renown, since it surpasses others of the same genre in terms of shape, design, and glaze.

Flasks were made throughout the Chosŏn period. However, during the first half of the period, nearly all were done in the Punch'ŏng tradition, and porcelain flasks are very rare. From the seventeenth century on, iron-painted porcelain flasks became common, and both white and blue-and-white porcelain flasks were made in great numbers in the eighteenth and nineteenth centuries. Beginning in the eighteenth century, two handles in the shape of squirrels were ordinarily placed opposite each other on the shoulder of the flask. Strings could then be passed through two holes located in the foot and be tied around the handles so that the owner could carry the piece and use it as a liquor container.

This piece is rather large for a flask, and not many of this kind have survived. One side is adorned with a painting of two figures engaged in conversation by a lake, whereas on the opposite side a single pavilion is drawn within a mountain landscape. This type of landscape motif became widespread during the course of the eighteenth century and can be found on jars, bottles, and water droppers. Judging from its shape and glaze, this particular flask was probably made in the second half of the eighteenth century. A similar example is in the Ewha Women's University Museum, Seoul. The foot was made separately and attached to the piece. The base is glazed.

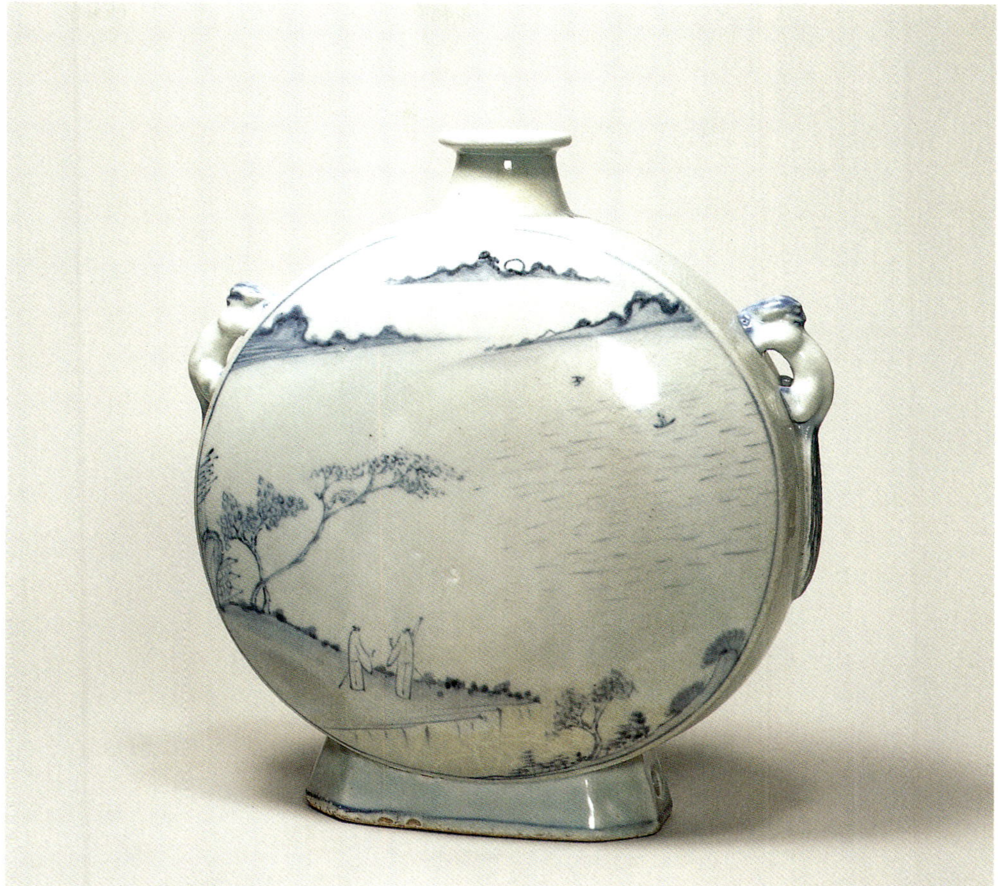

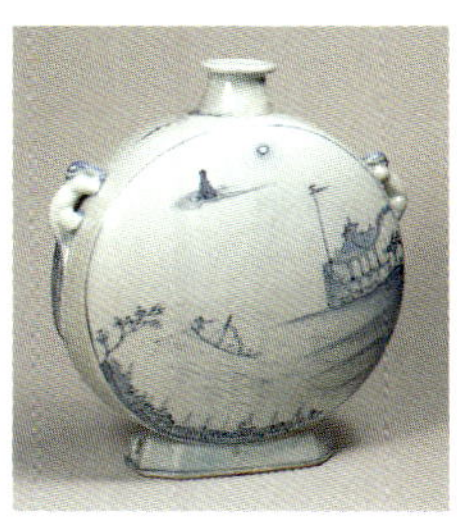

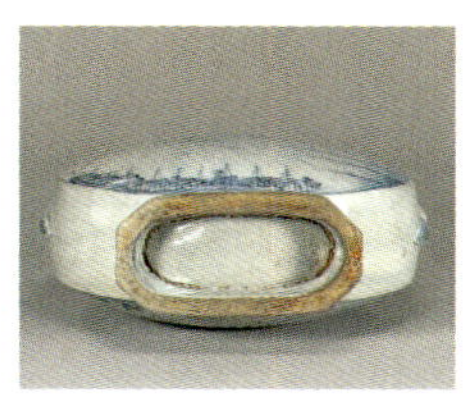

79

Two-handled flask with an underglaze landscape and human-figure design

Blue-and-white ware

Chosŏn dynasty, second half of the eighteenth century

30.3 cm

Acc. no. 20628

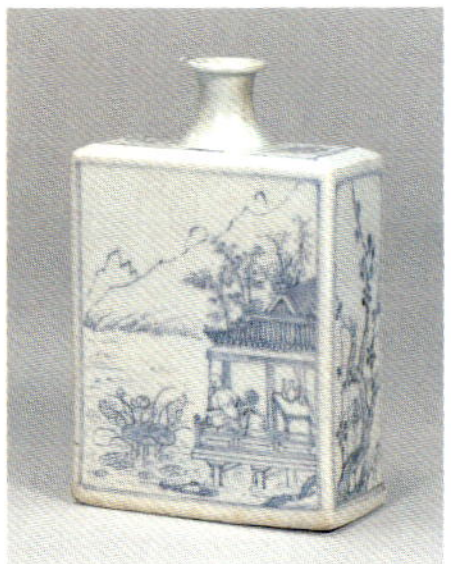

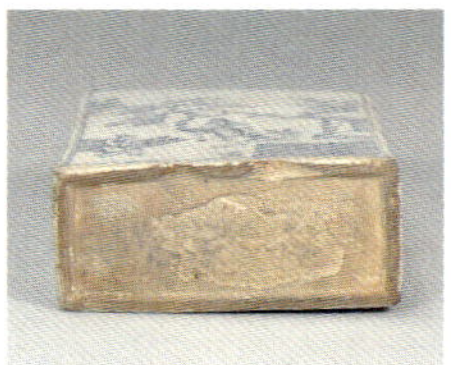

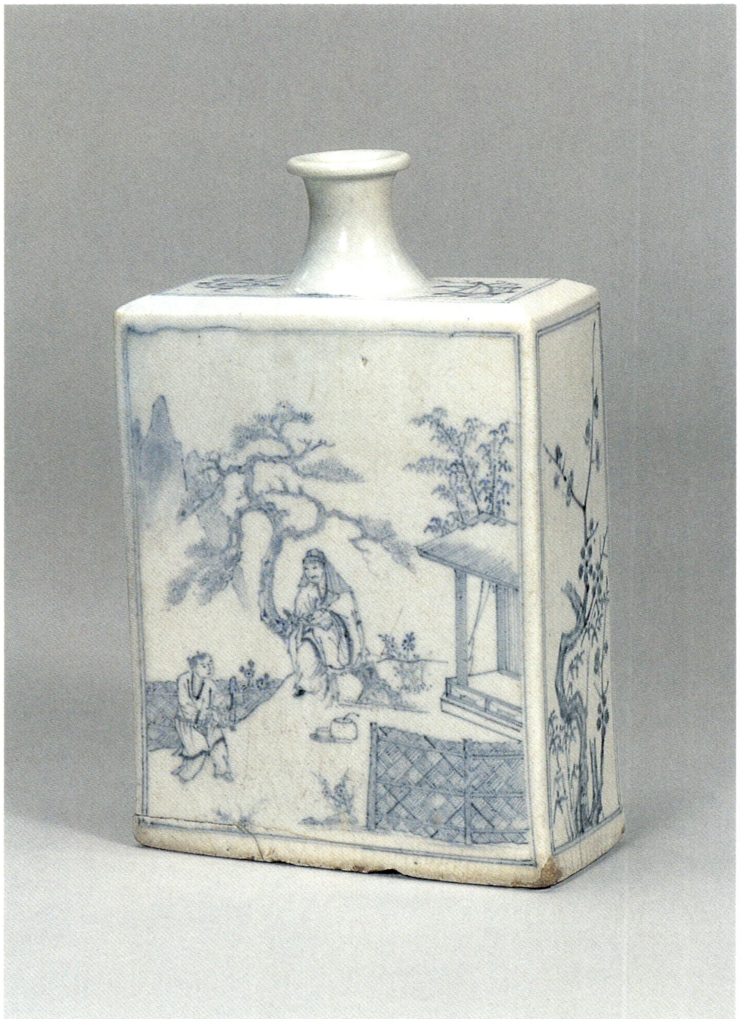

This is an unusually large example of a bottle done in a rectangular shape with flat, squared-off sides. It is also simpler than most such bottles, and does not include the squirrel-shaped handles that they normally feature.

The figural motifs on this piece depict traditional Chinese stories. On one side of the bottle a simply clad Daoist sits cooling himself in a pavilion overlooking a lotus pond, and, on the other side, a boy is seen handing chrysanthemums to an older man who rests under a pine tree. This sort of painting is unique and cannot be found on other similarly shaped flasks. During the reigns of Yŏngjo (1724–76) and Chŏngjo (1776–1800), which correspond roughly to the Qianlong epoch (1736–95) in China, Koreans drew much inspiration from Chinese scholarship, thought, and art. This bottle shows the extent to which Chinese taste had spread among the Korean literati. Originally, porcelains of this quality were produced for royal households, but, by the late eighteenth century, they had become popular among other classes as well. The shallow, flat base is trimmed and is not glazed.

80

Rectangular bottle with an underglaze landscape and human-figure design

Blue-and-white ware

Chosŏn dynasty, second half of the eighteenth century

22.9 cm

Acc. no. 20817

The design on this jar is very unusual for blue-and-white ware of the Chosŏn period. What seems to be a flowering clematis vine encircles the body, moving up and down in wave-like fashion so as to resemble a stylized scroll design. The upper and lower portions of this jar were thrown separately, and the lower edge of the floral wave touches the horizontal line that was created when the two pieces were fitted together. Ordinarily, jars were decorated with anywhere from two to four individual, isolated motifs; other examples of this sort of design, consisting of a single floral pattern horizontally encircling the entire piece, are virtually nonexistent. Perhaps this is a Korean improvisation on the theme of intertwined peony-scroll designs often found on blue-and-white ware in Ming China. The manner in which the flowers are painted, the placement of the leaves, and the flow of the vines all combine to create a feeling of lightness and rhythmic sensitivity. In addition to the clematis motif, a "seven-treasure" design appears at two places on the shoulder. This pattern commonly occurs on blue-and-white ware produced from the first half of the eighteenth to the nineteenth century, although it can be seen on some pieces dating to the fifteenth century. The glaze on this jar is a quiet blackish blue. The coloring, glaze, and design suggest that it was made in the first half of the eighteenth century. The concave hollow on the base is shallow, which is also typical of this period.

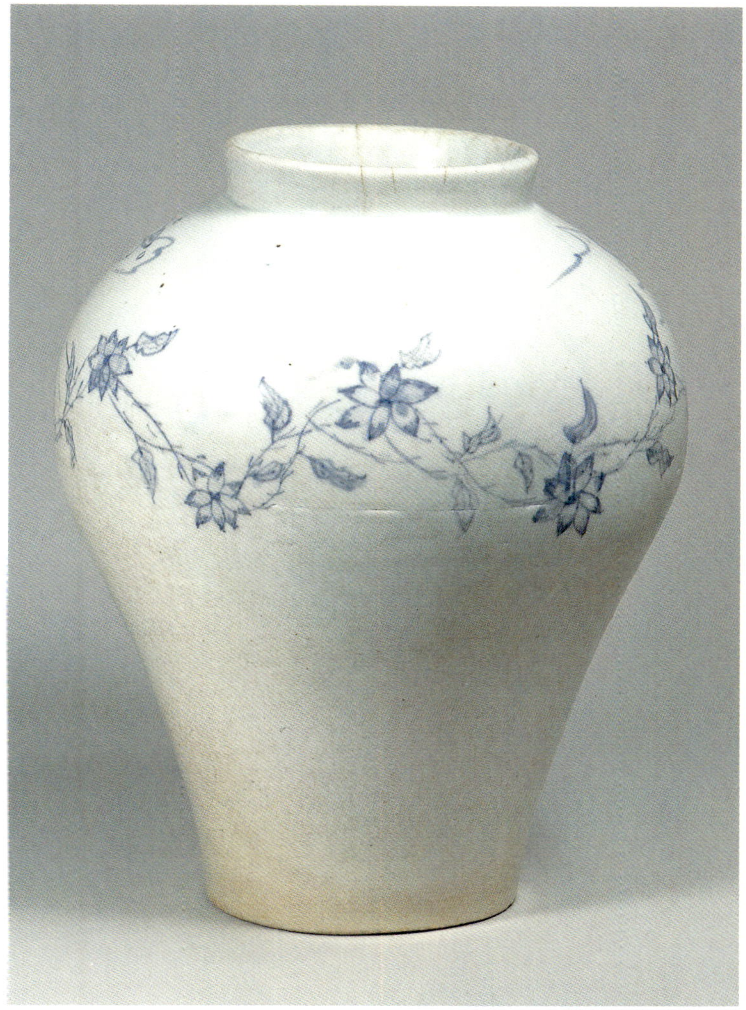

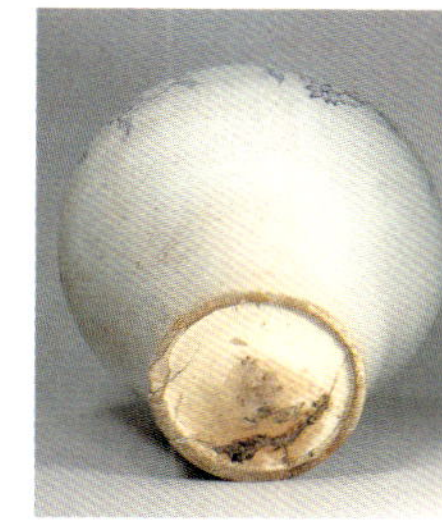

81

Jar with an underglaze clematis-scroll design

Blue-and-white ware

Chosŏn dynasty, first half of the eighteenth century

35.8 cm

Acc. no. 20178

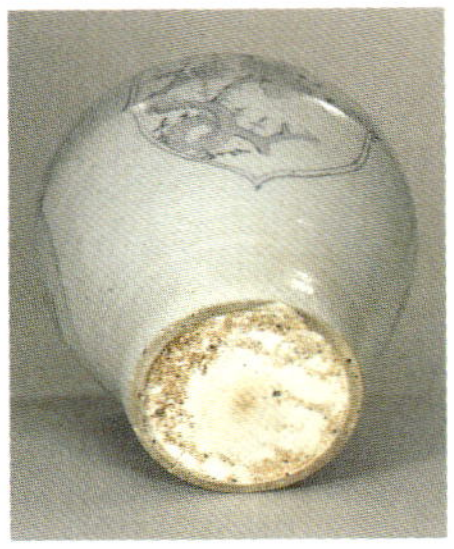

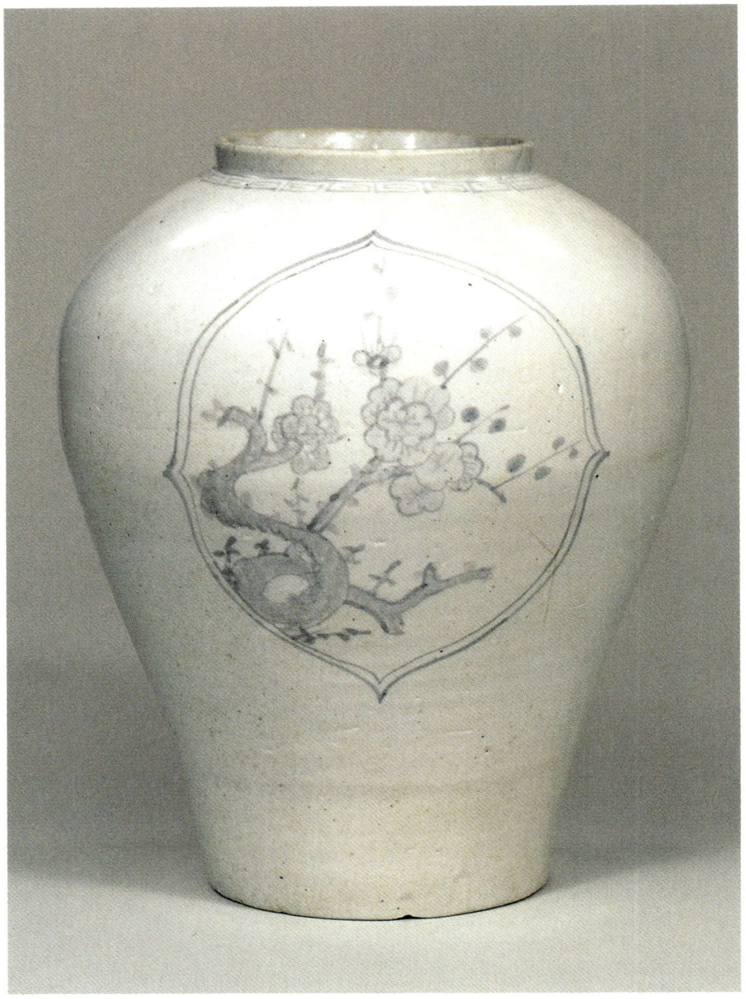

This jar is graced with classic proportions. Its contour line flows fluidly from the shallow mouth, sloping gently along the shoulder and down to the base. This profile, which was characteristic of pieces made in the early eighteenth century, later gave way to a silhouette that more closely resembles an *S*. Three "windows" done as outlines of water chestnuts punctuate the central portion of the body. Each frames a painting of aged plum branches drawn in a conventional, twisted style. The outlines of the plum branches and flowers have been sketched in fine lines in the front "window," and the darker shading of the bark helps to create a harmonious whole. The mood of the brushwork is rustic and naïve. Two similar "windows" appear on the other side of the jar, but there the brushwork falters somewhat more. Nevertheless, the soft blue tint of the cobalt compensates for the painting and produces a very tasteful piece. The key-fret pattern below the rim of the mouth is compressed and reserved, suited to the quiet atmosphere of this jar. The paleness of the blue perhaps results from the practice of diluting the valuable cobalt imported from China after the seventeenth century. Grains of fine sand are imbedded in the glaze covering the base.

82

Jar with an underglaze plum-tree design

Blue-and-white ware

Chosŏn dynasty, early eighteenth century

31.5 cm

Acc. no. 20430

This tall jar possesses the gentle overall form characteristic of jars of the first half of the eighteenth century. The wide, shallow mouth gives way to the gentle curvature of the shoulder, which gradually turns downward to the base The surface is pitted with slight irregularities, probably because the piece was not shaved smoothly enough after being thrown on the potter's wheel. They create a soft, rustic feeling.

Floral designs of fringed pinks, orchids, and daisies appear on three sides of the body. The lines of the leaves and stems are thin, spread equally to both sides, and flow smoothly upward. There are many brush strokes, but the overall impression is one of quiet gentility. This is another example of the "autumn-grasses" (Kor. *chucho*) motif found on eighteenth-century pieces and highly prized by the Japanese. During the nineteenth century, the brushwork tended to become thicker and the designs lost their compactness. Commonly, an "autumn-grasses" motif would be set off by drawing a horizontal line around the base and then placing the floral design above that line. No such line appears here, but the paintings are done with great care and skill. This piece, which is probably an early example of its type, was made in the first half of the eighteenth century at an official court kiln located in Kwangju. The base is glazed.

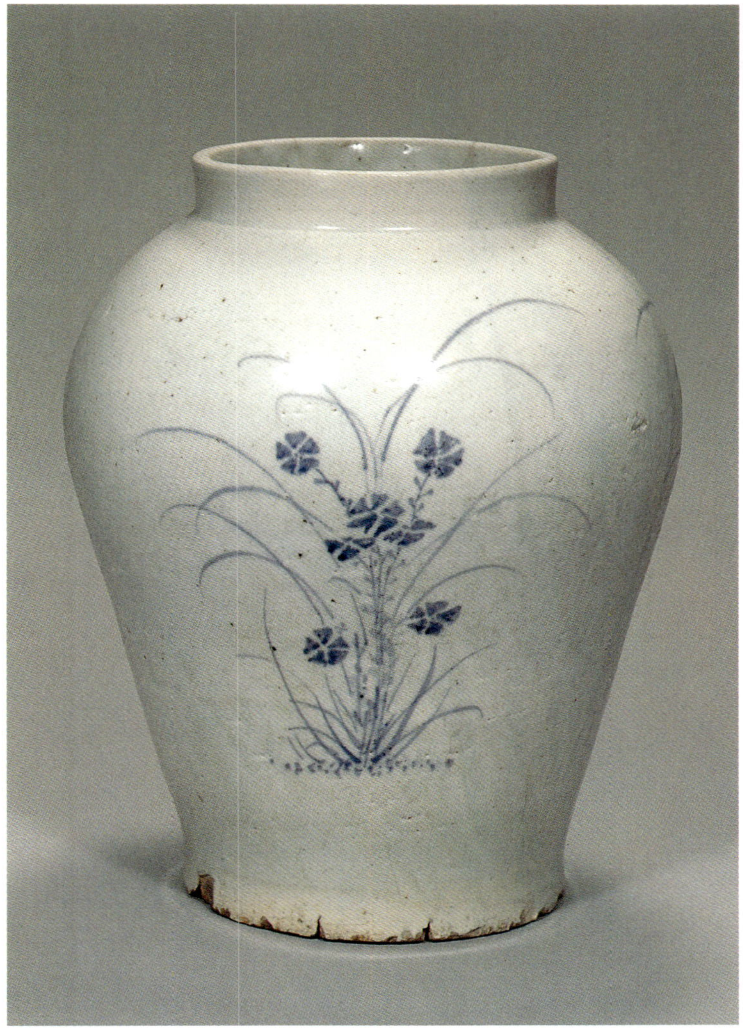

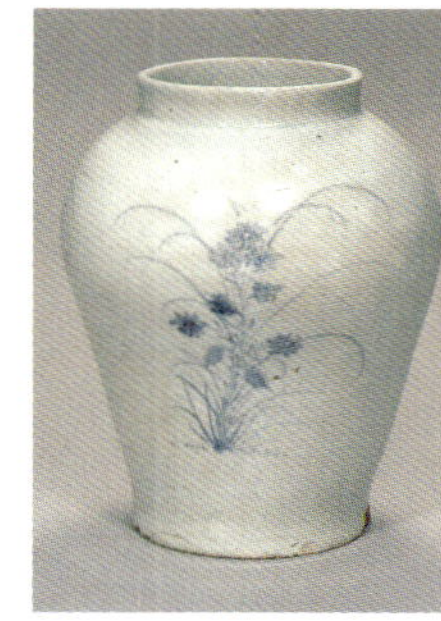

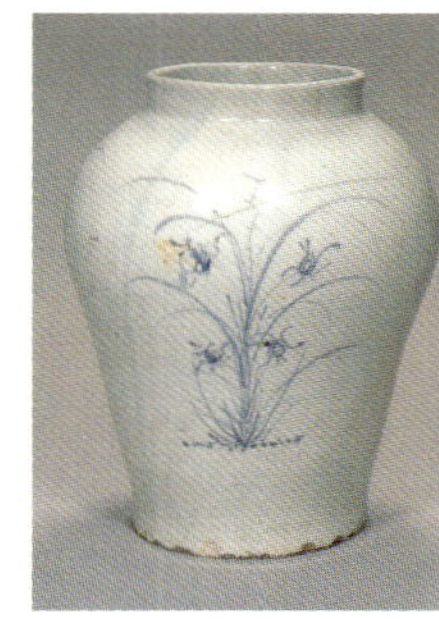

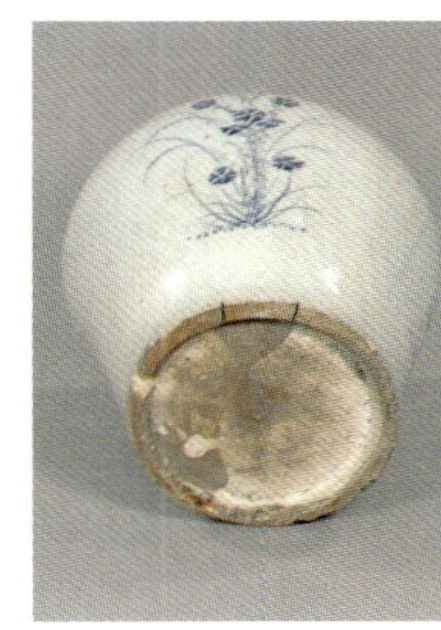

83

Jar with an underglaze floral-plant design

Blue-and-white ware

Chosŏn dynasty, first half of the eighteenth century

29.4 cm

Acc. no. 20601

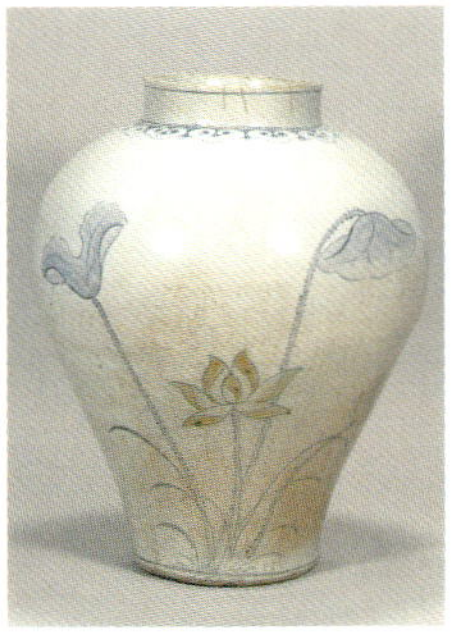

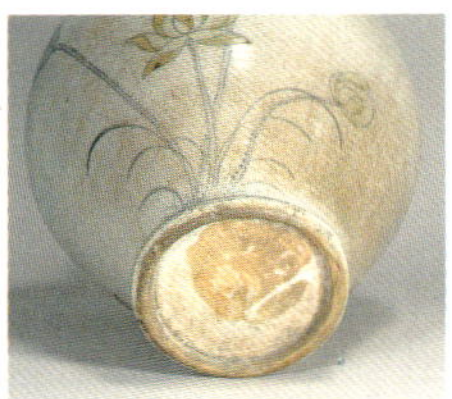

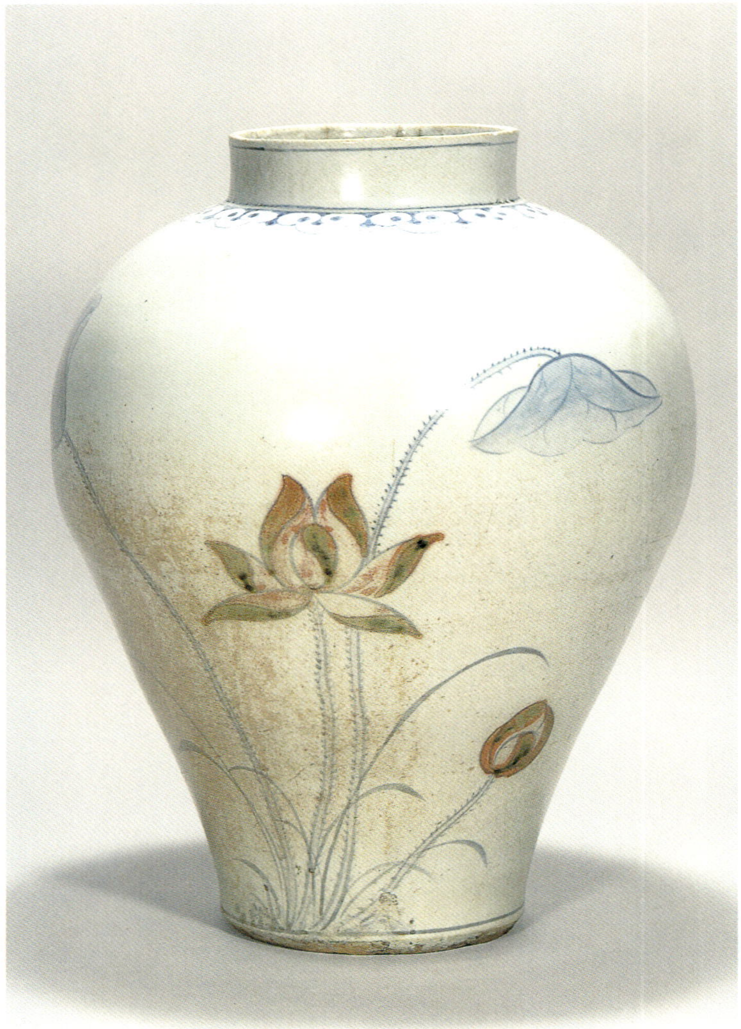

Centered on the body of this jar is a copper-red lotus flower, flanked on each side by a single blossom done in cobalt blue. One smaller red bud appears to the lower right of the main flower. The motif is repeated on the reverse side of the vessel. Although the names of potters and artists do not appear on ceramic works of the Chosŏn period, we can well imagine that this motif was drawn by a prominent artist who created a wonderful design by leaving the upper portion of the piece untouched.

Copper red is not a particularly good pigment to use on ceramics. It lacks stability and tends to fade into an unattractive green. But in this case, the green discolorations occurred during firing, and they impart a certain elegance and charm to the piece. This jar may have been made at an official kiln at Kumsa-ri, Kwangju, where cinnabar and iron pigments are believed to have been employed for decoration, although shards have not yet been discovered that would confirm this hypothesis. Many have praised this piece for its size and glaze, as well as for the refinement of its painted decorations. The base is glazed, but the glaze was partially scraped off.

84

Jar with an underglaze cobalt-blue and copper-red lotus design

Porcelain

Chosŏn dynasty, eighteenth century

44.6 cm

Acc. no. 21128

Gift of Eiichi Ataka

Only a few extant blue-and-white pieces date to the fifteenth century. This particular jar is known for its size and grace, for the intense severity of its paintings, and for the delicate coloration of the blue pigment. Its everted mouth is shallow and wide, and its contour line curves strongly outward at the shoulder and then slopes dramatically to the base. This shape is typical of Korean pieces produced in the first half of the Chosŏn period and differs from Chinese prototypes of the Ming period. The composition also contrasts with that of Chinese blue-and-white ware, the shoulders or bases of which are frequently adorned with bands of lotus petals or cloud-collar patterns.

The painting of plum and bamboo on this jar begins abruptly at the base and develops as it moves upward. The point from which to view the piece is clearly defined: the painting fills the front; and a large, blank space stretches across the back. In contrast to Chinese and Japanese practice, fifteenth-century court-appointed painters in Korea used ceramic space as if it were a canvas. But their names never appear on the ceramics. This piece is similar to a Punch'ŏng-style jar that has been assigned a mid-fifteenth-century date because it was excavated together with the birth record of Prince Wolsan (1454–?), a member of the royal family.

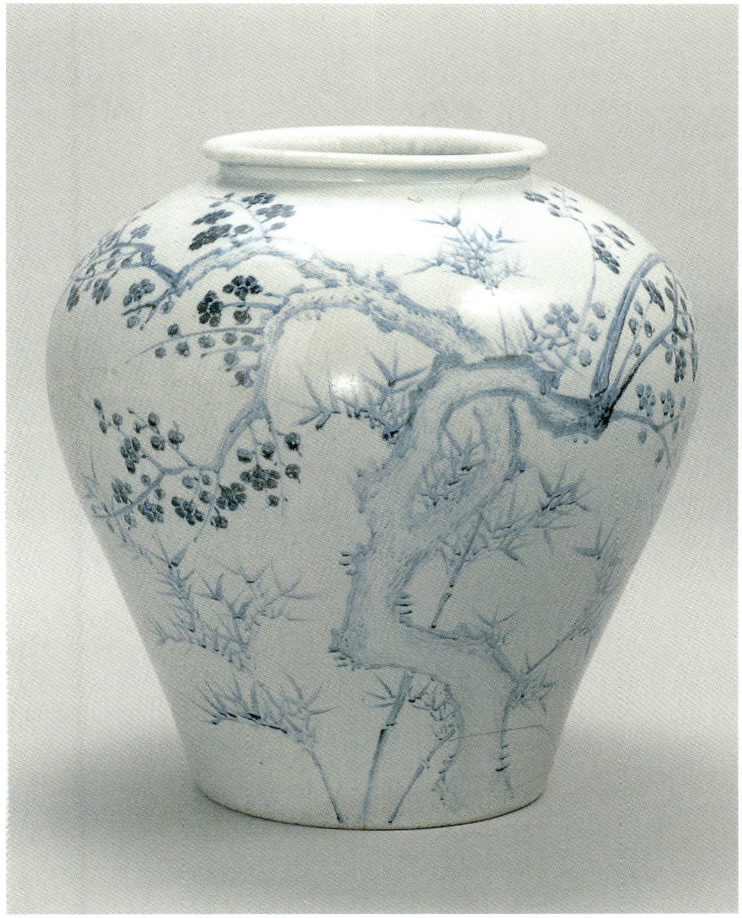

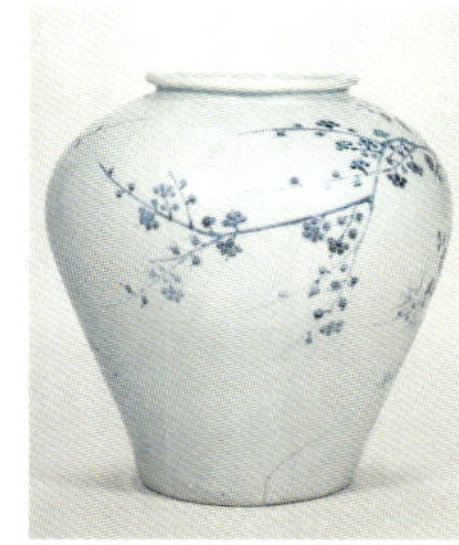

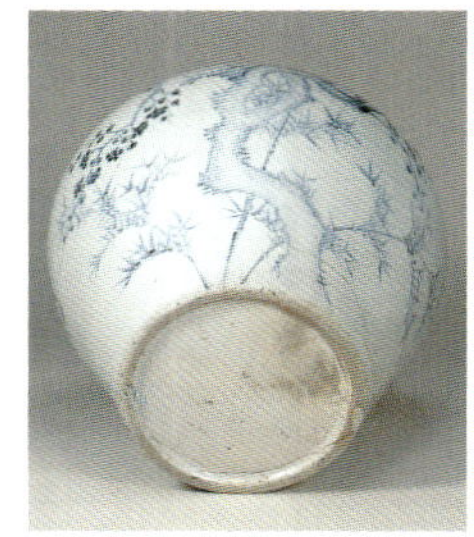

85

Jar with an underglaze plum-tree and bamboo design

Blue-and-white ware

Chosŏn dynasty, mid-fifteenth century

35 cm

Acc. no. 20325

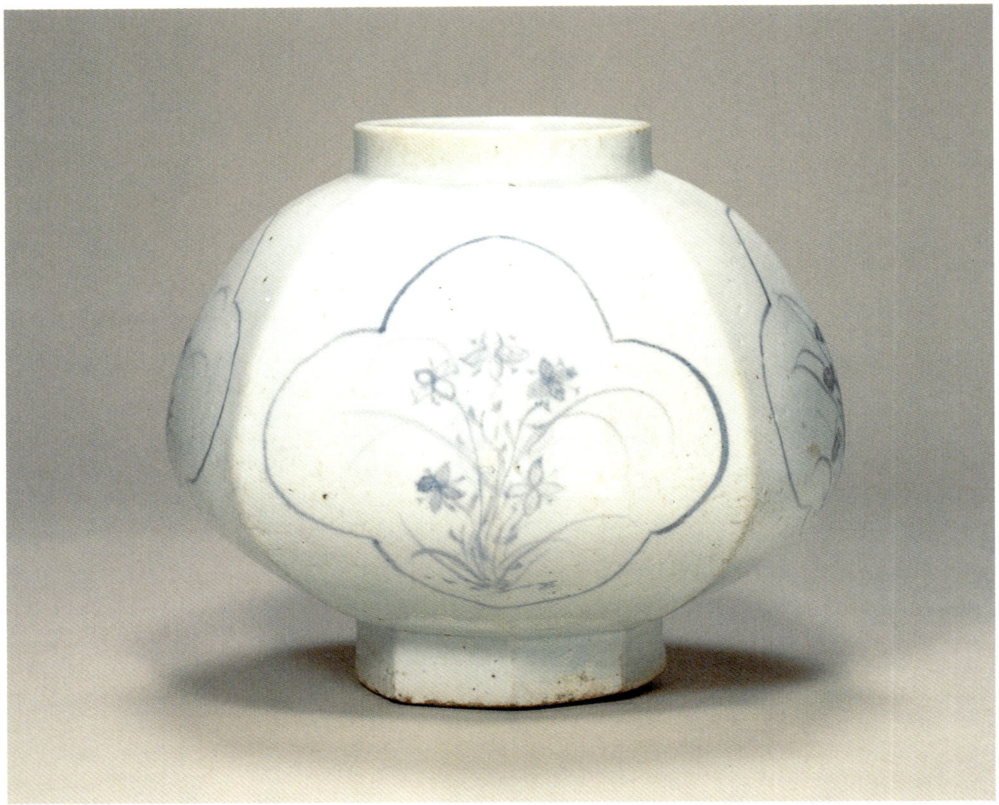

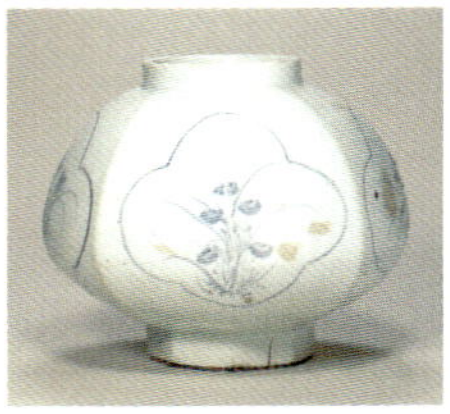

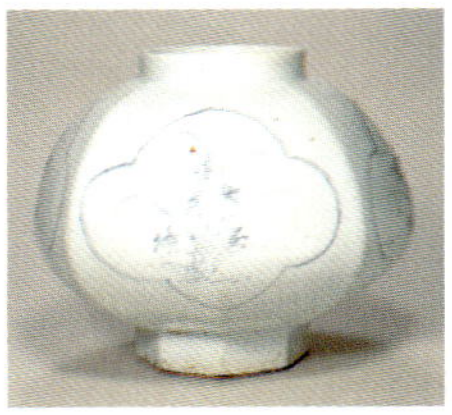

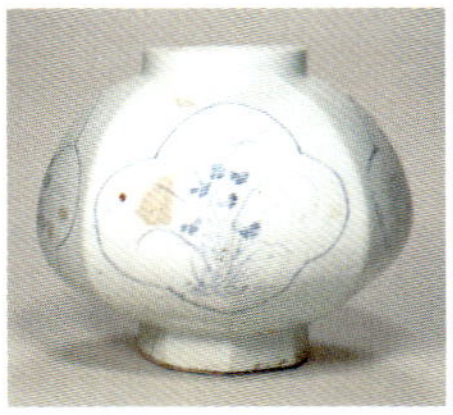

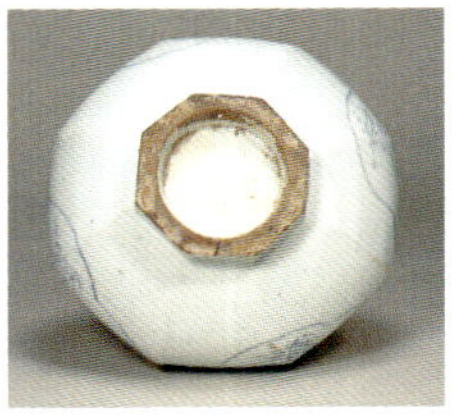

86

Beveled jar with an underglaze floral-plant design

Blue-and-white ware

Chosŏn dynasty, first half of the eighteenth century

23 cm

Acc. no. 20299

This is not a particularly large jar, but its shape creates a feeling of ample volume. Because of the manner in which the piece was hung while drying, the base is very sturdy and imparts stability to the whole. The jar was first thrown in the form of a spherical vessel, and the beveling was done with a spatula during the drying stage. Much skill was required to trim such large surfaces from top to bottom in a single stroke, especially since the curvature of the sides is so dramatic. Although many beveled bottles have survived to the present, only one other extant piece is similar to this example, perhaps because so much skill was needed to execute its complex construction.

The design consists of four cucumber-shaped "windows," each framing an "autumn-grasses" (Kor. *chucho*) scene of Chinese and autumn bell-flowers, meadow orchids, and fringed pinks. This style of painting is appreciated because it is so simple, yet so filled with sentiment. Its restraint and evocative mood are enhanced by the large expanses of milky white glaze, a color well suited to the blue of the drawings. This approach creates a settled feeling more often associated with earthenware than with porcelain. Indeed, many jars done in the "autumn-grasses" style are stained due to uneven firing or cracks in the glaze. Even the foot of this piece is beveled. It was trimmed with care and is glazed.

This jar was thrown as two pieces, and the shoulder expands outward to the horizontal line visible where the two pieces were joined. From that point, the contour line drops sharply to the base. The thin mouth has been flattened and resembles a sword guard, whereas the foot curves slightly outward. This combination of features is rather unusual.

The iron-painted design is equally distinctive. The tiger stands boldly, with front legs outstretched, head raised, and eyes wide open. Yet, his long eyelashes and exposed fangs almost reduce him to a caricature. Juxtaposed against this, the cloud in the shape of a fungus (Kor. *yongji*; Chin. *lingzhi*), the single lotus blossom, and the pair of herons standing back to back suggest a mood often found in Buddhist-inspired paintings. Perhaps the creator intended to express religious or political symbolism, although it is impossible to substantiate this suggestion. The jar's thick, opaque glaze contains some iron. A portion of the body near the base has been intentionally left unglazed.

No shards of pieces similar to this have been unearthed in the southern part of Korea, leading some to surmise that it was produced in the north. Some experts are of the opinion that this vessel was produced as early as the fifteenth century, but it seems safer to conclude that it was made in the seventeenth century, when iron painting reached its peak. This jar must be counted among the masterpieces of the Chosŏn period because of its individuality and strong composition.

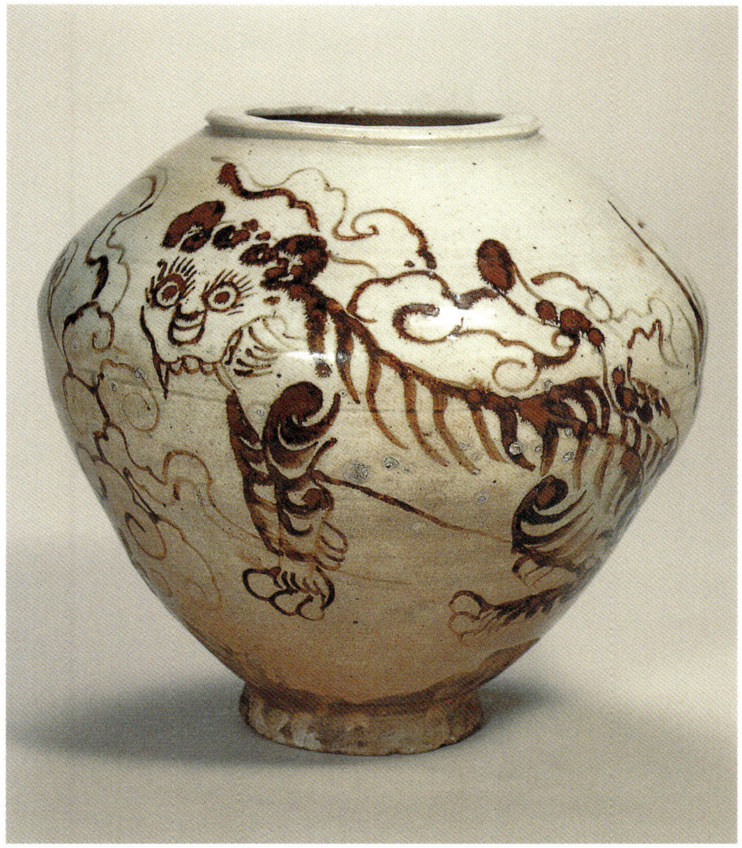

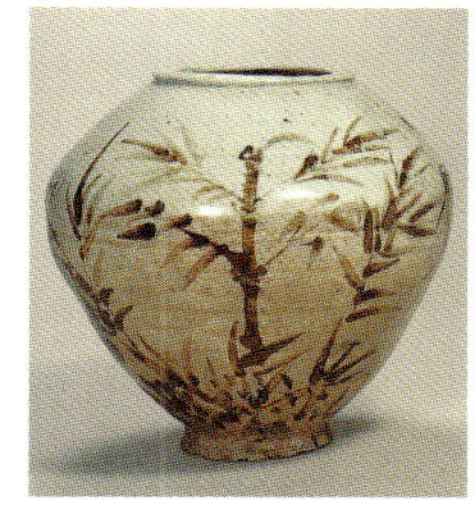

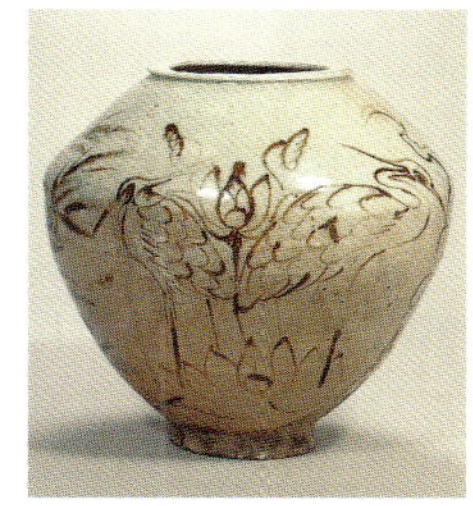

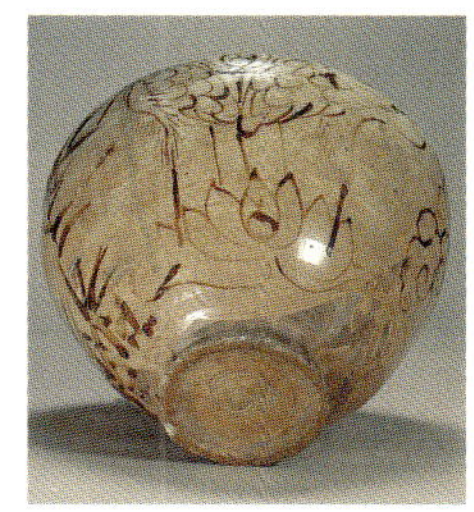

87

Jar with an underglaze iron-brown tiger and heron design

Porcelain

Chosŏn dynasty, seventeenth century

30.1 cm

Acc. no. 20546

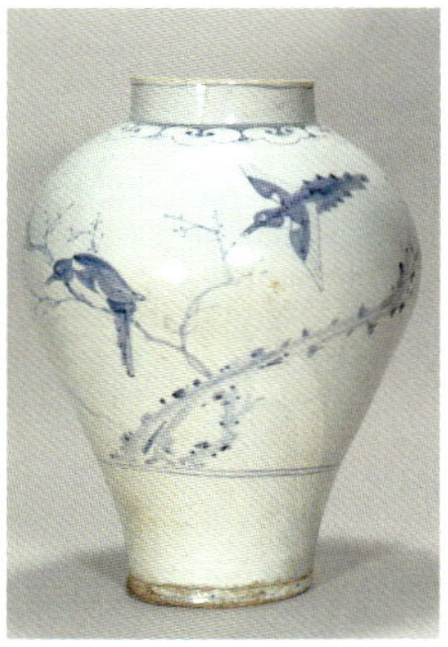

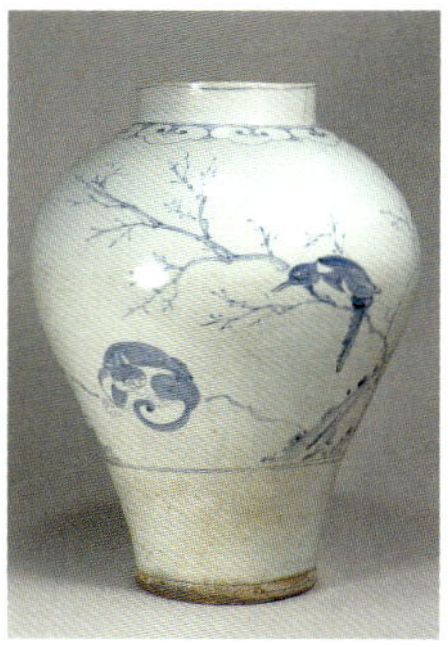

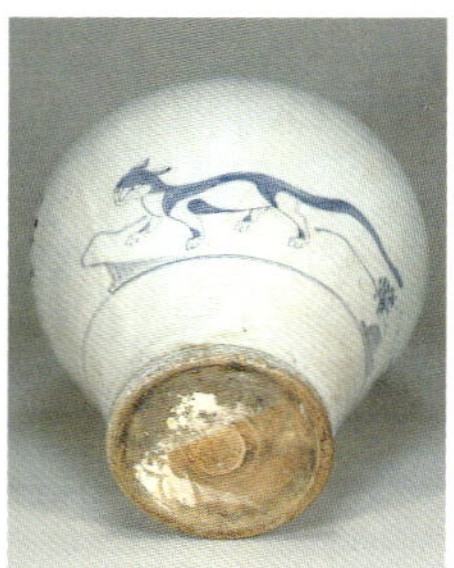

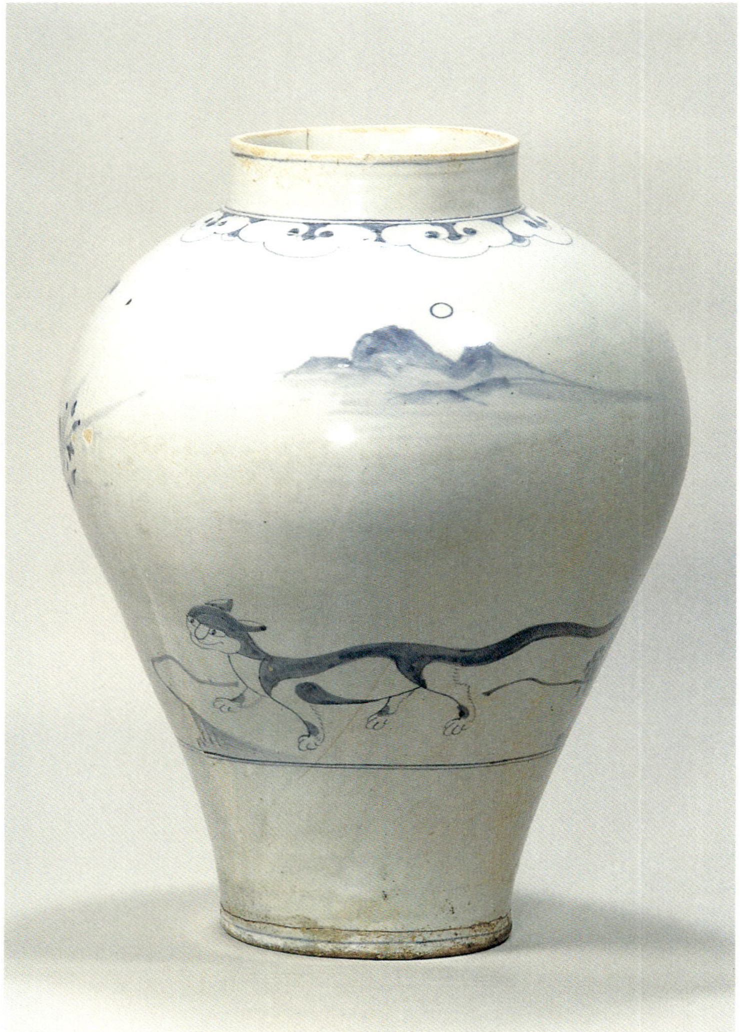

88

Jar with an underglaze tiger and magpie design

Blue-and-white ware

Chosŏn dynasty, second half of the eighteenth century

44.1 cm

Acc. no. 20182

This large jar has a tall, vertical mouth. The shoulder stretches out gently before tapering quickly to a narrow base, which ends with a slight outward flare. This is a very common shape for pieces fired at the official kilns in Kwangju during the middle and late eighteenth century. Also characteristic of these pieces are the two fine lines drawn around the lower hip to delimit the space given over to the imagery.

The motif of tigers and magpies, a common theme in Chosŏn-period folk paintings, is sometimes found on blue-and-white ware as well. Here, one tiger, drawn in a somewhat unusual, almost catlike fashion, is shown moving across some rocks on the front of the vessel. His sharp eyes fasten on a magpie, who is flying across the sky to call out a warning to a fellow bird perched on the branch of a dead tree on the opposite side of the jar. Crouching beneath the tree and eyeing that magpie is yet another tiger. The eyes of the tigers and magpies meet each other in a sequence that leads the viewer around the circumference of the piece. Ceramics of the Chosŏn period are decorated in a great variety of ways, but few tell a story as explicitly as this one does. It was probably produced in the latter half of the eighteenth century. The base is glazed.

This jar has a wide mouth and exceptionally wide shoulders. Its contour line then slopes inward quickly as it falls, so that the diameter of the base is considerably smaller than that of the mouth. As with other jars of this type, the upper and lower halves were made separately and the volume of the two parts is rather unbalanced. Despite this, the piece as a whole seems firm and stable. Three "windows," done as stylized, eight-sided flowers, are spaced around the upper portion of the body. Each frames a landscape painting probably based on the "Eight Views of the Xiao and Xiang Rivers" and the "Eight Views of Guandong," found so often on blue-and-white ware produced at the private Jingdezhen kilns in China during the late Ming and early Qing periods. The paintings on this piece are more simplified than their Chinese prototypes, however, and thus represent a reworking of those motifs to suit Korean sensibilities. All of the elements that make up the composition on the front of the jar—the moon, the bamboo thicket, the boat in the water, the distant mountain, and the cottage perched on a cliff—recede into the frame of the "window" as if drawn there by a magnet, leaving a large void in the middle of the composition. This treatment is reminiscent of the "autumn-grasses" (Kor. *chucho*) motif in that it also creates the quiet, evocative mood characteristic of blue-and-white ware produced in the first half of the eighteenth century. The base is glazed.

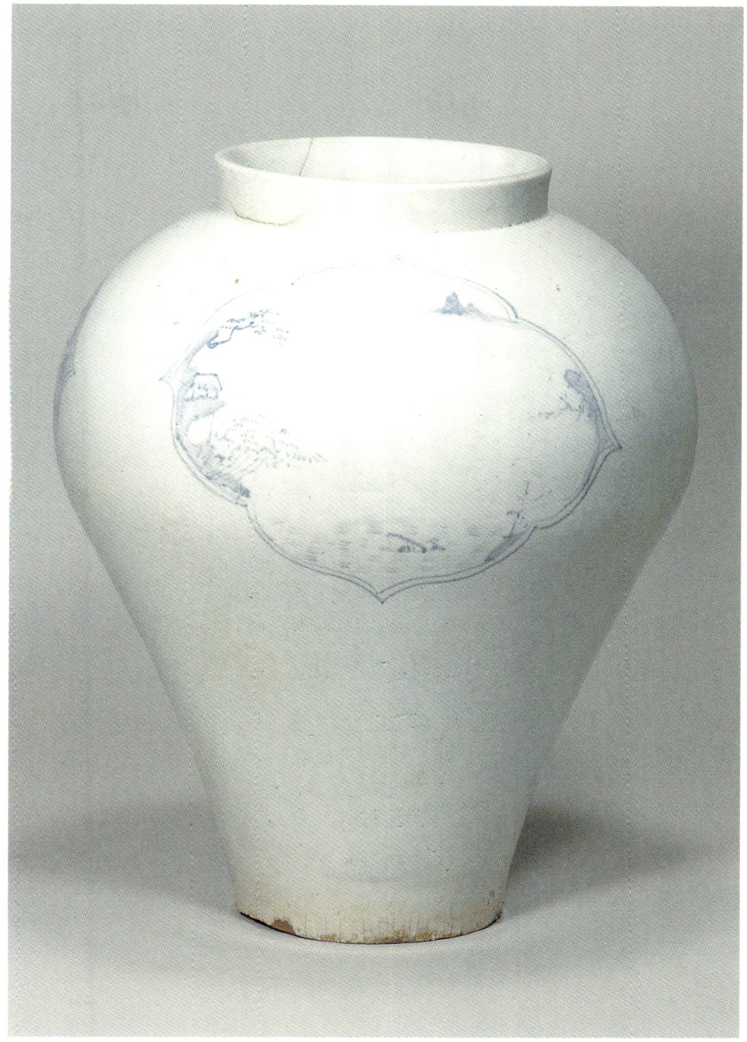

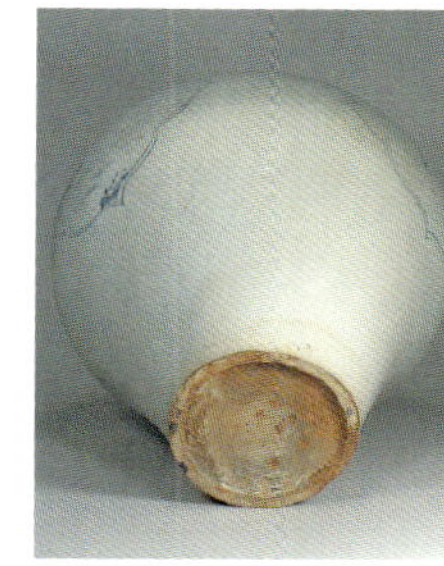

89

Jar with an underglaze landscape design

Blue-and-white ware

Chosŏn dynasty, first half of the eighteenth century

39.9 cm

Acc. no. 20502

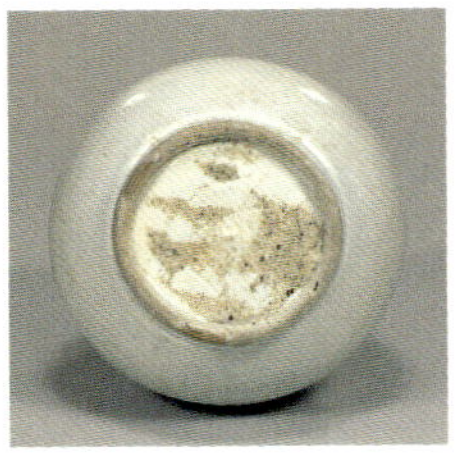

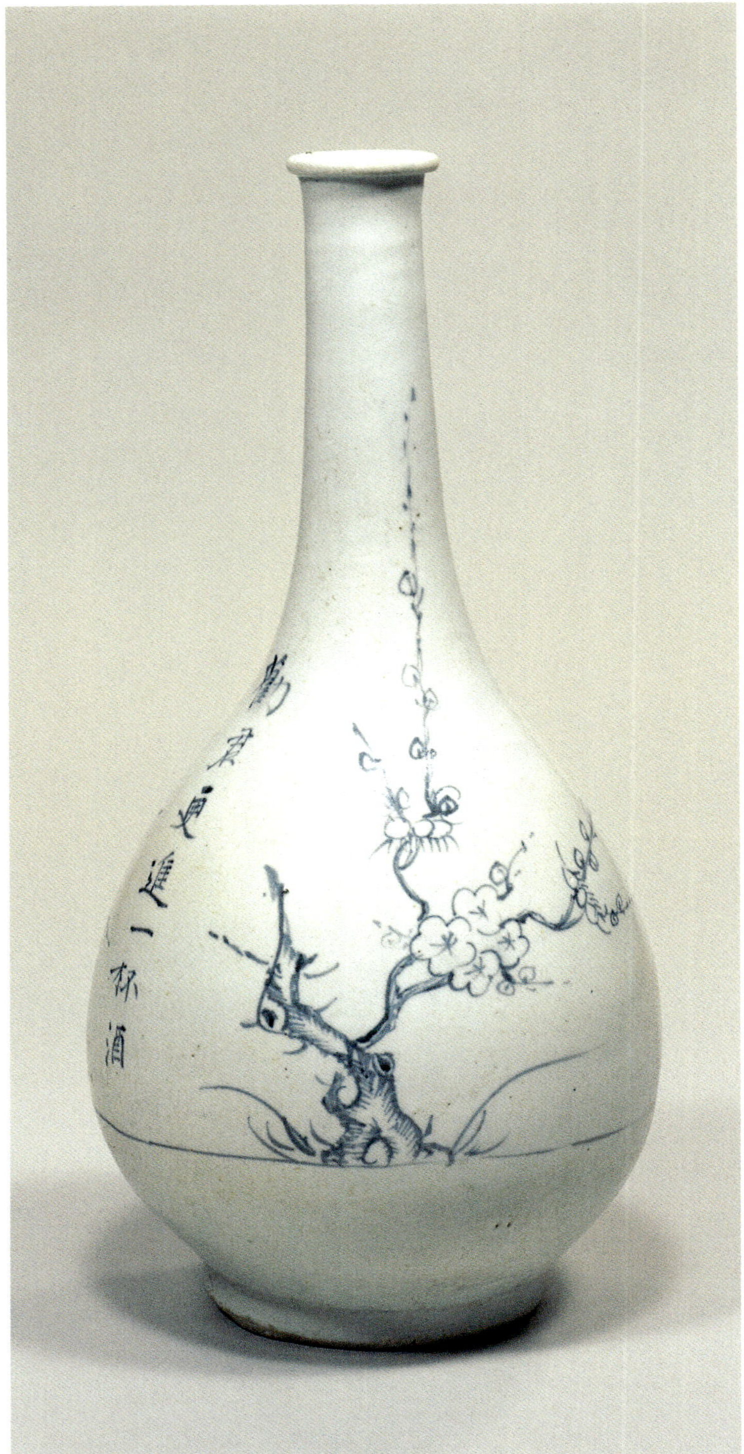

90

Bottle with an underglaze plum-tree design

Blue-and-white ware

Chosŏn dynasty, first half of the eighteenth century

36.2 cm

Acc. no. 20330

This is a slender bottle with a long neck. There are very few extant pieces of this height, which is about the same as that of the beveled bottle in no. 91. The rounded mouth is rolled back in an astragal-like shape (Kor. *okyŏn*; Jap. *tamabuchi*). The technique was used to prevent damage to the mouth and can be seen on many bottles made in the eighteenth century. This sort of slender silhouette, which does not swell out much at the base, is also characteristic of early eighteenth-century pieces, as is the pure white glaze lacking any bluish tint.

Plum trees adorn the center of the body. They are done carefully in thin, rapid brush strokes. A poem attributed to the Chinese poet Wang Wei of the Tang period appears between the two trees, indicating how interest in the Chinese literati gradually spread among Koreans during the eighteenth century. No suggestion of red tints the blue coloring, whose darkness gives an impression of having been produced by black Chinese ink. Grains of sand are imbedded in the glaze covering the base.

This is a large, eight-sided, beveled bottle. Imitating Chinese prototypes of the Yuan and early Ming periods, pieces executed in this style are rather common. The contour line of this bottle slopes gently downward, swelling out only slightly in its lower body. Numerous shards of similarly shaped, beveled bottles have been found at the Kumsa-ri kiln site in Kwangju, but very few of these come from blue-and-white ware.

Three "windows," done as stylized eight-sided flowers, frame pictures of pomegranates, peaches, and plums, respectively. All are fortuitous symbols that were especially popular among Koreans from the eighteenth century on. The brush strokes were carefully applied, and a special dotting technique was used to shade in the details of the leaves and branches. The blue coloring shows no evidence of a reddish tint, but almost seems as if it had been blended with black Chinese ink, another characteristic of blue-and-white ware produced in the first half of the eighteenth century. This piece was fired evenly, and the glaze has a bluish cast. It has a classic grace seldom found in similar examples.

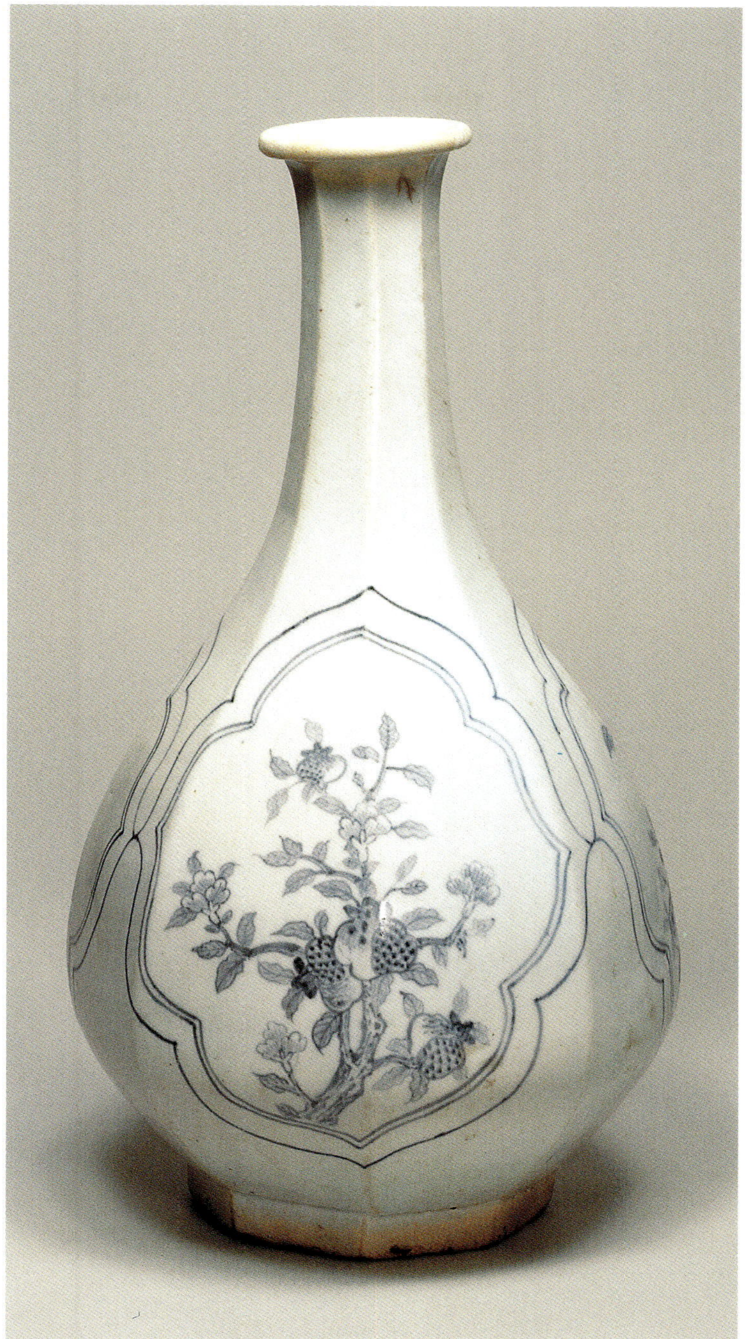

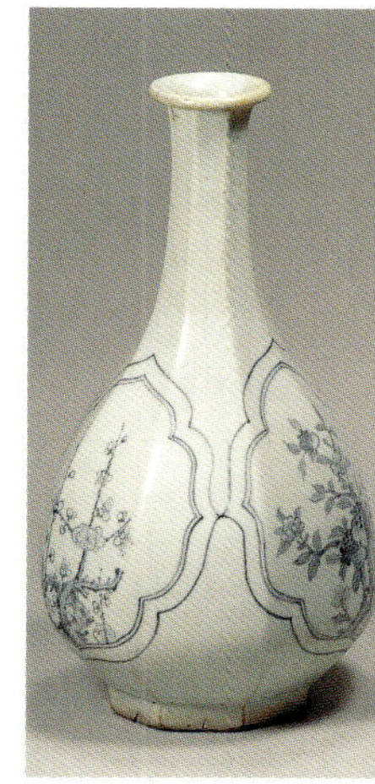

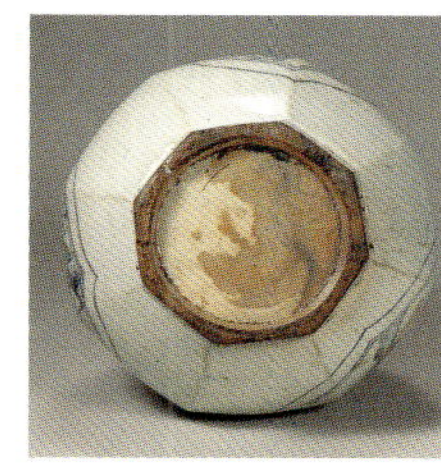

91

Beveled bottle with an underglaze pomegranate design

Blue-and-white ware

Chosŏn dynasty, first half of the eighteenth century

38.4 cm

Acc. no. 20119

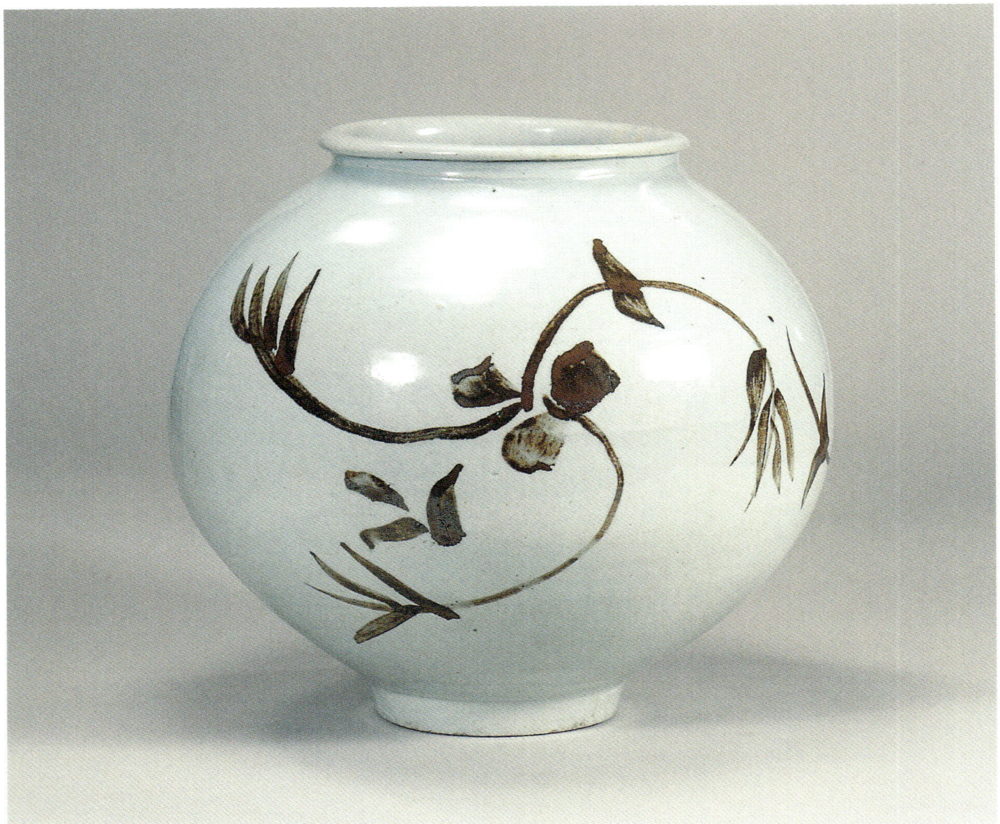

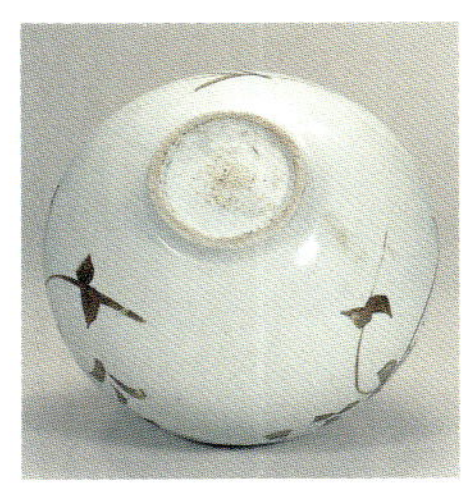

Iron-painted ware was produced in vast quantities at official government kilns in Kwangju, and then at rural kilns in Ch'ungch'ŏng, Kyŏngsang, and Chŏlla provinces, where many shards have been unearthed. Blue-and-white ware was also turned out in large quantities at the kilns in Kwangju during the seventeenth and eighteenth centuries, but, unlike iron-painted ware, production of underglaze blue-painted ware does not seem to have spread to other kiln sites. This was almost surely due to the difficulty of obtaining ample supplies of the blue cobalt used to make blue-and-white ware. The iron-painted ware of the Kwangju kilns far surpassed that of the rural kilns in terms of the quality of its glazes and the refinement of its decorative motifs. In addition to this jar, nos. 93, 96, and 97 were made at the Kwangju kilns; nos. 94 and 95 come from rural kilns.

This piece is adorned with a water plant that wraps itself around three sides of the body. The flow of the brush strokes and the harmonious placement of the design indicate that it was made at an official kiln in Kwangju. The jar is well balanced and fired. Some sand clings to the base, which is glazed.

92

Jar with an underglaze iron-brown water-plant design

Porcelain

Chosŏn dynasty, seventeenth century

27.3 cm

Acc. no. 20712

The glaze on this piece has a grayish tint, due to the presence of iron. Its overall shape and the composition of the decorative motif indicate that it was made at an official kiln in Kwangju. The oblique, inward cut of the low mouth is also characteristic of iron-painted pieces fired at the Kwangju kilns.

The jar swells out at the center in a curved shape known as an "abacus bead," a form typical of jars produced during the seventeenth century. One might well expect that the relatively small foot would produce a feeling of disequilibrium, but, in fact, it contributes to a sense of massiveness and stability. The powerful design suits the vessel's sturdy shape. With deliberate calculation, the painter employed bold brush strokes to draw the trunk of a bamboo plant across the center of the body in a single, broad sweep. The trunk and its leafed branches thrust upward, only to stop short of the shoulder. The vast, open spaces that are left make the jar appear larger than it actually is. To the right and left of the main branch, three thin bamboo twigs emerge from the foot of the jar. The iron pigment is of good quality and expresses the strength and vigor of many pieces produced in the Chosŏn period. The base is glazed and covered almost entirely with sand.

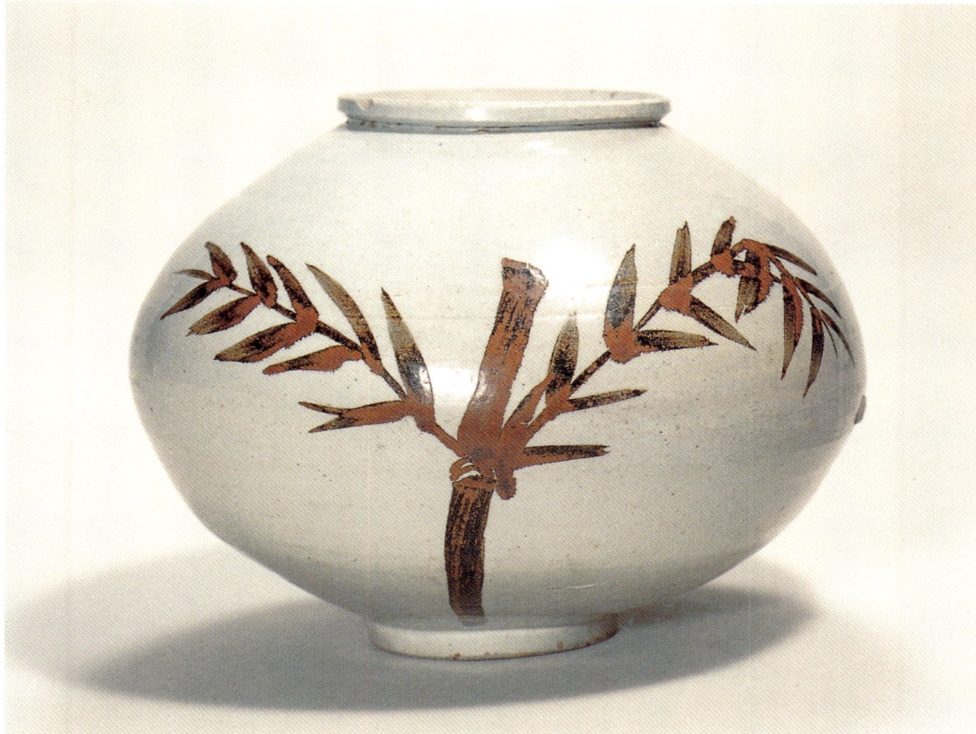

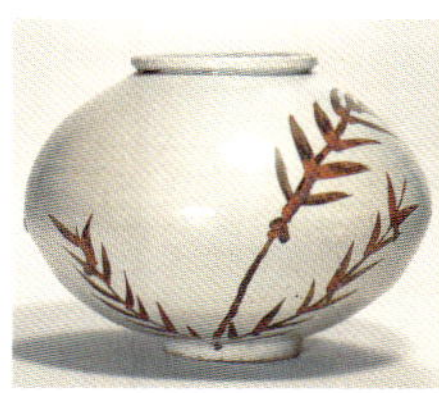

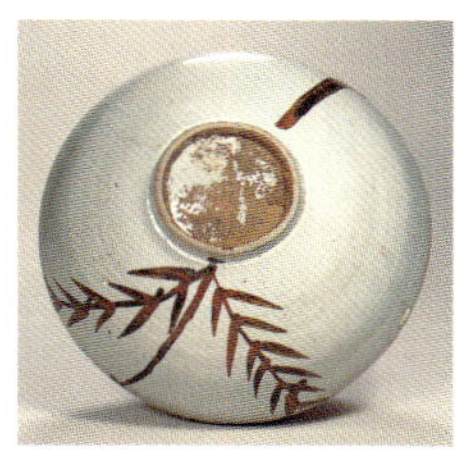

93

Jar with an underglaze iron-brown bamboo design

Porcelain

Chosŏn dynasty, seventeenth century

28.1 cm

Acc. no. 20228

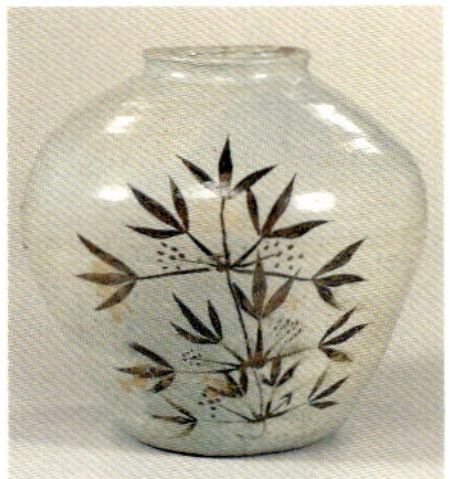

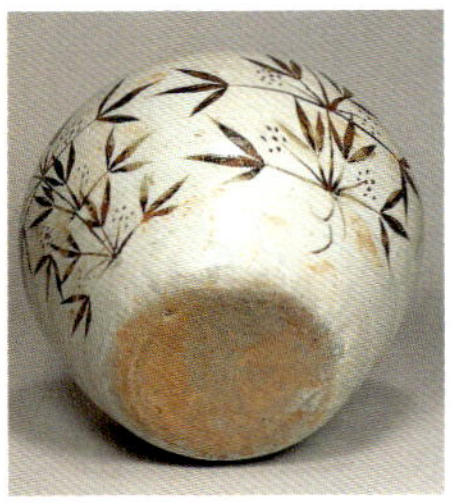

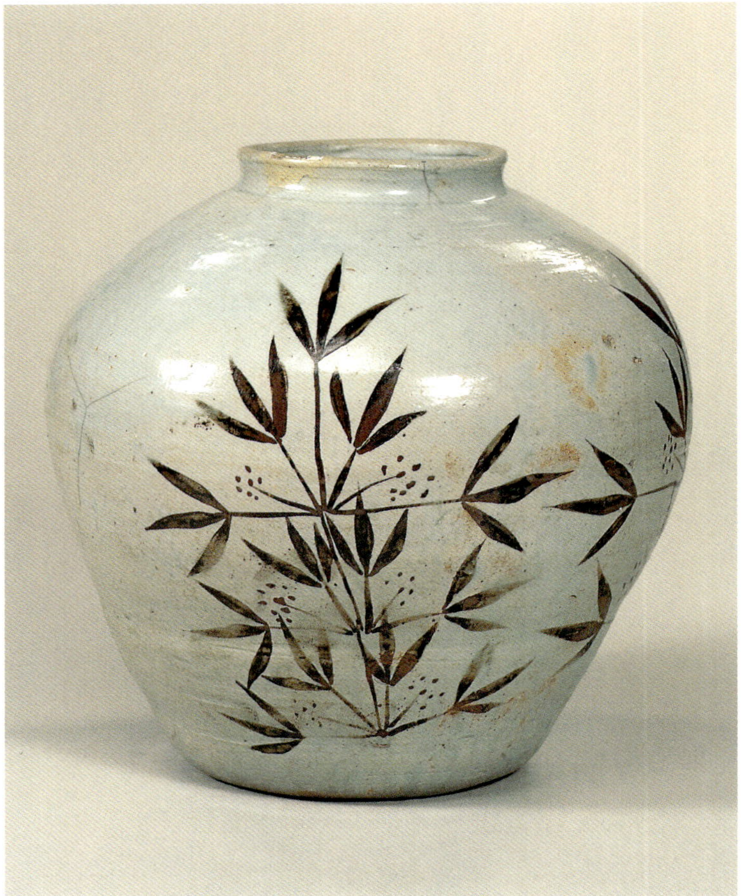

This vessel defies simple definition. The mouth cannot be categorized as raised, everted, or rounded in an astragal-like form (Kor. *okyŏn*; Jap. *tamabuchi*). The overall shape is equally unusual. The uneven slope of the shoulders was probably caused in potting when the piece shifted from the center of the wheel. The jar is also warped, and its silhouette changes depending on the angle from which it is viewed. The motif flows down to the lower right, keeping pace with the slant of the piece itself. There is a strong sense here that the potter simply let nature take its course. Because no standardized form was imposed on the clay, it almost seems as if the jar has its own organic bone structure. Iron-painted ware of this sort was produced in rural kilns in great quantity during the seventeenth century.

The design must represent flowering ginseng, whose root is popular as a medicine among Koreans. The dark iron-brown color goes well with the gray-white glaze. This is a very typical Chosŏn-period piece, and discovering its charm is one key to understanding ceramic ware of the time. The underside is indented and unglazed.

94

Jar with an underglaze iron-brown flowering-ginseng design

Porcelain

Chosŏn dynasty, seventeenth century

25.7 cm

Acc. no. 20252

Iron-painted ware could be produced in great quantities at rural kilns because of the ready availability of the paint, which was applied to the porcelain before glazing. This particular piece also has a high iron content in the clay, as well as in the glaze, which gives it a gray cast. The rather high mouth is everted, and the shoulder swells broadly outward and then narrows sharply as it approaches the base. The jar seems unbalanced, perhaps because there is no definite boundary between its mouth and body. No pieces of this type were produced at the official kilns in Kwangju. The motif under the glaze was executed in a hesitant hand that imparts a naturally simple mood to the scene, as if it were an innocent child's painting.

This jar cannot be considered a classic piece because of its contour and composition. Indeed, its maker never intended that it should be a perfectly proportioned vessel. The charm of such rural products lies in their lack of pretension. Their iron-brown colors are often not uniform; some are light, some dark. But for all of this, they almost always have a kind of simple harmony. One should not see these pieces as representing the lower standards of rural potters, but, rather, should appreciate them as expressions of the freedom and vitality of the countryside. It is impossible to pinpoint exactly where this jar was made, but similar shards have been unearthed at Koesan, South Ch'ungch'ŏng province. The base is glazed, and sand clings to one part of it.

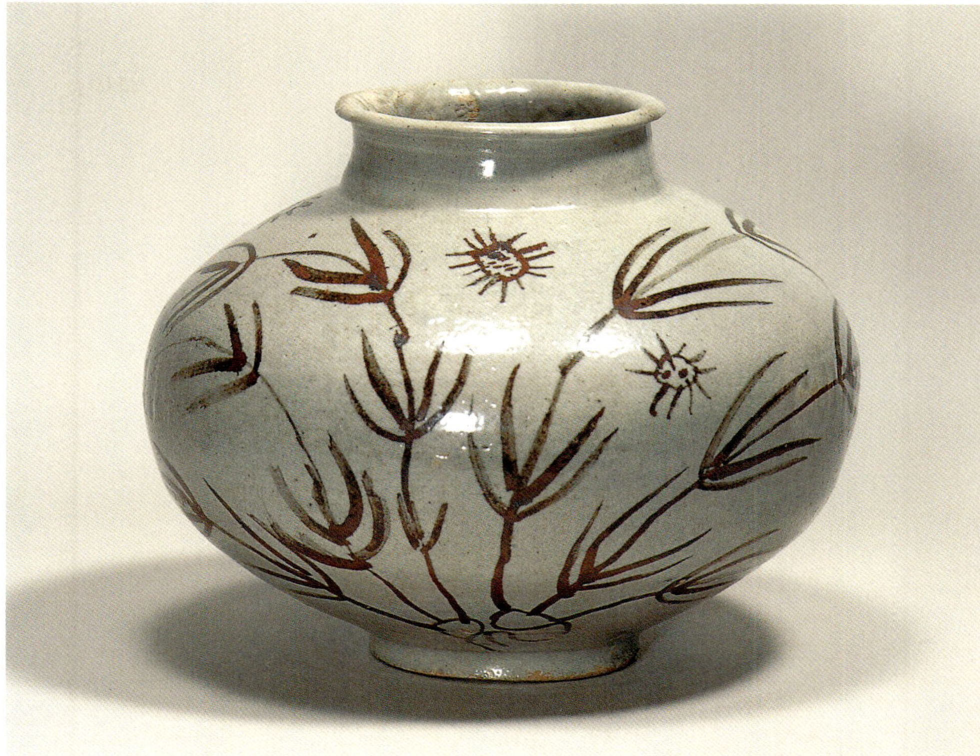

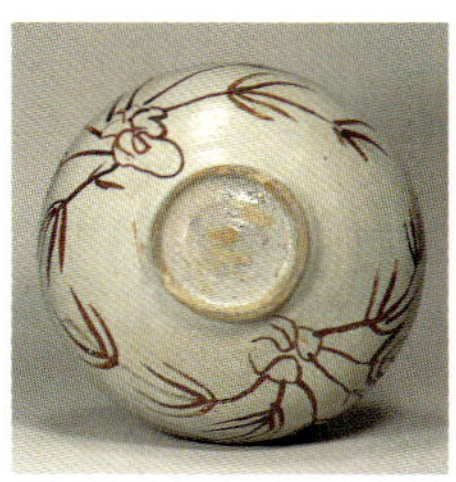

95

Jar with an underglaze iron-brown grass design

Porcelain

Chosŏn dynasty, seventeenth century

26.7 cm

Acc. no. 20069

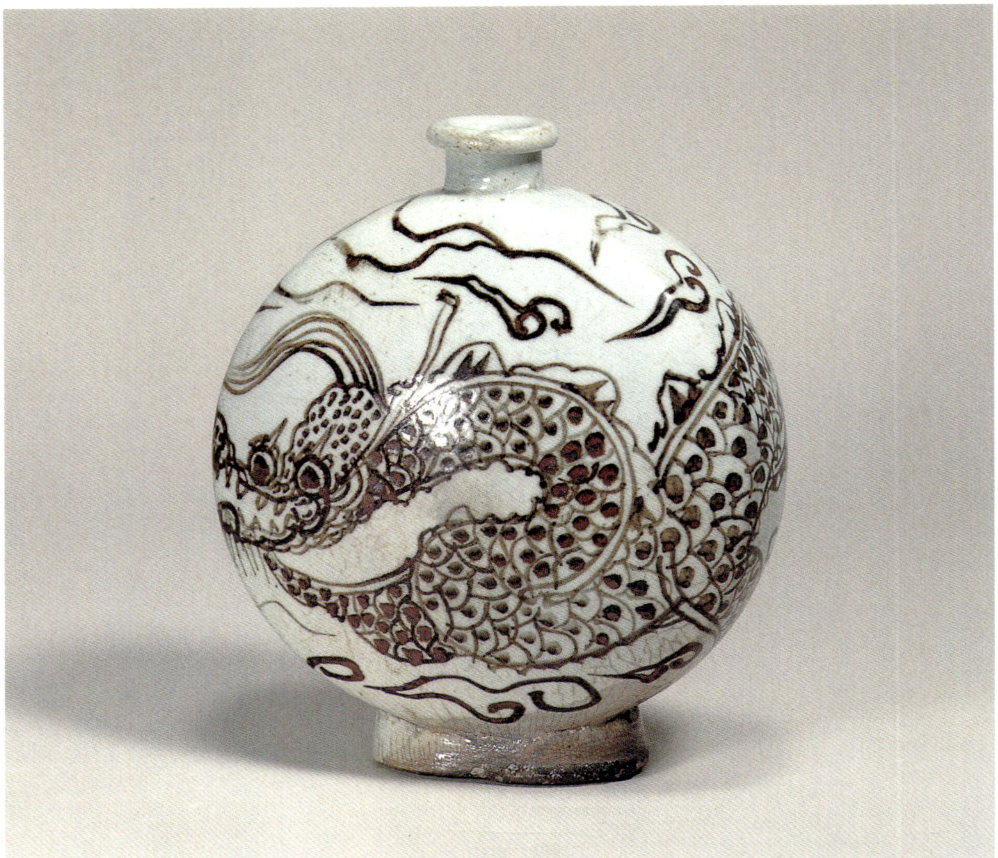

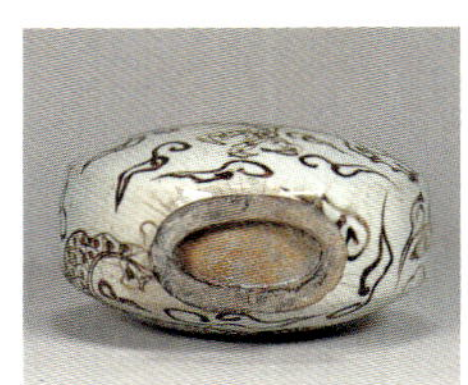

The iron-painted ware of the Chosŏn period includes vessels in the shape of flasks. The oldest porcelain flask, decorated with an inlaid floral design, was unearthed together with materials inscribed with the date 1466. No iron-painted flasks have yet been unearthed that date to the fifteenth century, and it is thought that most extant examples were made in the seventeenth century. Many such shards have been found at the official government kiln sites in Kwangju. Iron-painted flasks were made by fitting together two sections, each thrown in the shape of a dish, and then attaching the spout and foot. In the seventeenth century, the foot was usually made in an oval shape, but this gave way to a steplike shape in the eighteenth century. Typical motifs include plum trees, bamboo, grapes, and chrysanthemums, and some flasks are even inscribed with poems. The piece included in this exhibition is the only surviving flask with a dragon motif; the painting is done in the manner of a caricature and is based on examples from Ming China made during the Jiajing (1522–66) and Wanli (1573–1619) periods. The glaze and iron coloration are exceptional, and the base is glazed.

96

Flask with an underglaze iron-brown dragon design

Porcelain

Chosŏn dynasty, seventeenth century

21 cm

Acc. no. 20812

Although iron-painted jars survive in great numbers, this example is of an unusual size and form. Its shape resembles that of blue-and-white jars made during the second half of the seventeenth and the first half of the eighteenth century, indicating that it certainly was made at one of the official government kilns in Kwangju that, at the time, produced blue-and-white ware. A design of bamboo grass swaying in the breeze adorns the bottle. It was rendered with the sort of quick brush stroke that imparts a rich tone to flowers and birds. The leaves are not outlined, and the use of this "boneless" technique suggests that the drawing was executed by a literatus and not a potter. The glaze is milky, more opaque than on pieces fired at the higher temperatures that produce perfect calcination and a shiny finish. One cannot find jars with this kind of milky glaze among pieces made at official government kilns in China, since they were smashed when the firing was finished. Yet, because of their imperfect firing, jars such as this possess an especially soft, almost wet-looking glaze. In the Chosŏn period, ceramics were judged by a special aesthetic standard that permitted defects to exist and be appreciated. The base is glazed, and sand is imbedded on it.

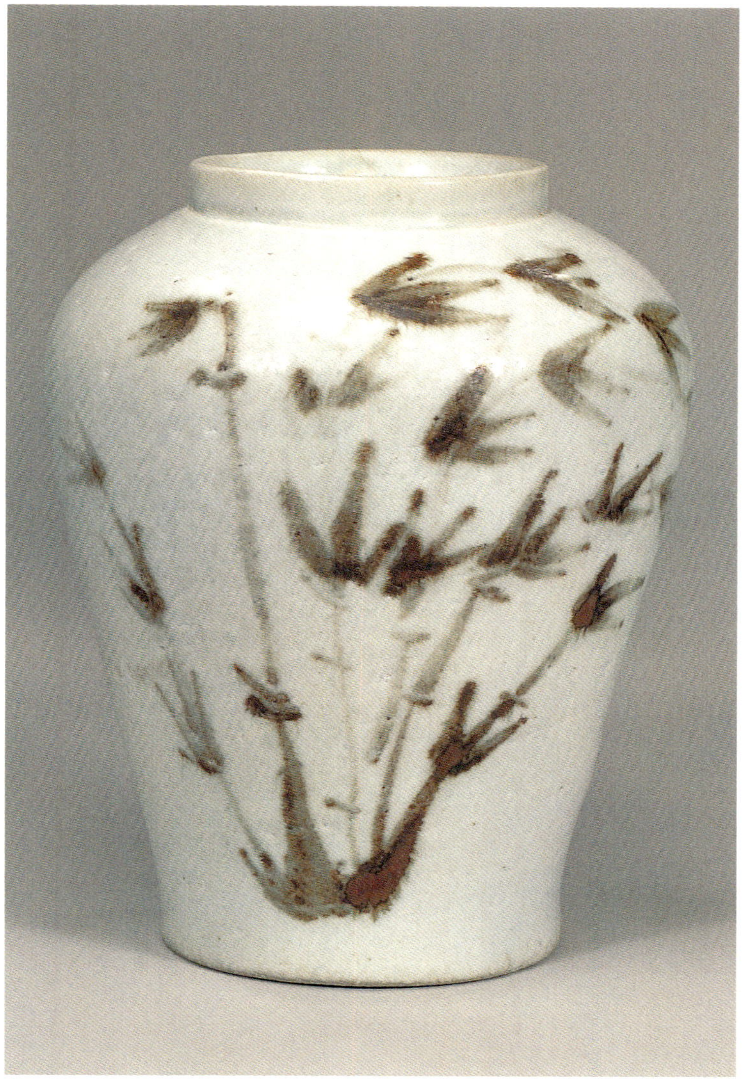

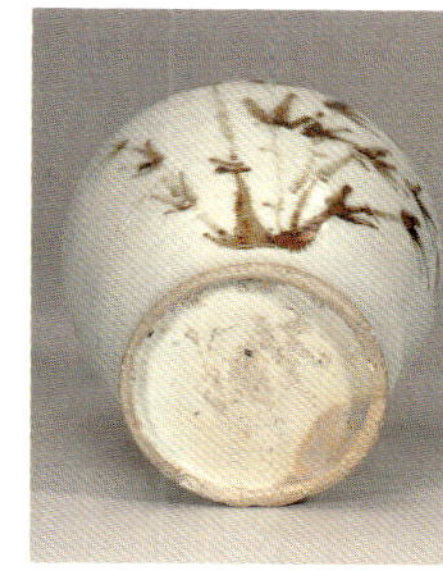

97

Jar with an underglaze iron-brown bamboo design

Porcelain

Chosŏn dynasty, second half of the seventeenth/first half of the eighteenth century

22.6 cm

Acc. no. 20444

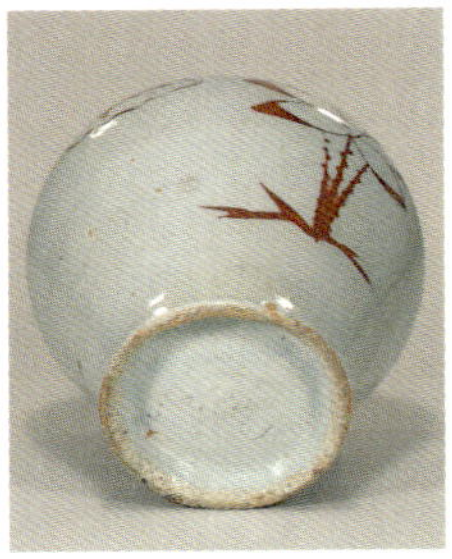

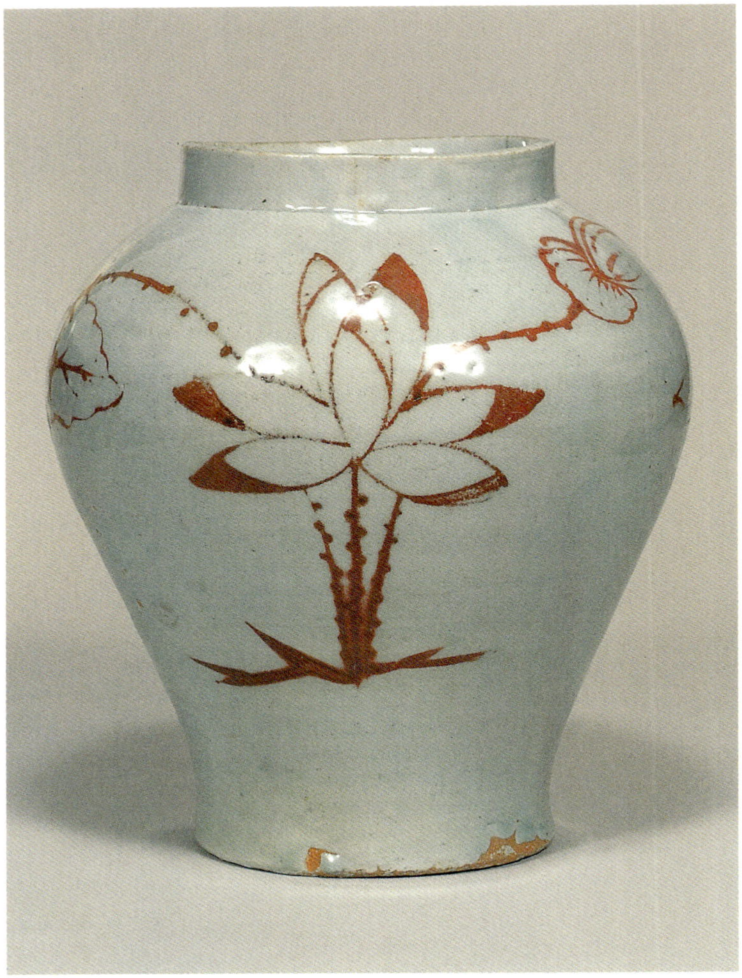

98

Jar with an underglaze copper-red lotus design

Porcelain

Chosŏn dynasty, eighteenth century

27.9 cm

Acc. no. 20748

Interestingly enough, underglaze painting with copper pigment on celadon was already known in Korea in the Koryŏ period. However, the technique of underglaze painting with copper pigment on porcelain was borrowed directly from China, where the practice began during the Yuan dynasty, rather than being developed from the earlier Koryŏ tradition.

Underglazed copper-red ware can be broadly divided into two classifications: those pieces made at the official government kilns at Punwon-ri in Kwangju, and those produced at kilns in the countryside. This jar belongs to the second category. The clay is not of particularly high quality, and the glaze contains a large concentration of iron. The treatment of the foot is also crude. These features would suggest that the piece was made at a rural kiln, as would the fact that no shards similar to it have been unearthed at an official Kwangju kiln. Nor has any rural kiln site been discovered where pieces such as this were made, so the whole matter of identifying the source for this jar remains somewhat of a mystery.

The central motif, which is a common one on underglazed copper-red jars, consists of a large lotus flower rising straight up from the base, with secondary stems branching upward to the right and left. Also typical is the red accent on the lotus petals. The drawing and the copper-red color are superior to that of most pieces done in this style. Copper-painted ware was produced throughout the latter half of the seventeenth century, but jars done in this style are thought to date to the eighteenth century. The base was shaved flat and is glazed.

This shape is characteristic of underglaze copper-painted jars. In particular, this kind of mouth, which slopes downward and outward to resemble a bamboo hat in profile, is rarely found on white porcelains or on blue-and-white or iron-painted wares. Indeed, the only blue-and-white piece with anything approaching the same overall shape has a phoenix motif. It is thought that this sort of mouth design was used so that a cloth or paper lid could be put onto the jar and secured with a piece of string. In any event, the overall shape of the piece suggests that it was used for some special purpose. Shards of underglaze copper-painted ware have not been discovered to date either at government kilns in Kwangju or at rural kilns.

The jar's principal motif consists of a crane poised on one leg, drawn very skillfully with a tiny brush. Thick, drooping pine branches, which are not outlined, provide the background. Many different sorts of motifs adorn underglazed copper-red jars, but this is the only known example of a crane and pine pattern. This jar exhibits such exquisite taste that it is difficult to imagine that it was made at a rural kiln. But the glaze and the style of the underside permit no other interpretation. The base is glazed.

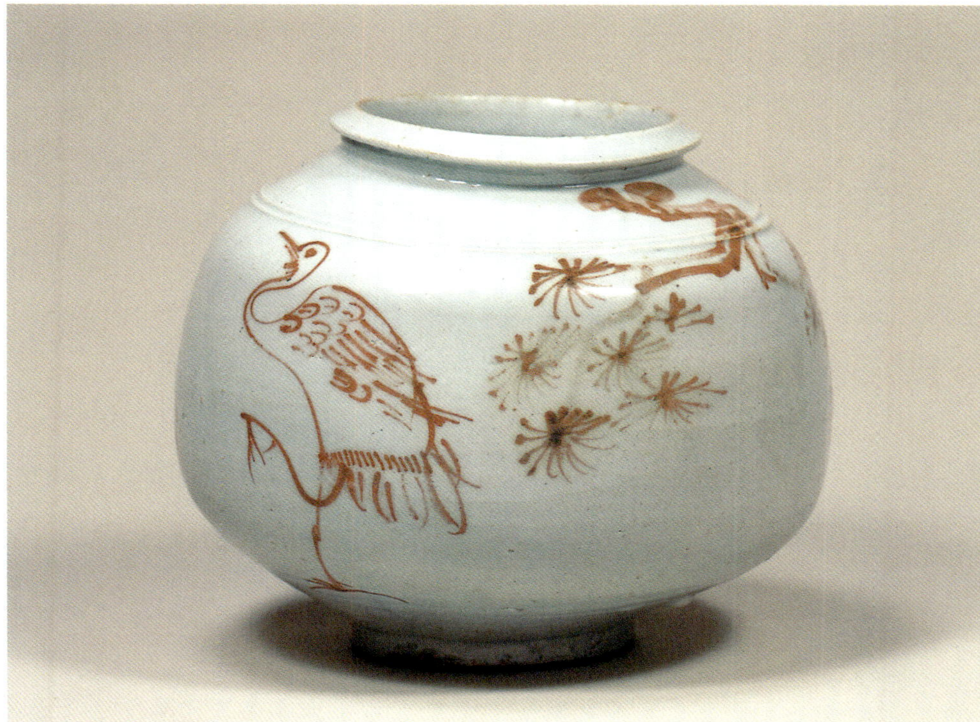

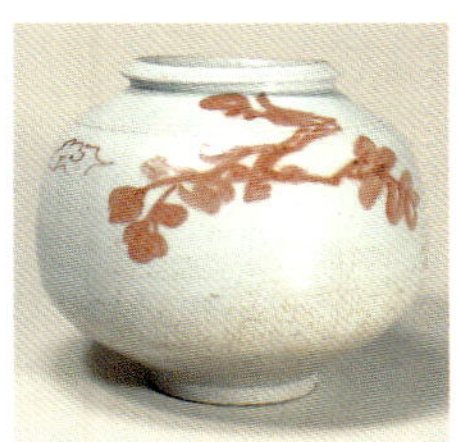

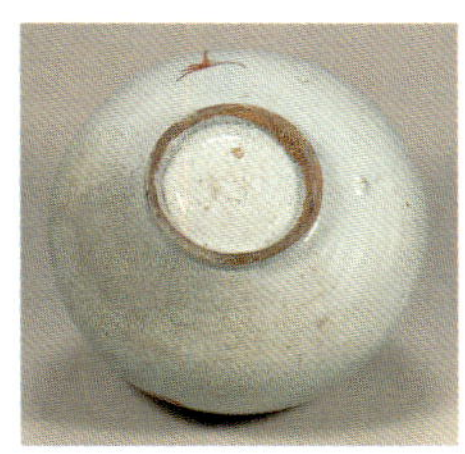

99

Jar with an underglaze copper-red crane and pine-tree design

Porcelain

Chosŏn dynasty, eighteenth century

24.8 cm

Acc. no. 20385

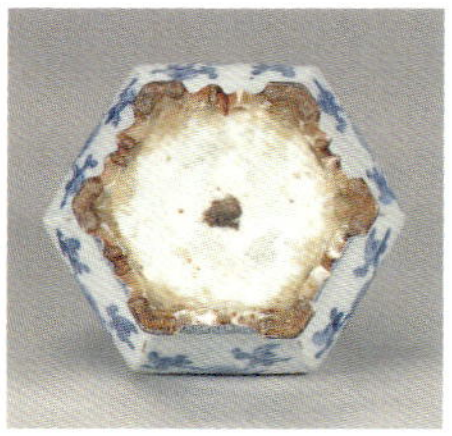

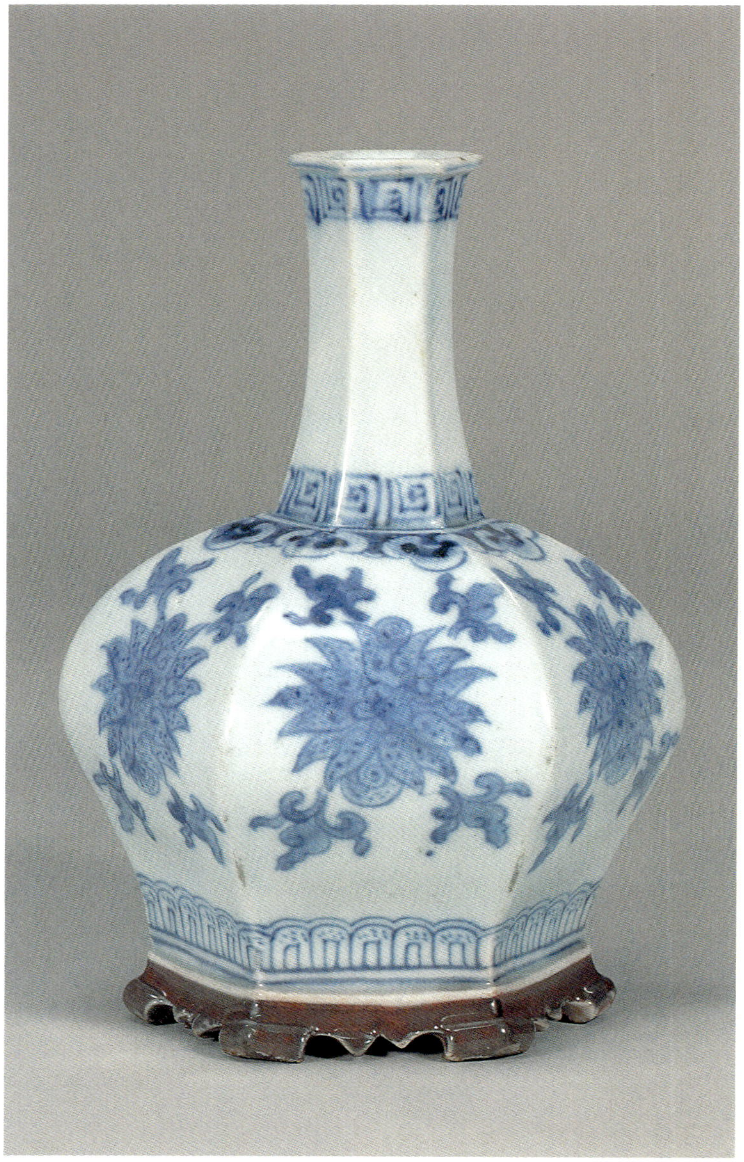

This beveled bottle has six sides. The middle portion of the body swells out boldly and is counterbalanced by the slightly shorter neck that rises above it. Although the shape of the bottle and its motif in themselves impart a sense of balance and stability, the potter enhanced that feeling by adding a cloud-shaped foot. He colored the foot maroon by applying an underglaze coat of copper red, his purpose having been to create the impression that the bottle rests on a small rosewood platform of the sort used in China to display ceramics.

A large floral blossom (Kor. *bosanghwa*; Chin. *baoxianghua*) is centered on each of the six sides, and bands done in stylized key-fret patterns appear at the top and bottom of the neck, while a cloud-collar design encircles the top of the shoulder and a band of lotus blossoms extends around the base. The bottle seems overly decorated, a tendency that became more pronounced during the course of the nineteenth century. The base is glazed.

100
Beveled bottle with an underglaze cobalt-blue floral design and copper-red-coated base

Porcelain

Chosŏn dynasty, nineteenth century

20.5 cm

Acc. no. 20818

The shoulders of this small, square bottle were trimmed diagonally in the typical manner. The square spout is tall and has thin walls, giving an impression of lightness. The paintings on the four sides are striking in their originality. Each depicts a scene of a faraway mountain with a riverbank in the foreground. The reflection of the mountain can be seen in the river, but, in each painting, the shape of the actual mountain and its reflection differs. Moreover, the perspective is distorted, since objects in the foreground are drawn to the same scale as those in the middleground and background. The technique is similar to that employed in Korean folk paintings of the Chosŏn period. The landscape elements were delineated without much regard for detail, whereas great care was used to elaborate on the minute nuances of the petals and vines. Because of this approach, the drawings are filled with an appealing poetic sentiment. This square bottle has the cheerful optimism of some ceramics of the latter half of the eighteenth century. The base is glazed.

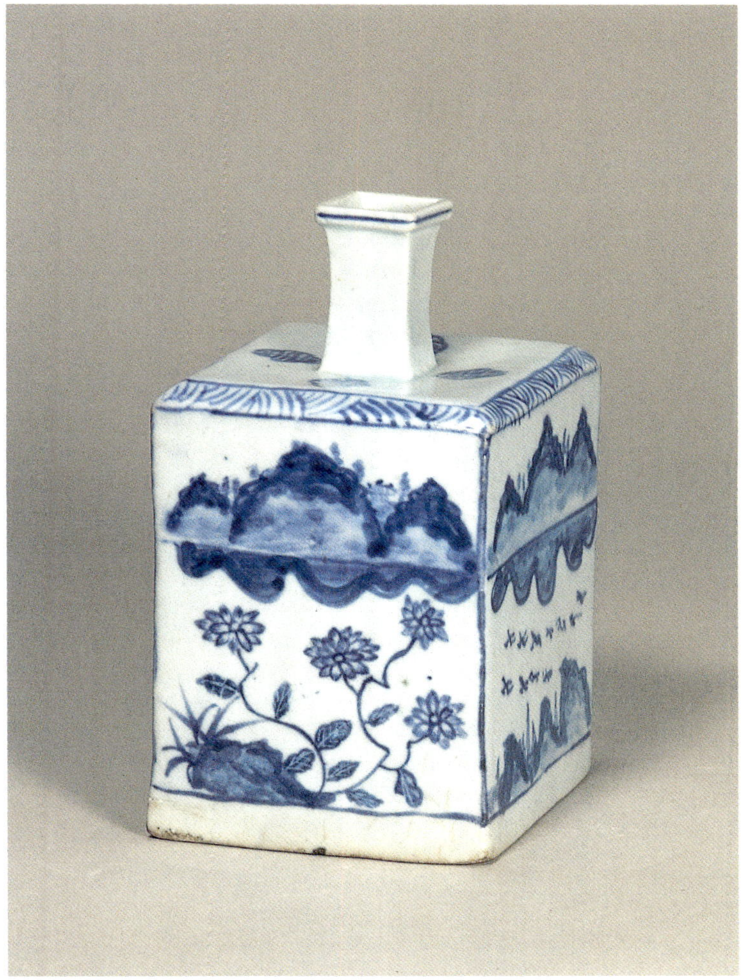

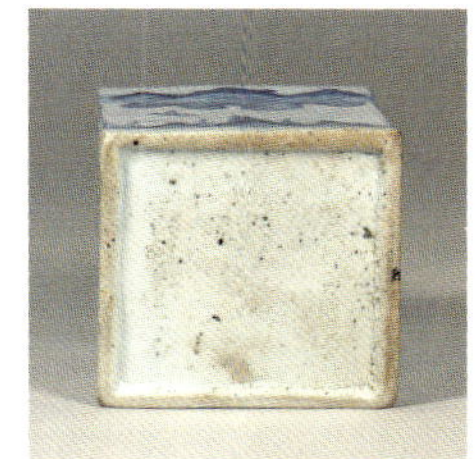

101

Square bottle with an underglaze landscape design

Blue-and-white ware

Chosŏn dynasty, second half of the eighteenth century

14.4 cm

Acc. no. 20985

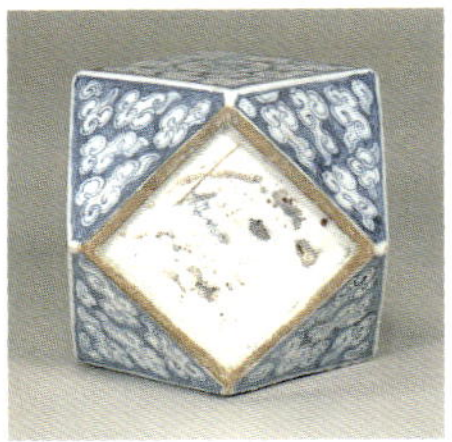

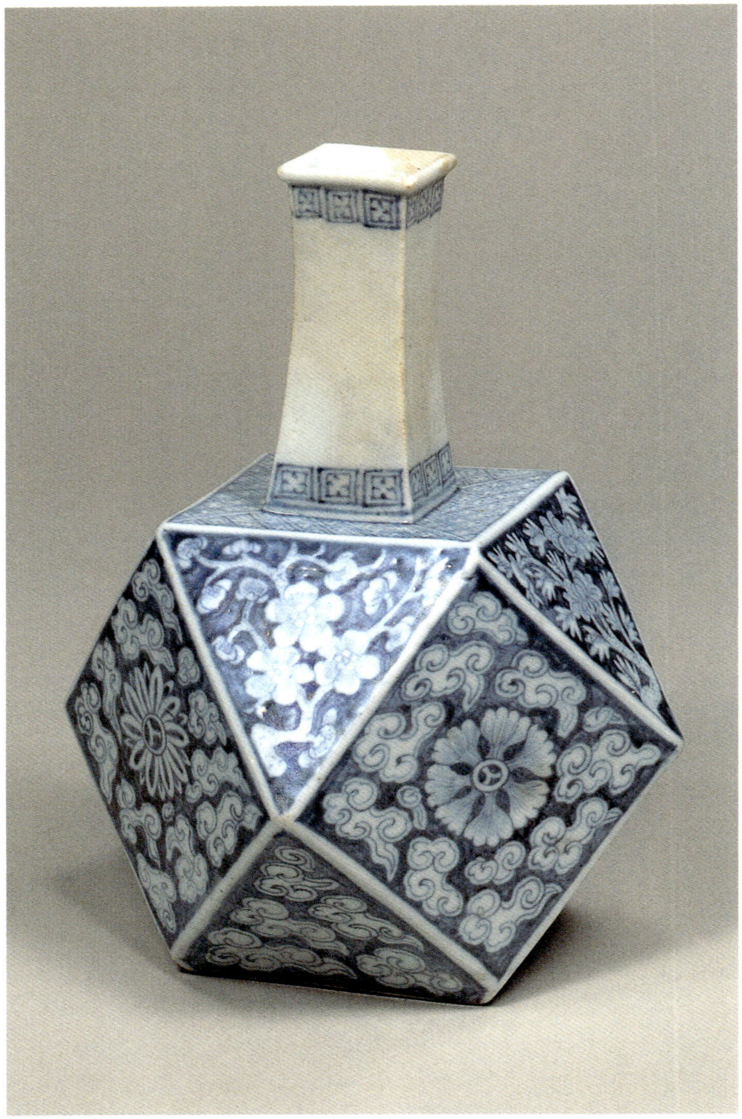

This bottle has a rather uncommon shape, even for a piece produced at the official kilns at Punwon-ri in Kwangju, which were known for their improvisations upon traditional designs. Made in a mold, the bottle is divided into twelve planes. The neck was formed separately and attached to the body before it had dried completely. Bottles executed in this complicated form are usually decorated in an intricate manner as well. On this piece, the artist left expanses of white within a bright blue background and then added motifs, done in blue, to those white spaces. Those motifs include plums, lotus flowers, peonies, peaches, and so forth, enfolded in clouds of fungus shape (Kor. *yongji*; Chin. *lingzhi*). The top and the bottom of the neck are decorated with a flowered lattice pattern set off effectively by an intervening space left in plain white. This bottle has a certain refinement, even though it was produced at a time when such pieces were overly decorated. The concave underside is shallow and glazed.

102

Faceted bottle with an underglaze floral design

Blue-and-white ware

Chosŏn dynasty, nineteenth century

19.9 cm

Acc. no. 20501

Ceramic brush holders were commonly included among the accessories found in the studies of literati. Bamboo, wood, and stone were also used to make brush holders, but, regardless of the material, their shape was invariably cylindrical. However, on some pieces, such as this one, the upper portion curls outward in a trumpet shape. These were used to store rolled-up paper documents and thus are sometimes called paper holders.

This holder is beveled on eight sides and is decorated with plum and bamboo motifs. These have been painted with cobalt blue and are delimited by a horizontal blue line around the base. The bamboo pattern resembles a fragment found at one of the official kilns at Kumsa-ri in Kwangju. The elegance of the paintings suggests that they were done by a court-appointed painter and not by the potter. Documents dating to the latter half of the fifteenth century indicate that such painters visited the Kwangju kilns each spring and autumn, but there are no extant documents that would confirm whether or not that practice continued into the eighteenth century. This paper holder probably was made in the first half of the eighteenth century, judging from its glaze, design, and color. The base is glazed, and grains of sand are imbedded in it.

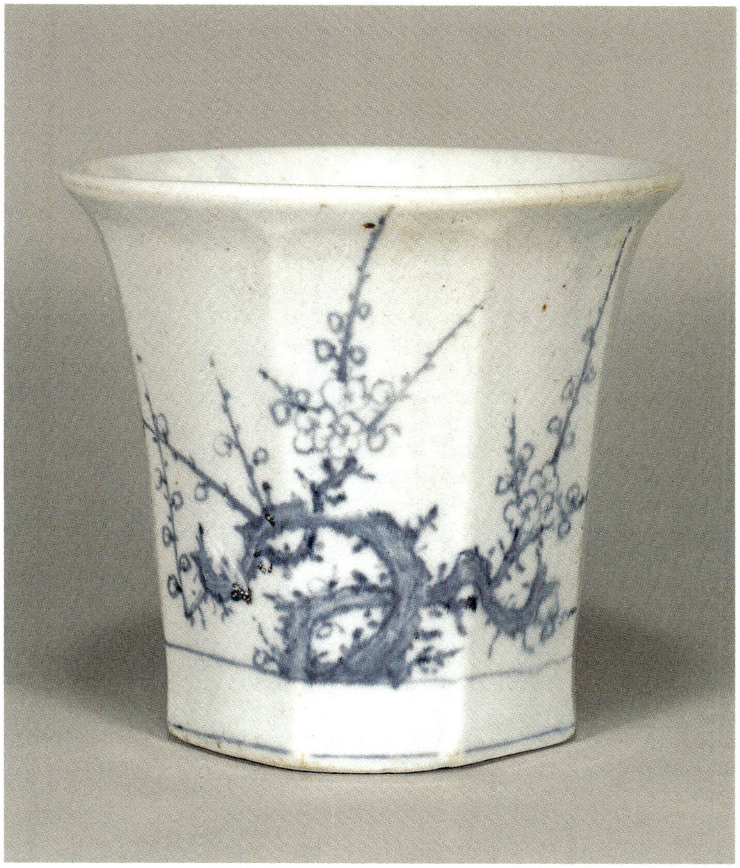

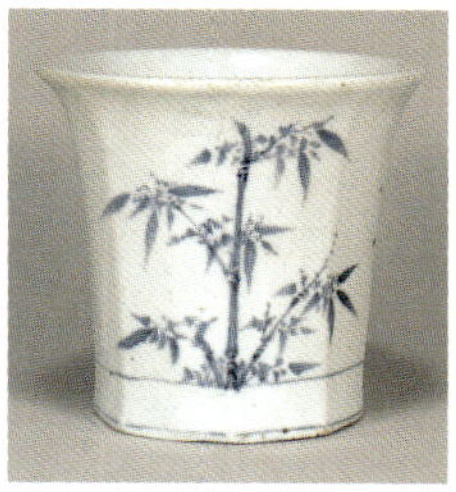

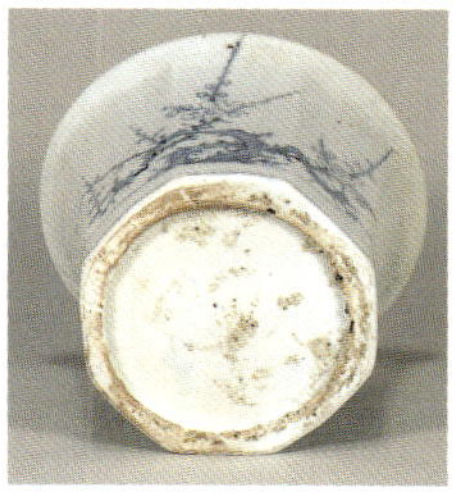

103

Beveled paper holder with an underglaze plum-tree and bamboo design

Blue-and-white ware

Chosŏn dynasty, first half of the eighteenth century

16.4 cm

Acc. no. 20674

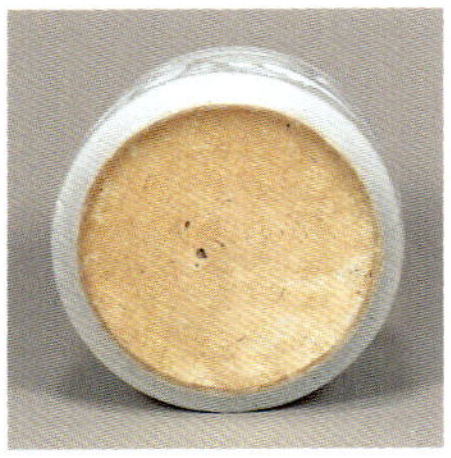

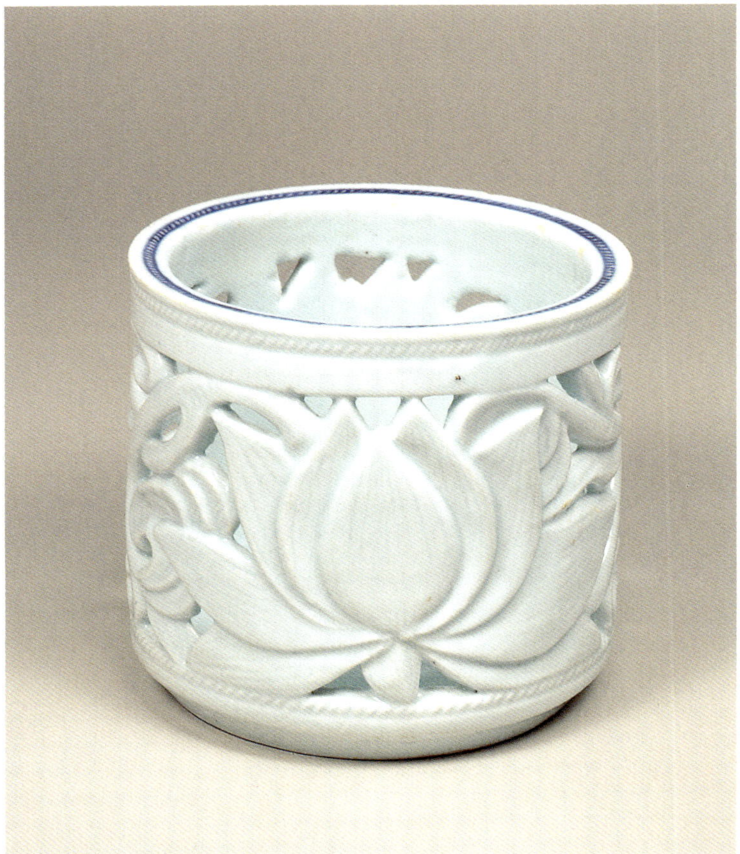

This is an unusually large brush holder. The openwork assumes the form of a lotus flower on each of the two sides of the holder, and these, in turn, are adorned with storks and flowers done in relief. Openwork brush holders and water droppers were produced in great numbers from the latter half of the eighteenth century at the official government kilns at Punwon-ri in Kwangju, and many still survive today. A relief resembling a braided rope encircles the top and bottom of the body, perhaps in imitation of the metal hoops used on wooden containers to prevent cracking and splitting. Another braided rope can be seen at the very tip of the holder, but it was painted on with blue pigment rather than being cast in relief, and it adds a nice decorative effect. The tint of the blue glaze and the hint of red indicate that this holder was made after 1752, when the official kilns at Punwon-ri were in production. The base is not glazed.

104

Brush holder with an openwork lotus design

Porcelain

Chosŏn dynasty, second half of the eighteenth century

14 cm

Acc. no. 20891

After the mid-eighteenth century, most brush holders were manufactured at the official government kilns at Punwon-ri in Kwangju, which opened as ceramic production was reaching its peak during the Chosŏn period. Produced in a great variety of styles, they might be incised, cast in relief, or feature openwork. Some were patched with porcelain clay; others were decorated with iron-brown or copper-red painting, and still others were done as blue-and-white ware. All utilized the pigments of the underglaze effectively.

This piece features an openwork design of fruiting grape leaves and vines, complete with a squirrel climbing on the vine. It is executed with skill and grace. To look at it is to appreciate more fully the high degree of refinement achieved in ceramics during the Chosŏn period. Reliefs of braided ropes, imitating metal restraining bands, decorate the top and the bottom. The glaze is bluish white, clear, and translucent; the base is glazed.

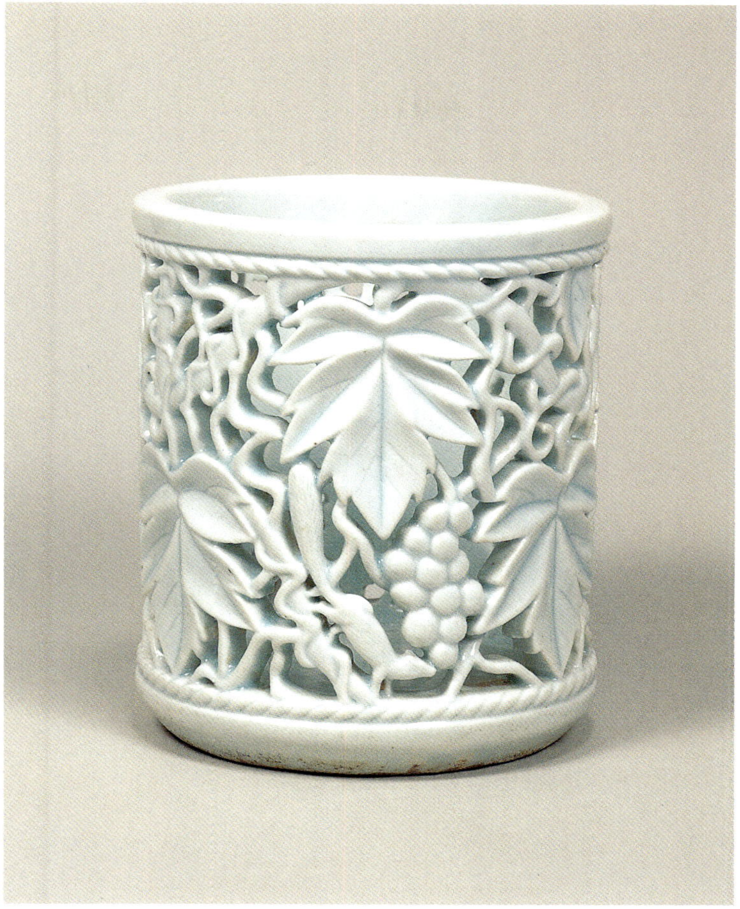

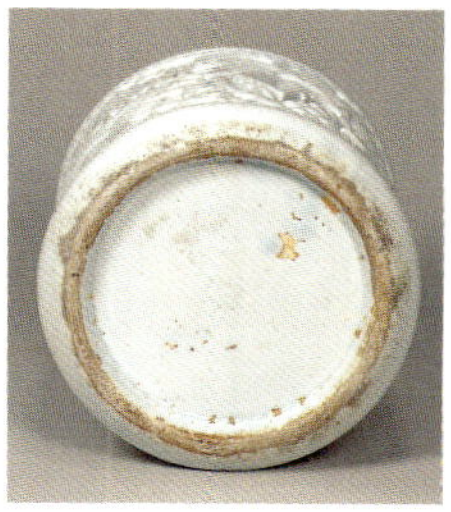

105

Brush holder with an openwork grapevine and squirrel design

Porcelain

Chosŏn dynasty, second half of the eighteenth century

15.1 cm

Acc. no. 20304

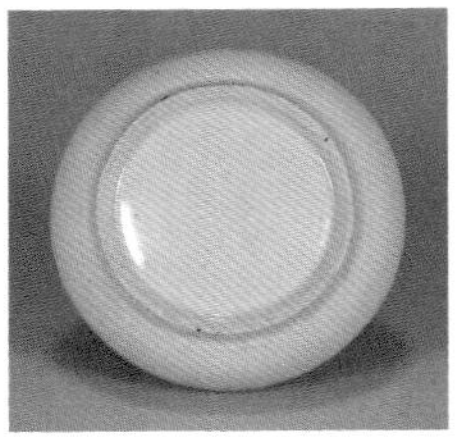

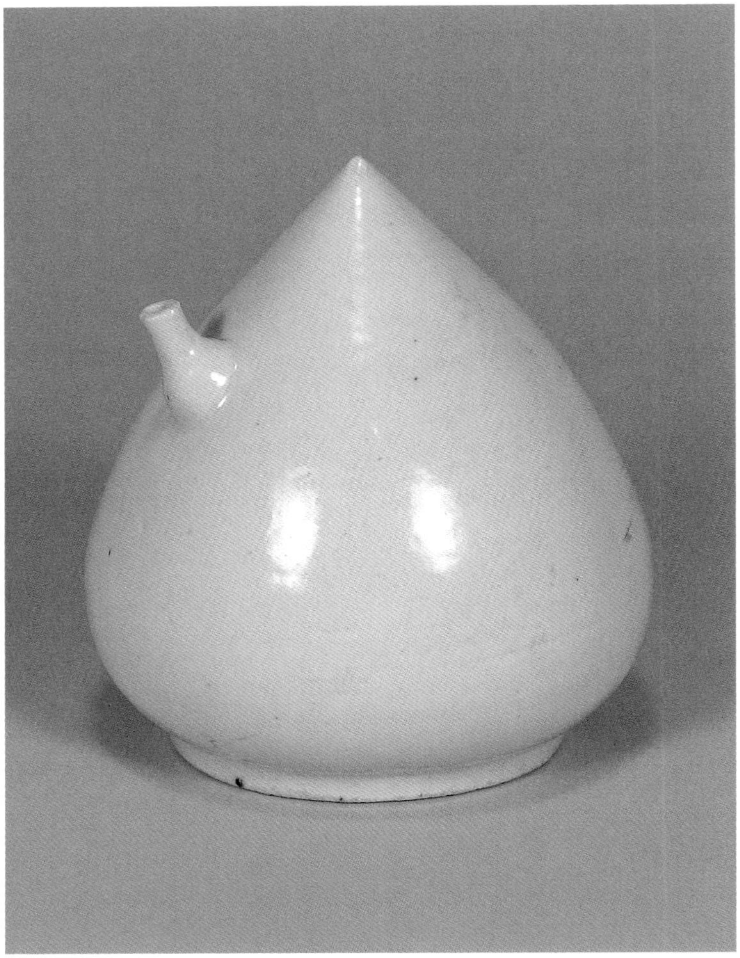

Water droppers were chiefly made in the latter half of the eighteenth and in the nineteenth centuries, and most examples that survive today date from that period. The production of water droppers, however, can be traced back to as early as the fifteenth century. They were done in the Punch'ŏng and porcelain traditions, although not with great frequency. On rare occasions, one even encounters a blue-and-white specimen as well. Many of the porcelain water droppers of the fifteenth and sixteenth centuries assume what is called a gem-stone shape, but it might be more appropriate to say that they resemble disks with elevated centers.

This water dropper has a very typical silhouette, although not many examples of that shape are extant today. The diagonal spout juts out sharply from the shoulder. Rather early examples like this one possess a certain tenseness of body contour that is suited to such a spout. One can observe the same clean, severe design in the spout as well as in the shape. The base was shaved with care and is glazed.

106

Water dropper in the shape of a lotus bud

Porcelain

Chosŏn dynasty, fifteenth/sixteenth century

11.8 cm

Acc. no. 21216

This water dropper is dome-shaped and has an outer shell with an openwork design of a lotus flower. A small container on the inside holds the water and is fastened to the outer shell at the base. Water emerges from a spout nestled in the stem of the lotus. Dome-shaped water droppers featuring openwork are very rare. The workmanship of this piece is excellent, and it has a powerful form. The key-fret pattern incised in a band around the base imparts a lively rhythm to the whole. It is thought that ceramics with this sort of simple, yet forceful, expression were made in the rather early stages of operation of the kilns at Punwon-ri. Potters there were increasingly interested in the decorative aspects of their work, and vitality of form was often sacrificed to such decorative dictates. This water dropper was probably made during the second half of the eighteenth century, when many excellent pieces were turned out at the Punwon-ri kilns. The slightly gray-green tint in the glaze is characteristic of the products of these kilns. The base is flat and is not glazed.

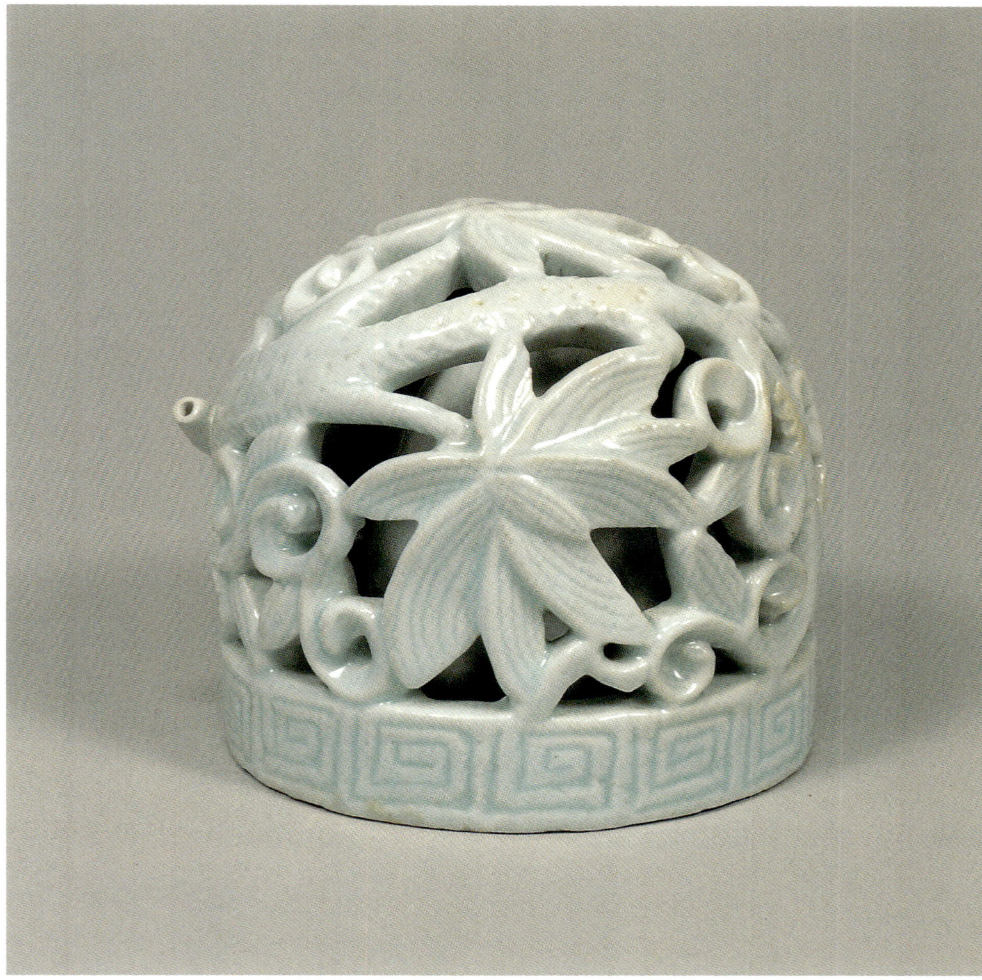

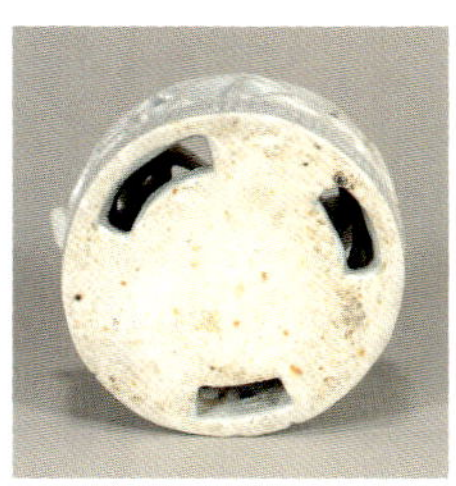

107

Water dropper with an openwork lotus-scroll design

Porcelain

Chosŏn dynasty, second half of the eighteenth century

9.5 cm

Acc. no. 20961

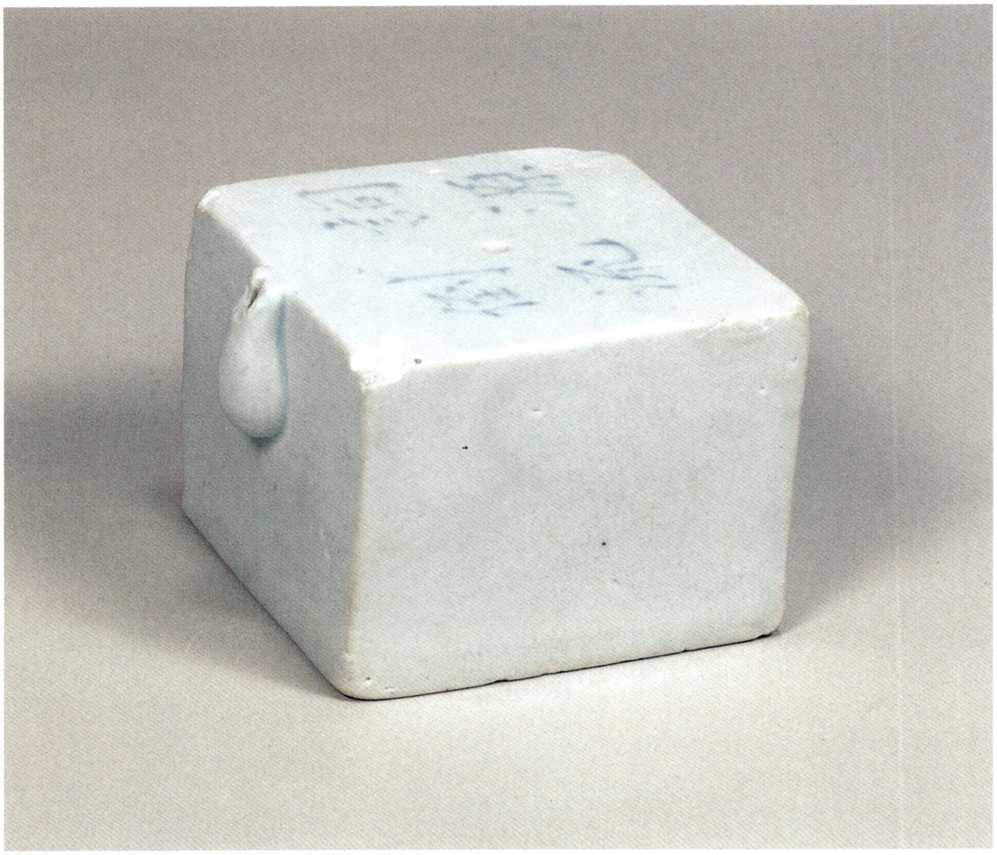

There was a strong tendency toward simplicity of design during the first half of the eighteenth century, due perhaps to the influence of Confucianism. Restraint was in fashion: simple, reserved shapes predominated; motifs tended to be small and minimal; and brushwork was executed with tiny strokes. This water dropper is representative of pieces done in that style. The body, a simple square shape to which a small spout has been attached, is warped and not perfectly proportioned. This denial of rigid uniformity apparently continues a tradition that dates back to the seventeenth century. A poem of four characters drawn in blue on the top constitutes the only decoration. The pale hue of the blue suits the vessel's mood of simplicity. The poem is from a dramatic piece dating to the late Yuan or early Ming period and alludes metaphorically to the ceaseless flow of water. The base is unglazed.

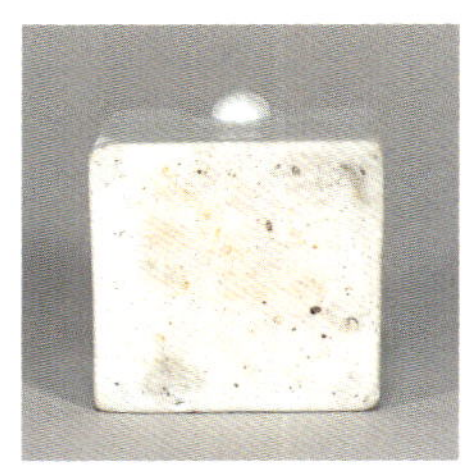

108

Water dropper with an underglaze inscription

Blue-and-white ware

Chosŏn dynasty, first half of the eighteenth century

6 cm

Acc. no. 20974

The tendency for ceramic wares made at the Punwon-ri kilns to be more elaborately decorated began at some point in the second half of the eighteenth century, perhaps reflecting a new vitality among the Korean people. This piece nicely illustrates that trend. The water dropper is encircled by a raised, narrow rim that separates the body from the base and draws attention to the foot. The top is not flat, but rather has a shaved, curved edge and a central area decorated with a lattice comprised of blue floral shapes. The square rim surrounding the lattice is adorned with a wavelike motif done in relief in white porcelain. Many water droppers have bottled-shaped spouts, but the spout on this piece depicts a squirrel. All of these elements come together to create a graceful, decorative effect. A banana-plant motif adorns three of the sides, and a design of brushes and brush holders fills the remaining side. The base is glazed, and one portion was shaved off.

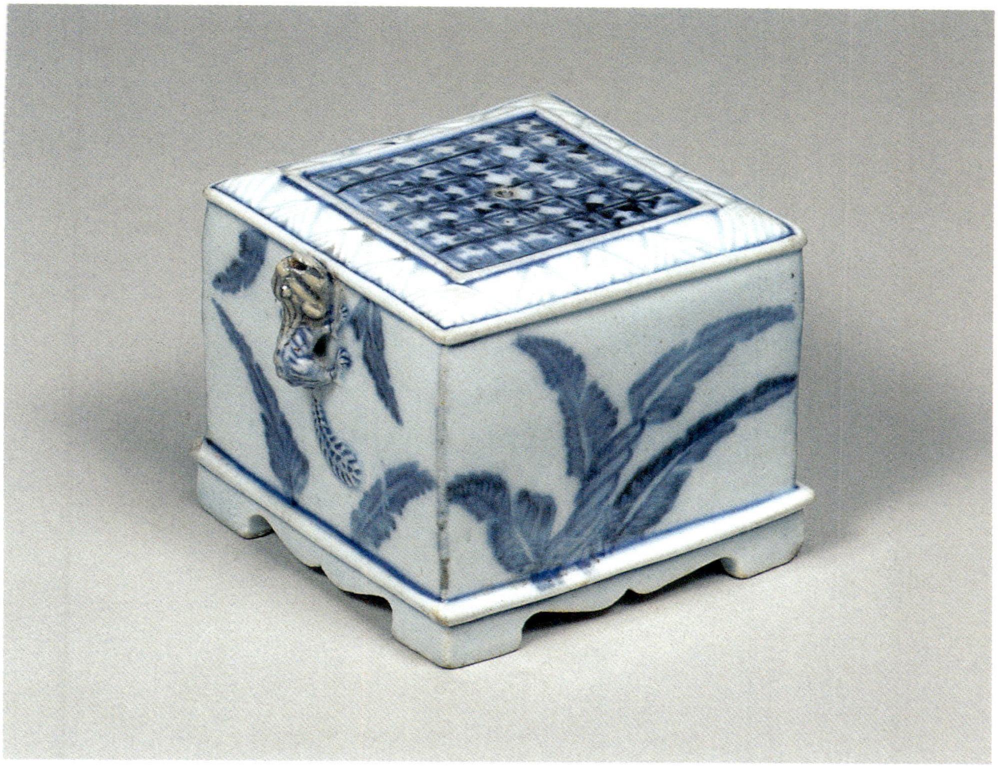

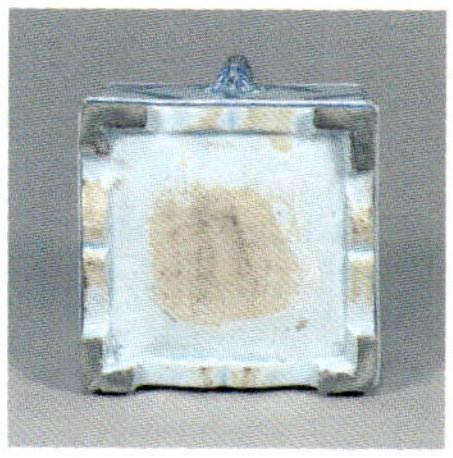

109

Water dropper with an underglaze banana-leaf design

Blue-and-white ware

Chosŏn dynasty, second half of the eighteenth century

7.6 cm

Acc. no. 21247

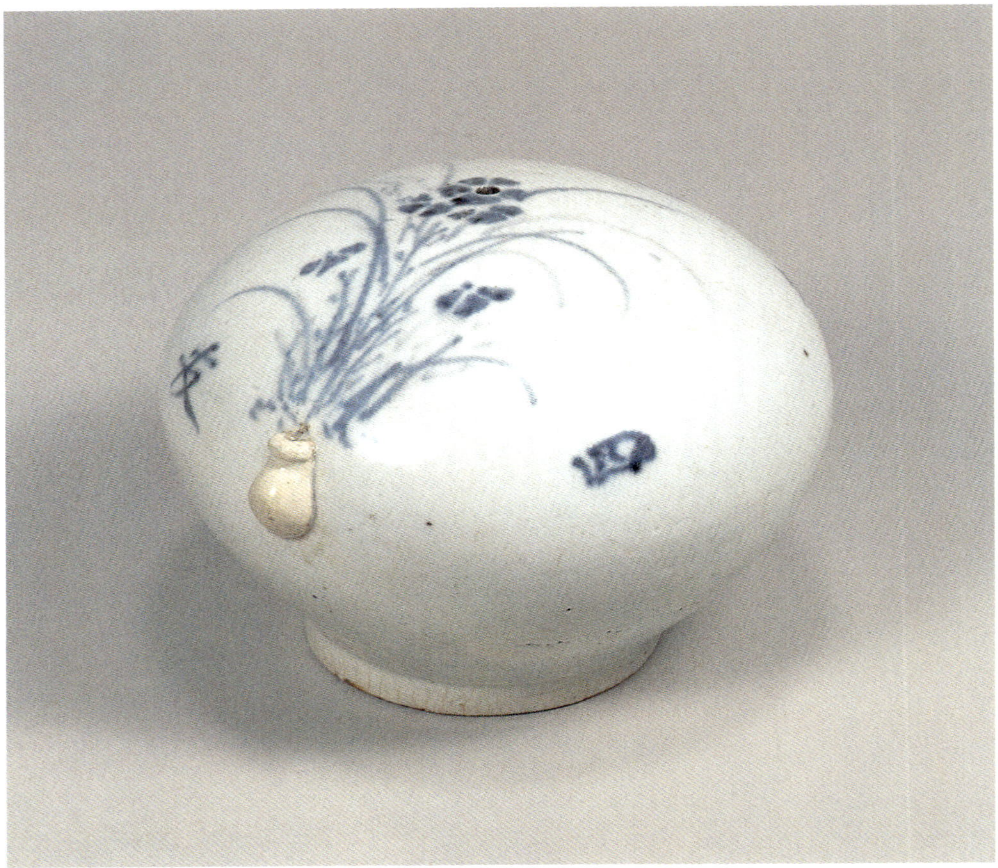

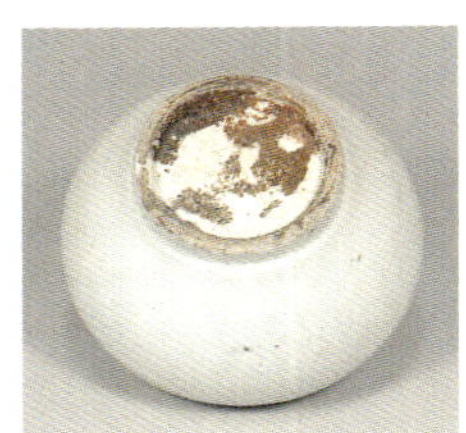

This water dropper resembles a bulging disk, a shape common for such pieces in the first half of the eighteenth century. On many examples, an extremely sharp turn inward toward the base is used to create a certain amount of tension and to suggest the feeling that a great deal of weight and bulk have been taken out of the lower part of the body. From the latter half of the eighteenth century on, however, this gave way to a more flattened, disklike shape.

A spout in the form of a half bottle emerges from the surface of the dropper. A floral motif of so-called China pinks (*Dianthus chinensis*) is painted in blue around the upper body. China pinks were a common motif during the first half of the eighteenth century, and those decorating this bottle follow the style of the time. A bee can be seen buzzing over the tops of the flowers, and the four characters for paper, brush, ink, and inkstone — the essentials of literary composition — appear around the perimeter of the floral design. The cobalt blue is very dark and without a tint of red, which is typical of its pigmentation on ceramic wares of the first half of the eighteenth century. The base is glazed, and grains of sand are imbedded in it.

110

Water dropper with an underglaze flowering-plant design

Blue-and-white ware

Chosŏn dynasty, first half of the eighteenth century

8 cm

Acc. no. 21176

This octagonal water dropper has an unusual construction in that its center is hollowed out to make a doughnut-shaped reservoir where water can be stored. The dropper holds less water than most, but it is light and easy to hold, rendering it highly functional. There are only a few extant droppers in this form, perhaps because they were so technically difficult to make. Motifs of plum trees and bees decorate three of the eight sides of the dropper. Around the top appear eight divination blocks, a popular motif in the middle and late eighteenth century. The spout is bottle-shaped. Spouts on water droppers made in the fifteenth and sixteenth centuries tend to project diagonally upward from the body. They broke easily, however, so that design eventually gave way to shorter spouts such as the one on this example. The glaze on this piece is milky, typical of the porcelains produced at the Kumsa-ri kilns, Kwangju, in the first half of the eighteenth century.

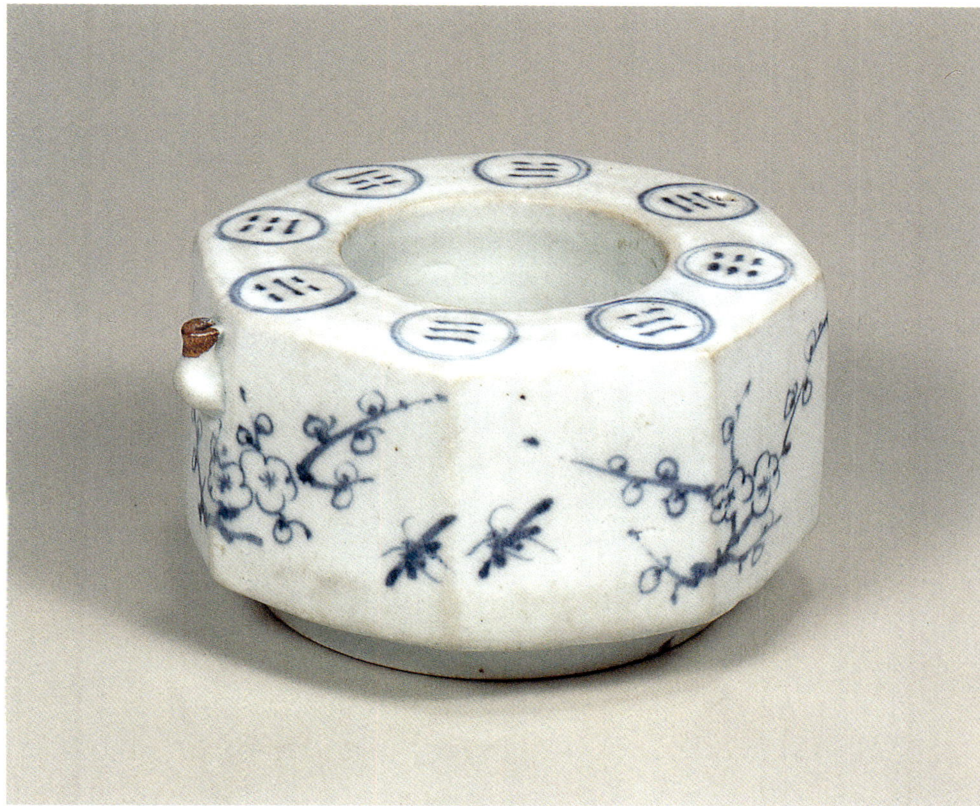

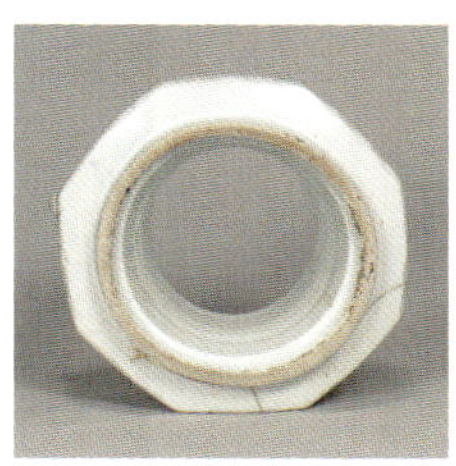

111

Water dropper with an underglaze plum-tree design

Blue-and-white ware

Chosŏn dynasty, first half of the eighteenth century

8.1 cm

Acc. no. 21223

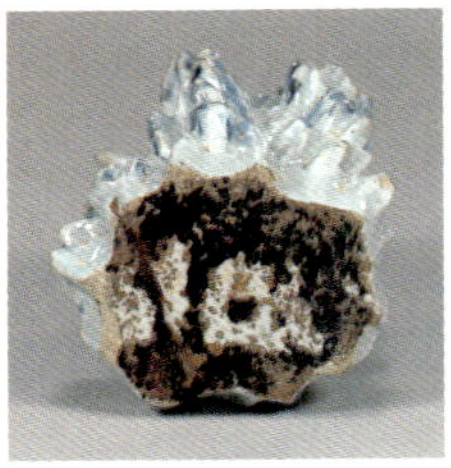

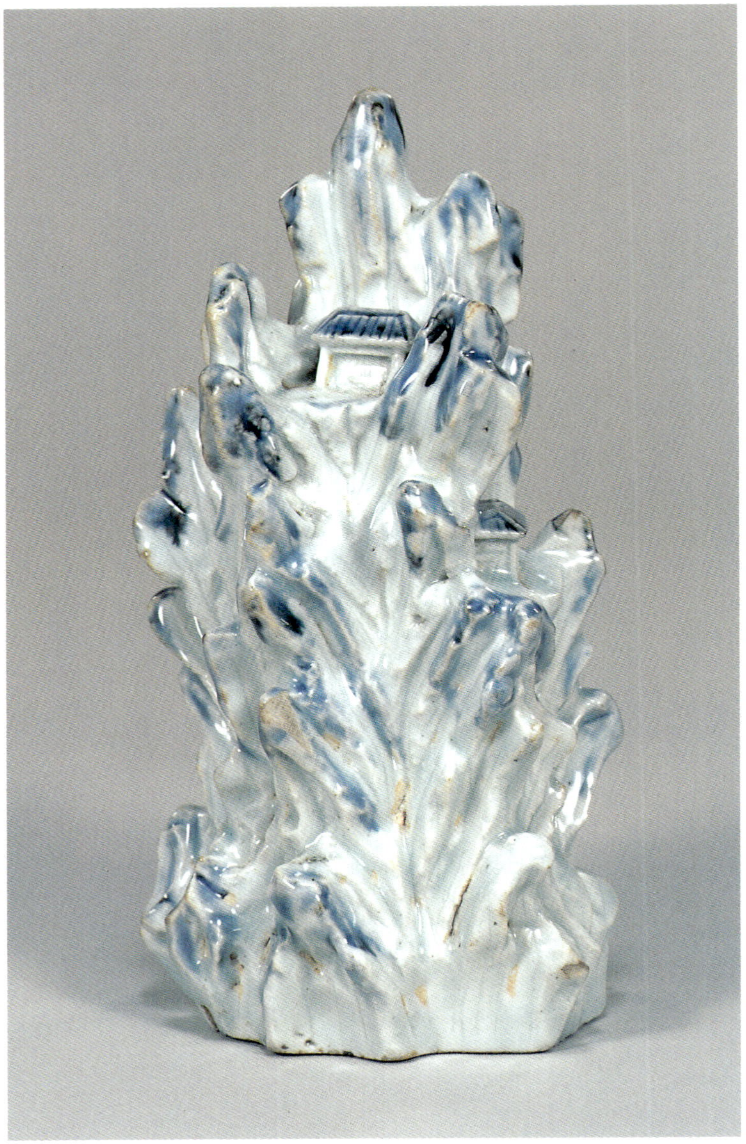

112
Water dropper in the shape of a mountain with an underglaze cobalt-blue coating
Porcelain
Chosŏn dynasty, first half of the nineteenth century
21.4 cm
Acc. no. 21161

During the Chosŏn period, some water droppers were done in the shape of animals, plants, personal residences, and mountains. This piece may represent Mount Kumgang, which was venerated as a sacred mountain. According to legend, more than one hundred temples and private sanctuaries were located on it. This dropper was probably made by piling a mass of clay around a cylinder and then forming the crags, ledges, and peaks on the surface. One temple is perched on a ledge, and another can be seen at the top of the mountain. Many water droppers representing Mount Kumgang are covered with copper-red or iron-brown paint that is distracting. This piece, however, was left white, except for the top of the mountain and the roofs of the temples, which have been painted cobalt blue, imparting a feeling of ethereality. An air hole can be found at the back of the temple on the peak, and a spout emerges from the right side of the temple on the ledge. The intense blue of the glaze indicates that the piece was made around the first half of the nineteenth century. The base is covered with glaze, which contains some black carbon sediment.

Very large water droppers probably served as decorative pieces rather than as functional objects, although one theory holds that they were made for use in schools and academies. This piece is the largest known octagonal water dropper. To shape it, clay was probably piled around a cylinder and then shaved to form the eight sides. The top and bottom were made separately and attached to the body. Adorning the eight sides are landscapes done in blue pigment, representations perhaps of the "Eight Views of the Xiao and Xiang Rivers." Grapevines with clusters of fruit and leaves decorate the top. Although the artist's name does not appear on the dropper, we can surmise with near certainty that the drawings were executed by a prominent painter, since the date and place of production are indicated. Adorning the top and side are a pair of squirrels whose details are defined with cobalt-blue and iron-brown pigment. The mouth of the male squirrel on the side of the piece forms an air hole, and a spout emerges from the left ear of the female squirrel on the top of the water dropper. The feet are in the shape of eight round pods, which were iron-painted to provide a balanced counterpoint to the squirrels. The base is covered with glaze. This piece has played an important role in helping to establish a chronology of production at the Punwon-ri kilns, for drawn in cobalt blue under the glaze on the underside are the words "Punwon" and "the sixth month," as well as two characters in the sexagenary cycle that correspond to the year 1783 in the Western solar calendar.

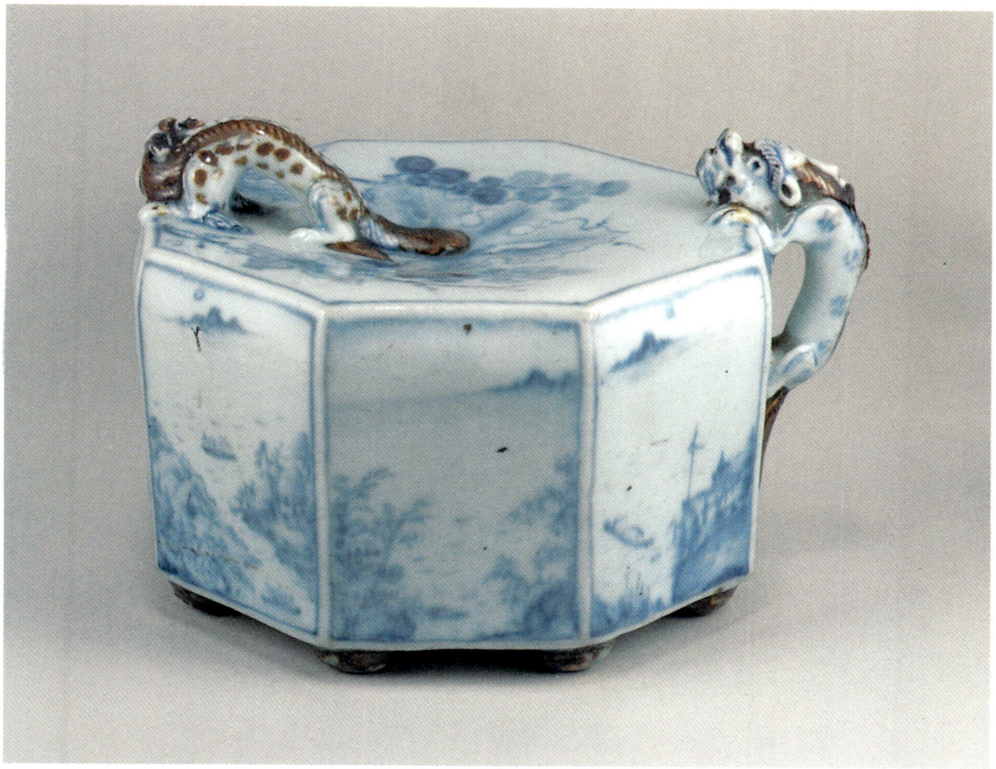

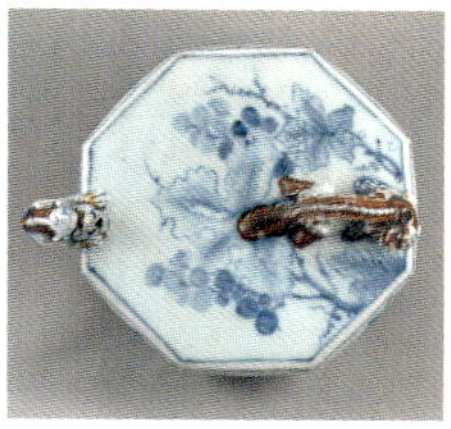

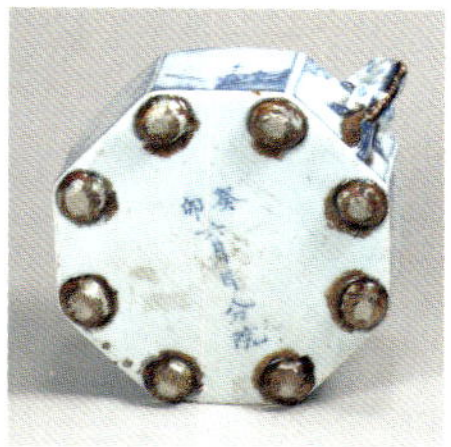

113

Water dropper with an underglaze landscape design

Blue-and-white ware

Chosŏn dynasty, eighteenth century, inscribed with the date 1783

12.8 cm

Acc. no. 21221

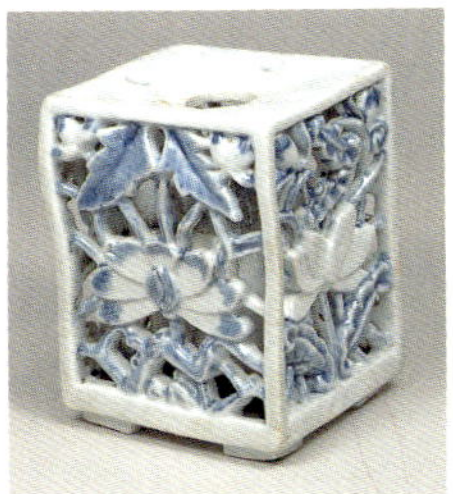

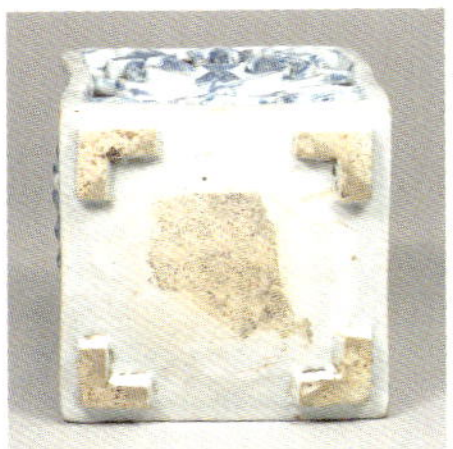

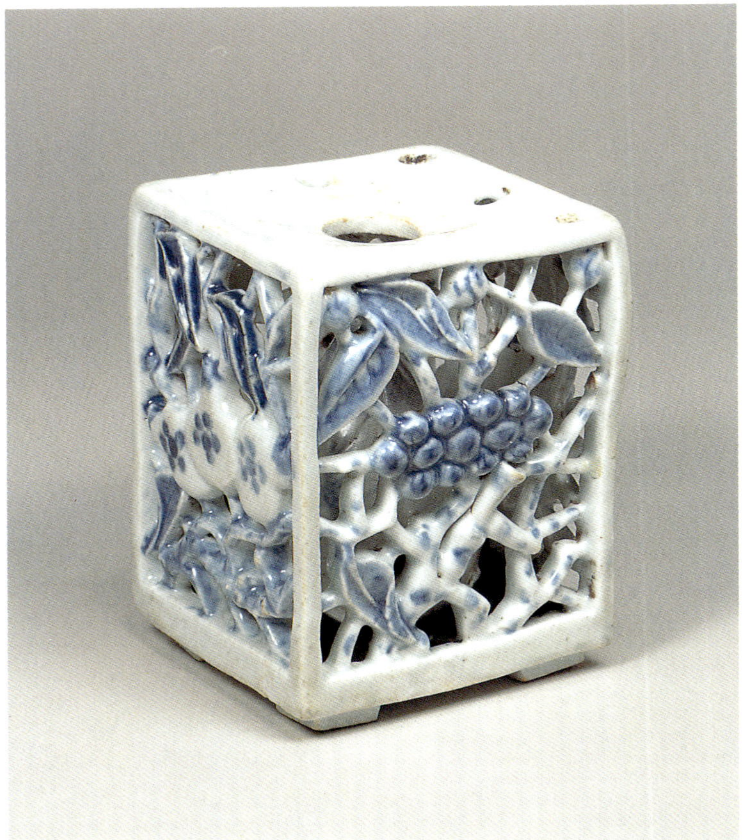

One category of water dropper combines within itself the additional function of a brush holder. This is one such piece. A peach-shaped water container nestles inside a rectangular outer shell. Nine holes in the top can be used to hold brushes. The four sides are done in openwork designs of peonies, lotus flowers, grapes, and peaches. The decorative effect is enhanced by exquisite details done in cobalt blue. The peach-shaped water container and the rectangular outer shell are connected at the bottom. Water is poured from the spout at the lower right-hand side of the grape cluster, whose main stalk provides a channel connecting the outer spout to the inner water reservoir. The piece is warped, and the openwork has been executed with little concern for precision. This imparts to the water dropper a sense of guilelessness and unpretentiousness. It is not, however, an inferior piece. Rather, the nobles who used these dropper-holders were fond not only of elegant and refined wares, but also of simpler pieces, which they felt embodied the flavor of literati attitudes. The base is glazed, and the body rests on four *L*-shaped feet.

114

Water dropper with an underglaze cobalt-blue coating on an openwork floral design

Porcelain

Chosŏn dynasty, nineteenth century

13.8 cm

Acc. no. 21184

The technology of Korean celadons

PAMELA B. VANDIVER

THAT KORYŎ DYNASTY celadon represents a peak of aesthetic achievement was recognized as early as the twelfth century by the Chinese.[1] From the beginning of the current century, Korean celadon has been a focus of Japanese and Western connoisseurship.[2] More recently, the publication and exhibition of ceramics excavated in Korea have brought some of the finest celadons to the attention of Westerners.[3] It is the reconstruction of the technology underlying manufacture of these celadons and the investigation of its relation to Chinese practice that are the subjects of this essay.

Celadon glazes consist of yellowish-, bluish-, or grayish-green glazes which have soft, lustrous, translucent appearances similar to jade in their depth and richness. They developed in Korea from a tradition of thin, yellowish- to brownish-green glazes called Nokch'ŏngja, which are often considered to have been based on southern Chinese Yue ware prototypes. Nokch'ŏngja is found in archeological contexts beginning in the eleventh century, and its production as common ware continued simultaneously with the manufacture of fine celadon for court and religious use. Koryŏ celadons began as molded, incised, and undecorated monochrome celadons, or Ch'ŏngja, during the first half of the twelfth century.

Innovations in decorative technology in the mid-twelfth century led to the production of inlaid celadon, or Sanggam Ch'ŏngja, in which the paste of the inlay was pressed into an incised or carved design. While manufacture of inlaid celadon continued through the fourteenth century, the inlaid tradition was transformed in the late thirteenth century, when a more rapid method of decoration using brushed and/or slip-inlaid decoration developed. Punch'ŏng ware, as this type of ceramic came to be known, evolved in the fourteenth century as a popular ware. The inlay was produced by stamping or roller impressing a pattern into the surface of the clay body; then the body was coated with slip which, when partially dry, was scraped to reveal the original surface design. Punch'ŏng, although usually not considered a type of celadon, developed directly out of the inlaid celadon tradition, as demonstrated by late thirteenth-century transitional pieces in the National Museum of Korea, Seoul, which have the color and inlay technique of inlaid celadon but are decorated with the stamped and rolled designs characteristic of the infilled technique of Punch'ŏng ware.

In China, green-glazed ceramics have been manufactured for about thirty-five hundred years, beginning in the middle Shang dynasty; the Chinese have produced a broad range of celadon types for approximately the last thousand years. In Korea, by contrast, celadon was made for a shorter period, extending from the eleventh century, in the Koryŏ period, through the fifteenth and sixteenth centuries of the early Chosŏn dynasty. A comparison of Chinese and Korean celadons reveals a high degree of consistency in the appearance of the Korean glaze, which can perhaps be attributed to its limited duration. Another factor that may have contributed to the continuity of the Korean celadon glaze is Korea's compact geography. Green-glazed ceramics of various types evolved at widely separate kiln sites in northern and southern China, which were sometimes separated by distances of thousands of miles. Different raw materials, craft practices, and technological methods used at these regional production centers accounted for variation in the character of the wares. In Korea, however, production of celadon was concentrated in the southern half of the peninsula, to the south and west of the Koryŏ capital of Kaesŏng (slightly north of the present capital at Seoul), in an area extending over a distance totaling only several hundred kilometers.

The geographic proximity of the Korean peninsula to China (map, p. 14) facilitated contact with the centers for Chinese ceramic production, and some scholars now propose that Korean ceramic practice may have influenced Chinese ceramic technology.[4] Certainly, the converse belief that Chinese ceramics at points influenced the development of Korean ceramic technology is widely held.[5] For example, it is generally believed that methods for applying a green glaze to stoneware first entered Korea from China by the tenth century.[6] Dating for Korean shapes is also based on the chronology of Chinese wares.

The purpose of this study is to investigate the physical basis for the visual appearance of Koryŏ celadon (for examples, see nos. 1–23) and to reconstruct empirically the materials and processing constraints that produced such objects. Comparison of the technology of manufacture of Korean celadon with that of celadon production in China hopefully will shed light on the development and maintenance of the conservative Korean technological tradition. Finally, by comparing the technology of Korean celadons with that of Chinese celadons, affinities between Korean and certain Chinese celadon wares will yield information helpful in defining distinct areas of possible technological interaction between the two cultures.

"Deconstructing" the underlying technology of visual appearance

Determing why celadon glazes look the way they do involves understanding the interaction of light with the internal glaze structure just below visual detection.[7] Analysis of the microchemical and microstructural elements (i.e., elements below the limit of resolution of the unaided eye, about 0.1 mm) enables us to work backward, "deconstructing" the visual appearance of the glaze to define its physics, chemistry, and technology. To accomplish this goal, I use the approach and methods of modern materials science; that is, I analyze the relationships among residual structure, composition, and properties through measurements and then establish what processing behavior is sufficient and necessary to produce them. This approach clarifies the underlying technology and the chemical and physical behavior responsible for the visual appearance of the glaze.[8]

The special grayish tint of the Koryŏ celadon glaze, which distinguishes it from Chinese celadon glazes, results from the presence of about 0.5–0.75% manganese dioxide dissolved in the glaze.[9] The presence of iron oxide, which is about 1–2% of the glaze solution by weight, is responsible for its green color, as is the case also in Chinese celadon glazes (fig. 1). The translucency of the Koryŏ glaze is produced by heterogeneities in the glaze layer just at visual detection but easily distinguished using a scanning electron microscope in backscattered mode, which registers variations in composition as a gray scale with low atomic number elements being darkest and high atomic number elements being lightest (fig. 2). These heterogeneities are identified as partially dissolved quartz and quartzite particles ranging in size from 5–100 microns, with most in the 20- to 30-micron range, and as fine, agglomerated black particles of magnetite and ilmenite (an iron titanate), in the same size range (figs. 2a, 2b). The largest of these black-and-white particles are visible on close examination of the surface of the glaze. The white quartz particles, which appear gray in the micrographs, create translucency by bending and scattering light, thus increasing the path length of light through the glaze. Even though partial melting has slightly decreased the volume of the particles, it is clear that they were unusually finely ground during preparation of the glaze; the multiple-grained quartzite particles have a large surface area and scatter light internally among the grains of the particles, thus adding to the glaze's translucency. Clumps of cubic, black magnetite particles, present as submicron cubes that appear white in the micrographs, act to absorb light and are responsible for the darker, grayish areas. The dispersal of these particles results in a shimmering light-and-dark texture that adds considerable depth to the glaze. This method of alternating light scatterers with light absorbers is not found in Chinese glazes.

The translucency achieved by the presence of these heterogeneities is further enhanced by small, spherical bubbles of less than 100 microns (or a tenth of a millimeter) and the network of fine cracks, or craze lines, which serves to brighten the glaze through internal reflection of the incoming light, sometimes giving the optical illusion of interior illumination. Crazing results when particular glaze formulations, especially those high in lime, contract on cooling in the kiln so much more than the underlying clay body that the tensile strength of the glaze layer is exceeded. The low-silica, high-calcia composition of the Koryŏ glaze is at the limit of durability for stoneware; thus, crazing derives from the poor fit of the glaze and clay body. Many Korean celadon glazes have weathered to an iron-stained brown color as the result of corrosion along craze lines through which water easily penetrates, leading to hydration of the glaze and leaching of the alkalies, which cause slight dimensional changes and yet further crazing.

Although crazing is usually considered a flaw because the cracks are permeable to liquid, its overall effect is to enhance the shimmering surface of the Koryŏ celadon glaze. The two surfaces of each craze line reflect and scatter incoming light, and specular reflection radially scatters light from the surfaces of the bubbles to form areas and networks of locally concentrated light that reinforce the stippled texture of the glaze as shade and shimmer when strong illumination is present. In the Korean celadon glaze, the volume of these inclusions and bubbles is about 10%, but there are areas of some glazes where up to 20% may occur. Quartz particles and bubbles are common in Chinese green glazes, but the quartz is not as finely ground. The control of quartz-particle size was not essential for translucency in Chinese celadons. Crazing, however, is only seen in southern Chinese wares such as Yue and Guan, which also have a high-lime glaze composition.

FIGURE 1

Compositional range of Korean celadons.

	MEAN	RANGE
SiO_2	60	(54.3–64.3)
Al_2O_3	12	(10.9–16.7)
CaO	19	(8.8–27.0)
MgO	2.5	(0.4–3.2)
K_2O	2.5	(1.7–5.7)
Na_2O	0.75	(0.3–3.0)
FeO	1.0	(0.8–2.7)
MnO	0.5	(0.1–0.9)
TiO_2	0.1	(0.05–0.4)
P_2O_5	0.8	(0.2–1.4)
SO_2	0.3	(0–0.9)

Innovation in ceramic technology

The special gray-green color, translucency, and light-and-dark stippled texture of Korean celadon were further enhanced by black-and-white inlaid designs beginning sometime before 1159, the earliest date for a tomb yielding inlaid celadon.[10] These inlays, which were a Korean innovation, were not made of white kaolinitic and dark ferruginous clays, as is commonly believed and as is the case with the inlays of some modern Korean replicas from the Ichon Valley south of Seoul that I have tested. Instead, the white inlays are composed of quartz particles sintered together with small amounts of glass and a potassium-rich clay that has been partially transformed to mullite during firing.[11] In figure 3a, a cross section of an inlay shows that a deep incision of about 2 mm was made into the body of the pot while it was still somewhat wet but able to hold its shape. The groove was filled with the white inlay, which the cracks show shrank during drying. When the glaze layer was applied, the glaze materials filled in the crack (fig. 3b). Figures 3a and 3c illustrate the porous nature of the white inlay at increasing magnification, as well as the presence of individual glass-coated quartz particles.

To produce the black inlays required the development of a separate technology. These inlays, which are composed mostly of glaze materials,[12] are much more dense than the white ones. Figure 4a shows that, with the exception of the crystals in the black inlay, the gray level, and thus average composition, is almost the same in the glaze and the inlay. In figures 4b and 4c, the contrast has been increased to show the crystalline inclusions in the inlay, which consist of magnetite, ilmenite, quartz, and anorthite (a calcium-rich feldspar visible in figure 4b at a higher magnification). The cubic particles of magnetite (fig. 4c) are the chief black colorant. The anorthite particles, which grow during an extended, high-temperature firing, are commonly seen as long, needlelike particles in the inlay but only rarely at the glaze-body interface. Since these particles are not found near the black impurities in the clay body, the celadon body was clearly not prefired prior to glazing, for example, by a biscuit firing; this observation is consistent with East Asian ceramic practice, which traditionally involves a single firing. Therefore, the presence of anorthite indicates that the black inlays had been prepared through a separate heat treatment, namely, fritting, a heating and grinding process, prior to being applied. Since black magnetite combines with excess oxygen during firing to form red hematite (commonly seen as red rust on iron), prefiring of the inlay was used to surround the magnetite with inert glass and thus prevent this change.

To achieve technical success, such that the inlay on the body of the vessel could be seen through the glaze, required tight processing control of glaze translucency. Thickness, consistency of composition, and control of particle size were important factors in producing the desired effect. At an appropriate glaze thickness of about 0.2 mm (measured on the horizontal axis), the inlay is seen slightly indistinctly through the glaze; at 0.3 mm, the pattern is obscured. At a depth of 0.1 mm, which is characteristic for eleventh-century celadon and Punch'ŏng ware, the glaze is virtually transparent, and the inlay and sometimes even the clay body are clearly visible through the glaze. This evidence shows that precise control of processing parameters was practiced by Koryŏ-period potters.

Conservatism and change in ceramic technology

The question of the continuity or development of Koryŏ ceramic technology over time can be evaluated by noting the changes in glaze micro- and macrostructures (fig. 5). During the Koryŏ period, glaze thickness remained fairly constant for large and small vessels. The glaze thickness and microstructure produced optimally translucent and inlaid effects as the amounts of quartz and extremely finely ground black particles were increased in comparison with eleventh-century celadon. Inlays, although compositionally the same, tended to become shallower and not as sharp at the edges and their materials not quite as pure toward the end of the Koryŏ dynasty. In pieces of late thirteenth- and fourteenth-century date, the components of the black-and-white inlays sometimes intermix so that the colors become somewhat muddy or mottled.

During the early Chosŏn dynasty, the technology of celadon changed to promote the use of stamped and impressed Punch'ŏng as a more rapidly produced common ware.[13] Porcelain replaced celadon at court, while fine, plain, translucent celadon continued to be made as a Buddhist ritual ware for use on altars. Thus, the change in the function of Punch'ŏng ware required alterations in ceramic technology to accommodate large-quantity workshop production of common ware. Although the glaze composition of celadon and Punch'ŏng remained the same, there were some significant differences. The glaze thickness of Punch'ŏng ware decreased to about half that of inlaid celadon. With only half the glaze raw materials used in manufacture, most of the translucency of the glaze was lost. In fact, the presence of fewer and larger bubbles and the occurrence of quartz and magnetite particles responsible for the trans-

FIGURE 2

Microstructures of the polished glaze-body cross section of a maebyŏng *fragment. (a) Irregular quartz and quartzite inclusions and spherical bubbles in the glaze (140x). (b) Partially dissolved quartzite particle and bubbles (750x). (c) Unusual anorthite growth (in gray) at the glaze-body interface (950x).*

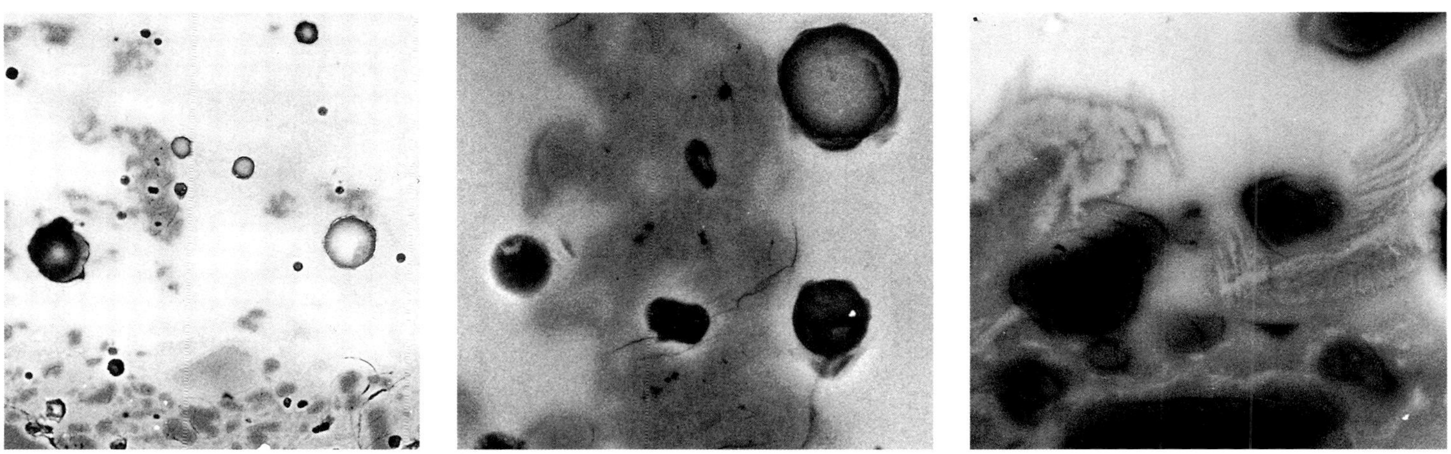

a *b* *c*

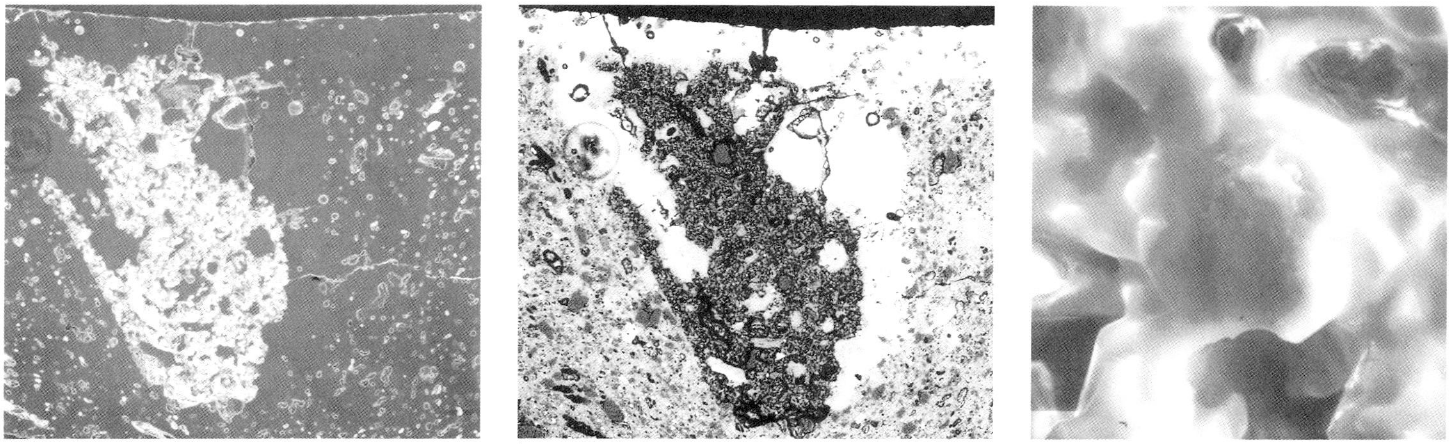

a *b* *c*

parent and glossy character of most Punch'ŏng glazes results from the more limited attention given to glaze preparation as raw materials were more coarsely ground and poorly mixed (fig. 6). Many of these heterogeneities are larger in scale than those in Koryŏ glazes and can be seen with the naked eye. In addition, black inlay was no longer used, and the white inlay contained considerable clay. Instead of being incised or inlaid in carefully drawn patterns, decoration was applied using the more rapid techniques of slip painting of rolled or stamped impressions, which eliminated the time-consuming precision and draftsmanship of inlay.

The conservative nature of Korean celadon technology is demonstrated by the composition of its glaze, which was remarkably constant over a five-hundred-year period that began in the early twelfth century.[14] Composition of the Koryŏ glaze is closer to that of the southern Chinese tradition than it is to the northern Yaozhou tradition (figs. 7 and 8). In terms of relative proportions of potassia, calcia, and alumina, the Koryŏ and Punch'ŏng glazes consist of about 60% SiO_2 and are grouped near the Yue, Longquan, and Guan wares (fig. 8).[15] With few exceptions, this consistency of composition and structure was achieved through knowledgeable control of processing. If the glaze were too thick or underfired or the quartz particles too coarse or too many, the inlaid pattern would not show through the glaze. If the glaze were too thin or overfired, it would appear transparent rather than rich and unctuous. Developing and maintaining this narrow envelope of physical and chemical parameters supports the conjecture that the potters expanded their technology empirically by analogical reasoning, conservatively manipulating only a few variables at a time to produce particular innovative effects within the craft tradition. Once developed, this technology remained constant until changes in taste at court created the new demand for white porcelain.

Within this tradition, artisans found ways to impress their patrons by producing exciting visual effects. By making minor changes in processing, for example, potters used and controlled crazing as an intentional decorative element in many of the finest Koryŏ celadons. In the earlier molded or incised celadons, the pattern of craze lines on the surface is random. In the white inlays, however, the pattern of crazing follows the linear pattern of those inlays because of the large difference in thermal contraction; in the case of the black inlays, no crazing pattern is observed because of the close

FIGURE 3

Microstructures of a white inlay. (a) Topographical view of a cross section of white inlay about 2-mm deep (30x). (b) Cross section of a white inlay (in gray) with the white, crazed glaze filling in the crack around it (31x). (c) Close-up of the inlay showing quartz particles (915x).

FIGURE 4

Microstructures of a black inlay. (a) Cross section of a black inlay showing crystals in the inlay (46x). (b) Enlarged view of the top of the inlay showing anorthite needles, ilmenite, and fine white crystals of magnetite (128x). (c) Close-up of cubic magnetite crystals (in white; 1320x).

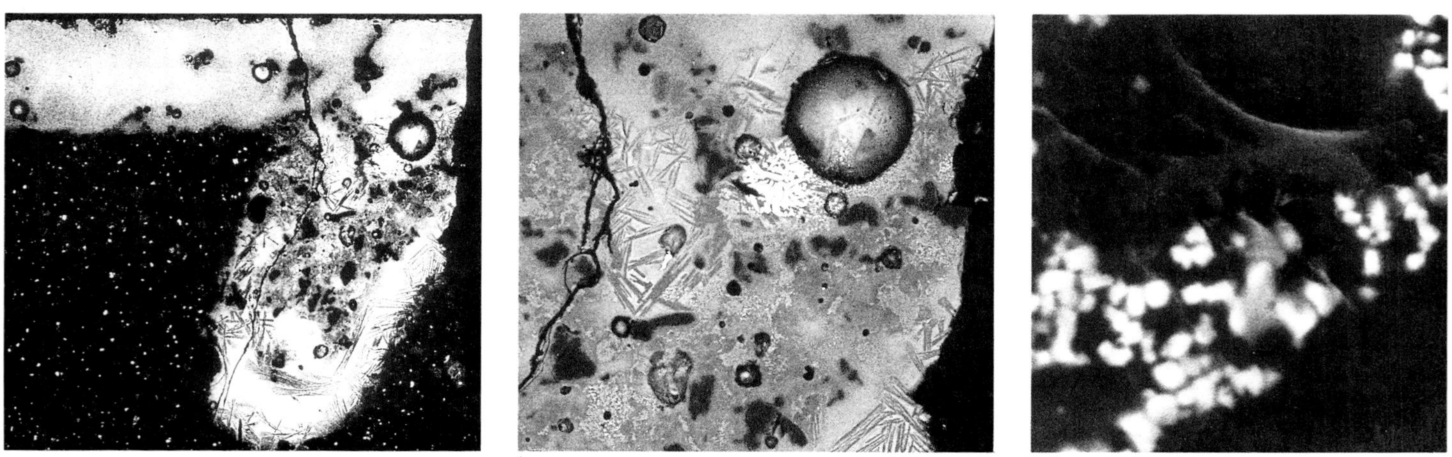

a *b* *c*

match of composition and thus of thermal contraction. Sometimes, a craze line does occur outside a large patch of black inlay about 0.5 cm into the background glaze and acts to outline that inlay. Craze lines also tend to splay at the ends of white inlaid lines, becoming subtle continuations of the inlaid lines penetrating into the background of the glaze and concentrating light at their tips. As light focuses around the inlaid pattern, it creates the illusion that the inlay floats above the level of the glaze. Craze lines serve as a texture that connects one decorative motif with another nearby element. Since the visual effects of inlaid designs were enhanced by craze lines, they were used to connect decorative elements, and many design motifs were clearly selected intentionally on the basis that their edges would splay or spread out (e.g., clouds, birds, and flowers) rather than enclose the pattern.

The development of craze lines can be predicted and controlled. Crazing occurs during the cooling cycle in a specific range of temperature, when the glaze has contracted sufficiently more

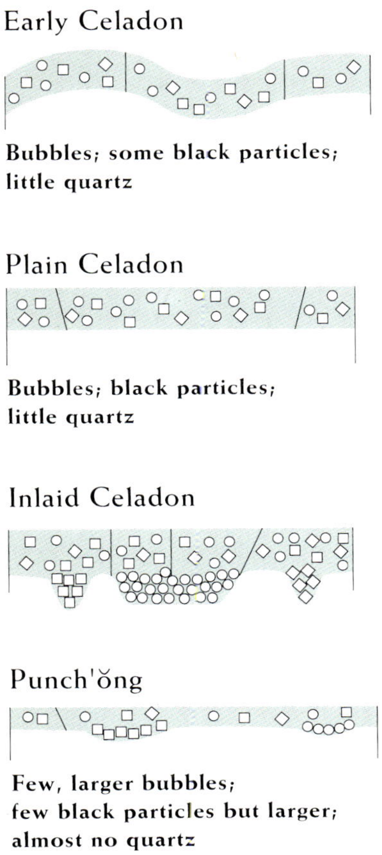

FIGURE 5
Sketches of glaze-body cross sections at about 60x showing differences among plain, inlaid, and Punch'ŏng celadons.

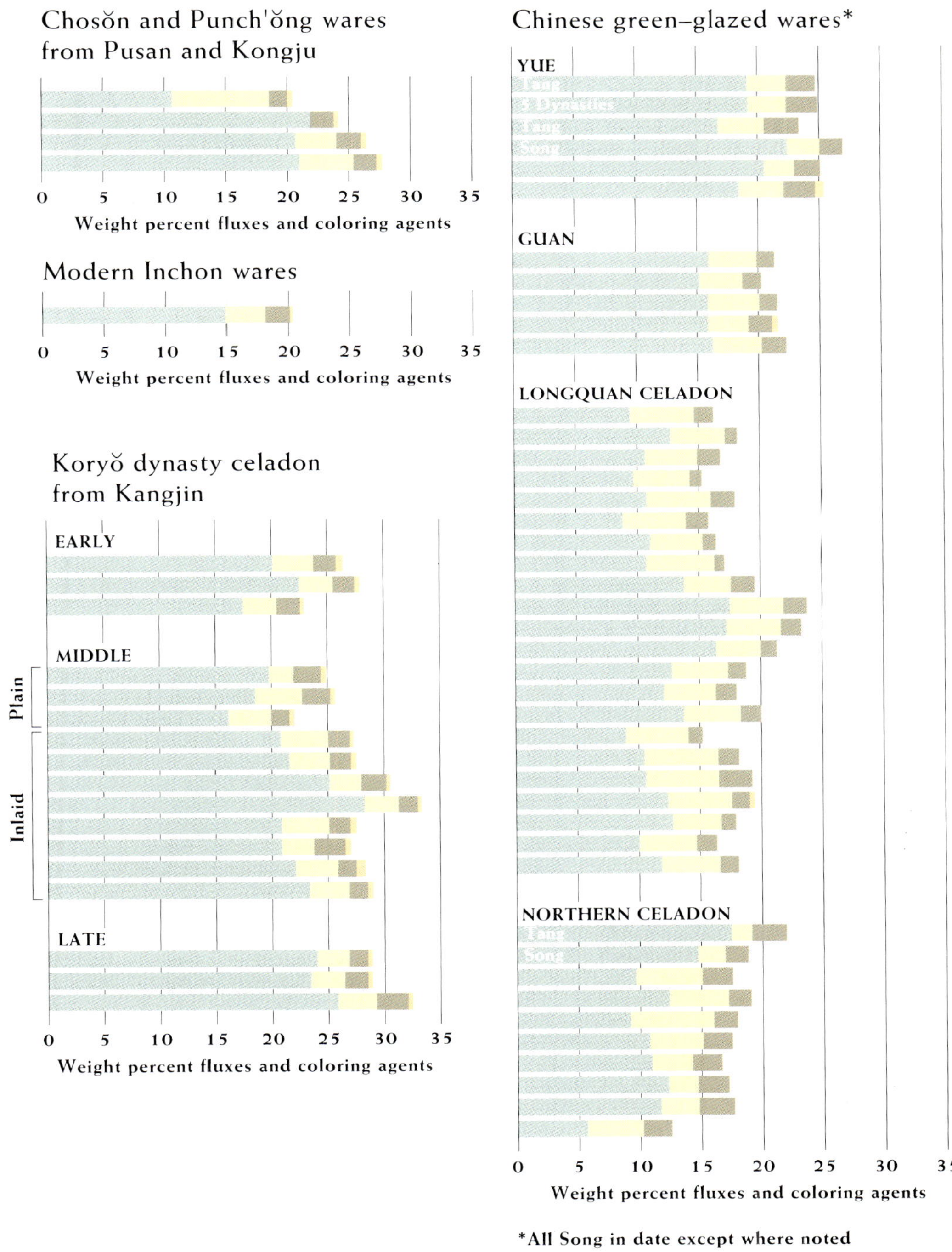

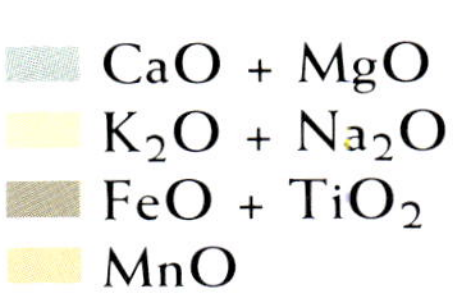

FIGURE 6
Histogram of glaze compositions showing weight percents of fluxes and coloring agents. (a) Korean green-glazed wares. (b) Chinese green-glazed wares (all Song in date, except where noted).

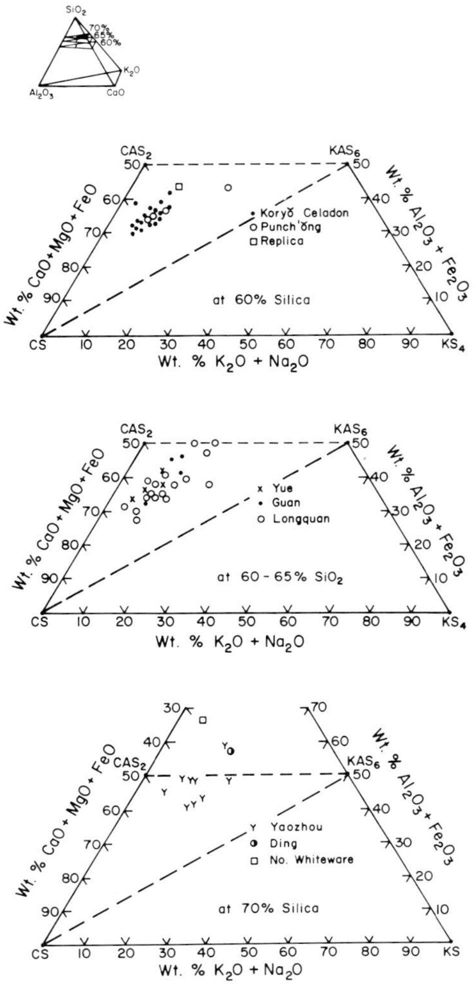

FIGURE 7

Three sections at differing silica contents in the SiO_2-Al_2O_3-CaO-K_2O quaternary which differentiate various Korean and Chinese wares.

than the body that its tensile strength is exceeded and T-shaped cracks form. During the time when the kiln temperature decreases from about 960°C (the point at which the glaze ceases to flow and begins to act like a brittle, solid glass) to 573°C (when the quartz undergoes a transformation and contracts about 2% in volume), the high-lime Koryŏ glazes shrink more than the quartz-containing clay body and much more than the quartz inlays. The glaze cracks where the greatest misfit occurs, that is, over the quartz-rich inlay; nevertheless, there is sufficient misfit to produce crazing over the rest of the body, which is generally initiated from cracks over the quartz-rich inlay. The 2-mm depth of the black inlay, as well as its compositional similarity to the body, makes crazing less likely in those areas than in the glaze over the body. Slowing the cooling of a test specimen in this temperature range results in less crazing; however, Koryŏ kilns were of the rapidly cooling dragon, or hill-climbing, type; and, thus, cooling was a function of kiln structure and not amenable to direct control.

About 270 Koryŏ kilns have been identified.[16] All but one are reduced versions of the hill-climbing variety known from southern China, rather than of the rounded single-chamber type common in northern China. They are small, about 1 meter across and 5–17 meters in length, with a firebox at the base, smoke holes at the top, and stoke holes along one side. This efficient type of kiln was fired sequentially from the bottom, with the result that the hot exhaust or flue air preheated the pots in the area above the region being fired; as firing commenced at the next level up, the pots in the lower region were rapidly cooled by the incoming air, while the incoming air flowing over the cooling pottery was preheated prior to combustion at a higher level in the kiln. Late Koryŏ kilns, unlike most Chinese kilns, were divided into separate firing chambers to permit better control of firing at higher temperatures in their individual sections.

The view that Koryŏ celadons were fired at a relatively low temperature of about 1100°C to 1150°C, at the low end of the firing-temperature range for Chinese glazes, and were cooled quickly, as were southern Chinese wares, is based on two findings: the rare occurrence of such crystals as anorthite and the complete lack of wollastonite and cristobalite, all of which nucleate and grow during slower glaze firings; and the low melting temperature of the well-fluxed, high-lime glaze composition. Anorthite, commonly found in Chinese green and white glazes of the Song and Yuan periods, adds to the translucency of most Chinese celadons; it forms a white layer between the body and the glaze that acts as a white ground for the glaze. During firing, anorthite requires lengthy exposure at peak temperature to nucleate and a sustained period at a lower temperature to grow. In Koryŏ celadon, anorthite is rare (fig. 2c), and in Punch'ŏng ware it is not found at all. Only when a Koryŏ celadon sample is held at 1100°C for a period of eight hours does it grow into the shape of these etched needles (fig. 8).

Overall lack of a white anorthite layer in Korean celadon, as well as the manganese content of the glaze, renders the celadon grayer in color. In addition, it also provides an indication that Korean firing practice allowed for a brief, relatively low-temperature heat treatment and more rapid cooling than did Chinese kiln firing. This belief is reinforced by the Korean kiln design—the firing temperatures of the hill-climbing kilns tended to be lower, the cooling tended to be more rapid, and the pottery remained at peak temperature for a shorter time because kiln size was smaller than that of northern Chinese kilns of approximately the same period.

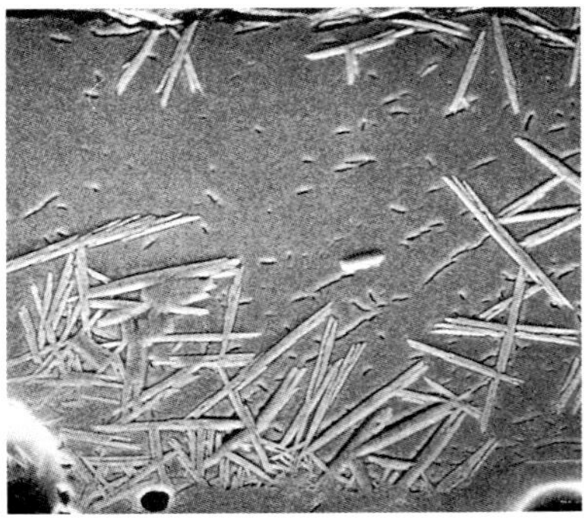

FIGURE 8

Growth of anorthite in a refired shard held for eight hours at 1100°C.

Conclusion

Korean potters learned empirically to control the technical aspects of ceramic manufacture to create outstanding, jadelike celadons and to implement innovative new processes, such as fritting, to produce particular visual effects. They operated within a narrow range of technical parameters in which they controlled color chemistry, the purity of their raw materials, the application of inlays, the grinding of glaze materials and preparation of glazes, the optimal glaze thickness for translucency and visibility of the inlay, the development of crazing patterns through control of the glaze composition and inlay designs and the rapid firing and cooling of their relatively small hill-climbing kilns.

Within this conservative celadon tradition, the development of a new market and new function dictated technological adapation of techniques used for court ware to the manufacture of common ware in the fifteenth century. The attendant development of aesthetic conventions was based on the continuity of technological constraints, indigenous aesthetic preferences, and simplification of processing raw materials.

The technology of Korean celadon manufacture is most closely related to that of the southern Chinese rather than to that of the northern centers of production. The range of firing temperatures for Korean celadon, its glaze composition and microstructure, and the design of its kilns all come closest to southern Chinese models.

NOTES

In acquiring samples, I gratefully acknowledge the help of Marian Mitchell and Lee Jeong; Rha Sunwha of Ewha Women's University, Seoul; Dr. George Kuwayama of the Los Angeles County Museum of Art; and Louise Cort of the Freer Gallery of Art, Washington, D.C. Karen Andelman assisted with sample preparation; Dr. Eugene Jarosewich of the Mineral Science Probe Lab, National Museum of Natural History, Smithsonian Institution, collaborated on the chemical analyses; and Harold Dokstadter printed the microstructures.

1. Indianapolis Museum of Art, *Ice and Green Clouds: Traditions of Chinese Celadon*, exh. cat. by Yutaka Mino and Katherine Tsiang (1986), no. 65, p. 166; and G. St. G. M. Gompertz, *Korean Celadon and Other Wares of the Koryŏ Period* (London, 1963), pp. 33–45.

2. London, Victoria and Albert Museum, *Catalogue of the Le Blond Collection of Korean Pottery*, cat. by Bernard Rackham (1918); James M. Plumer, ed., "Korean Ceramics," *Far Eastern Ceramics Bulletin* 9, 3–4 (1957); Gompertz (note 1); idem, *Korean Pottery and Porcelain of the Yi Period* (New York, 1968); Goro Akaboshi and Heiichiro Nakamaru, *Five Centuries of Korean Ceramics: Pottery and Porcelain of the Yi Dynasty*, trans. June Silla (New York and Kyoto, 1975); Cologne, Museum für Ostasiatische Kunst, *Seladon-Keramik der Koryŏ-Dynastie, 918–1392* (in German and in English), exh. cat. by Choi-Bae Soontaek (1984); Cambridge, England, Fitzwilliam Museum, *Korean and Chinese Ceramics from the 10th to the 14th Century*, exh. cat. by Margaret Medley (1975).

3. Seoul, National Museum of Korea, *Pottery Arts of Korea*, Korean Arts Series 1 (slides, booklet, and tape), prepared by Chŏng Yang-mo (1976); *Sekai tōji zenshū* (Catalogue of world ceramics), vol. 18 (Tokyo, 1978); Seoul, National Museum of Korea, *Selected Treasures of the National Museum of Korea* (in Korean and in English) (1977); New York, Asia Society, *The Art of the Korean Potter: Silla, Koryŏ, Yi*, exh. cat. by Robert P. Griffing, Jr. (1968); Seoul, National Museum of Korea, and San Francisco, Asian Art Museum, *5000 Years of Korean Art*, exh. cat. (San Francisco, 1979); London, British Museum, *Treasures from Korea: Art through 5000 Years*, exh. cat. by Roger Goepper, Roderick Whitfield et al. (1984).

4. Kim Yŏng-wŏn, "A Comparative Study of Koryŏ and Chinese Ceramics," *Kogo Misul—Art and Archaeology*, Hankuk, Misulsa Hakhoi, Seoul (Art Historical Association of Korea) 171–72 (Dec. 1986): 73–90; Chŏng Yang-mo, "A Study of Koryŏ Celadons," in ibid.: 1–18; Yun Yong-i, "A Study of Koryŏ Celadon Kilns," in ibid.: 59–73.

5. See, for example, Feng Xianming, "Ancient Korean and Iranian Ceramics Excavated in China," in *Chinese Silicate Society, The Second International Conference on Ancient Chinese Pottery and Porcelain: Its Scientific and Technical Insights* (abstracts) (Beijing, 1985), pp. 109–10.

6. Indianapolis (note 1), no. 1, pp. 28–29.

7. Pamela Vandiver and W. D. Kingery, "Song Dynasty Jun (Chün) Ware Glazes," *Bulletin of the American Ceramics Society* 62, 11 (1983): 1269–79; idem, "Composition and Structure of Chinese Song Dynasty Glazes from Longquan," *Bulletin of the American Ceramics Society* 63 (1984): 612–16; idem, "Variations in Microstructure and Microcomposition of Pre-Song, Song, and Yuan Dynasty Glazes," in W. D. Kingery, ed., *Ancient Technology to Modern Science*, Ceramics and Civilization, vol. 1 (Columbus, 1985), pp. 181–223; idem, *Ceramic Masterpieces* (New York, 1987).

8. A corpus of research on Chinese glazes utilizing this methodology already exists: see Li Jiazhi, Chen Xianqiu, Zhang Fukang, Guo Yanyi, Chen Shiping, et al., *Scientific and Technical Achievements in Ancient Chinese Pottery and Porcelain* (Shanghai: 1984), esp. pp. 211–29, for firing-temperature measurements; Jiazhi Li, "The Evolution of Chinese Pottery and Porcelain Technology," in Kingery, ed., *Ancient Technology* (note 7), vol. 1, pp. 135–62; Fukang Zhang, "The Origin and Development of Traditional Chinese Glazes and Decorative Ceramic Colors," in ibid., pp. 163–80; idem, "The Development of Ancient Chinese Glazes and Decorative Ceramic Colors," in Shanghai Institute of Ceramics, *Scientific and Technological Insights on Ancient Chinese Pottery and Porcelain*, ed. Academica Sinica (Beijing, 1986),

pp. 46–54; Guozhen Li and Lingxiang Guo, "The Development of Chinese Celadon and Its Influences," in Kingery, ed., *Ancient Technology* (note 7), vol. 3 (Columbus, 1986), pp. 129–52.

To date, however, only three technical studies of Korean celadon glazes have appeared: Kurt Hangst, "Results of Mineralogical Examination of Koryŏ-Period Celadon Shards," in Cologne (note 2), pp. 233–47; P. B. Vandiver, L. A. Cort, and C. A. Handwerker, "Variations in the Practice of Ceramic Technology in Different Cultures: A Comparison of Korean and Chinese Celadon Glazes," in P. E. McGovern, ed., *Cross-craft and Cross-cultural Interactions in Ceramics* (Westerville, 1989), pp. 347–88; and Richard Newman, "A Compositional and Microstructural Study of Korean Celadon Glazes of the 11th to 15th Centuries," in P. B. Vandiver, J. Druzik, and G. S. Wheeler, eds., *Materials Issues in Art and Archaeology II*, Materials Research Society Proceedings, vol. 185 (Pittsburgh, 1990).

9. Shards from Kangjin were tested by me; since then, I have tested shards from twenty other excavated sites. These results agree with those of Hangst (note 8).

10. Indianapolis (note 1).

11. Mineralogically, X-ray diffraction indicates the presence of quartz and minor amounts of glass and mullite. The major phases present in the white inlays by volume are porosity, then quartz; the minor phases are glass and mullite. Mullite can form at temperatures as low as about 970°C. In the black inlays, the major phases present are glass and magnetite; the minor ones are quartz, ilmenite, and porosity.

12. For the composition of black inlays, see note 11. These inlays can actually range in color from gray to black.

13. Punch'ŏng is generally considered a household ware but has also been found in the tomb of Prince Onyong, thus testifying to its use at all levels of society (Seoul, National Museum of Korea, display captions, Nov. 1987). Similarly, fine-quality Koryŏ celadon has an analogue in popular ware. The plain green-glazed ware known as Nokch'ŏngja was once considered an experimental precursor to fine Koryŏ celadons, but recent excavations have shown that this ware was produced simultaneously with green-glazed court ware for household use (Rha Sunwha, personal communication, Nov. 1987).

14. The SiO_2 at about 60% (in a range of 58–65%) is quite low for East Asian glazes; Al_2O_3 is 11-15%, K_2O 2–5%, MgO about 2.5%, Na_2O about 0.75%, FeO-Fe_2O_3 about 1%, and MnO_2 about 0.5%. Further analyses from other sites have corroborated that a narrow range of compositions was employed. The calcium oxide (CaO) concentration varies across the glaze cross section, decreasing near the body according to the temperature and duration of firing, which affects the extent of interdiffusion of constituents at the interface of glaze and body. In one exception, an overfired waster (shard 6 in fig. 3a), the body has warped, but the glaze is mat, due to its low calcia content.

15. Three different sections are plotted throughout the silica, alumina, potassia, and calcia quaternary phase diagram in fig. 10. Note that pure silica is on top, and the relative proportions of potassia, calcia, and alumina are shown as corners in each of the truncated triangles. Through similarity of valence, titania has been grouped with silica, alumina with Fe_2O_3, potassia with soda, and calcia with MgO and FeO. One modern celadon replica and one Punch'ŏng sample are outliers, while the Yaozhou, Ding, and northern Chinese white wares plot nearest the 70% SiO_2 section and are rich in alumina.

The results cited in figs. 9 and 10 are based on recent analyses by five different researchers (those by Li and Guo and by Zhang for China and those by Hangst and by Vandiver for Korea, all cited in note 8, with numbers retained from their original publications).

16. Cologne (note 2), p. 30.